Organoid Technology: Disease Modelling, Drug Discovery, and Personalized Medicine

Edited by

Manash K. Paul

Department of Radiation Biology and Toxicology
Manipal School of Life Science
Manipal Academy of Higher Education
Manipal, Karnataka, 576104, India

&

Department of Pulmonary and Critical Care Medicine
University of California,
Los Angeles (UCLA)
USA

Organoid Technology: Disease Modelling, Drug Discovery and Personalized Medicine

Editor: Manash K. Paul

ISBN (Online): 978-981-5238-69-3

ISBN (Print): 978-981-5238-70-9

ISBN (Paperback): 978-981-5238-71-6

First published in 2025.

need for a court order if at any point you breach any terms of this License Agreement. In no event will any delay or failure by Bentham Science Publishers in enforcing your compliance with this License Agreement constitute a waiver of any of its rights.

3. You acknowledge that you have read this License Agreement, and agree to be bound by its terms and conditions. To the extent that any other terms and conditions presented on any website of Bentham Science Publishers conflict with, or are inconsistent with, the terms and conditions set out in this License Agreement, you acknowledge that the terms and conditions set out in this License Agreement shall prevail.

Bentham Science Publishers Pte. Ltd.
No. 9 Raffles Place
Office No. 26-01
Singapore 048619
Singapore

Email: subscriptions@benthamscience.net

CONTENTS

,Kaplana Mandal, Shalaka Wahane, Muhammad Nihad, Anubhab Mukherjee Bharti Bisht, Chrianjay Mukhopadhyay, Bipasha Bose and *Manash K. Paul*

Jyotirmoi Aich, Sangeeta Ballav, Isha Zafar, Aqdas Khan, Shubhi Singh, Manash K. Paul, Shine Devrajan and *Soumya Basu*

PREFACE

Organoids are miniature in vitro 3D models that mimic the near-physiological structure and function of the respective tissues and organs. Organoid bioengineering is a transdisciplinary approach that uses stem cells' capacity to self-renew, differentiate into several lineages, and self-organize into organoids. Using organoid bioengineering, scientists have employed induced pluripotent stem cells (iPSCs), embryonic stem cells (ESCs), and tissue-resident adult stem cells (ASCs) to generate these tiny tissue replicas. Several research teams have developed endodermal, mesodermal, and ectodermal organoids by manipulating stem cells in vitro. Numerous organoids may now be created, including those of the kidney, brain, lung, colon, intestine, breast, retina, and liver. Given the gap between animal-based models and human disease pathology, a paradigm change was required to simulate human diseases accurately. The 3D human organoid platform provides an unmatched opportunity to develop better models and get a more profound knowledge of human pathophysiology. Organoids provide information on human disease-related processes, such as disease-specific signaling alterations, cell-cell interactions, therapeutic target identification, therapeutic screening, and discovery.

Organoid technology has been used to model diseases across different organ systems, drug screening, and regenerative medicine. Recent advances, including the development of the novel organoid platform, engineering organoid complexity, disease modeling, introducing pathological aspects together, and drug discovery, have provided a ray of hope for human-specific therapeutic discovery. Patient-derived tumor organoids may be created from individual patients, and biobank, and used for therapeutic screening and personalized treatment. Significant progress is made toward large-scale organoid production. Novel technologies like high-resolution 3D imaging, genome editing, hybrid culture techniques, single-cell transcriptomics, microfluidics, organ on a chip, 3D printing, nanotechnology, and other cutting-edge technologies facilitate the development of physiologically accurate human disease models.

Significant numbers of animal-based preclinical studies leading to human clinical trials fail due to safety or efficacy concerns, raising the question of whether animal experiments can truly aid in the development of effective therapies and at what cost. Numerous alternatives, such as in vitro human-specific 3D microphysiological systems such as organoids and microfluidic-based organs-on-a-chip that closely mimic human physiology and architecture, have enabled cutting-edge animal-free research. Although optimistic, these technologies have not yet attained their summit, and a complete substitution of animal-based experiments may take decades. Scientists believe that addressing the obstacles of appropriate validation, proper standards, development of protocols for large-scale production, funding, regulatory rules, and ethical issues may enhance the use of human biology-based models, thereby improving the lives of humans and animals. In an endeavor to reduce animal dependence, in late December 2022, President Joseph Biden signed a law that novel therapeutics no longer require animal experimentation. After eight decades of medication safety regulation, this long-awaited action could help end animal experimentation and make therapeutic interventions that are tangible and effective. The purpose of these chapters is to shed light on the developing resources addressing the concepts of organoids and disease models, including cancer. This book focuses on organ-specific organoids and disease modeling.

Manash K. Paul
Department of Radiation Biology and Toxicology
Manipal School of Life Science
Manipal Academy of Higher Education
Manipal, Karnataka, 576104
India

&

Department of Pulmonary and Critical Care Medicine
University of California, Los Angeles (UCLA)
USA

List of Contributors

Anubhab Mukherjee   Esperer Onco Nutrition Pvt Ltd, 4BA, 4th Floor, B Wing, Gundecha Onclave, Khairani Road, Sakinaka, Andheri East, Mumbai, Maharashtra, 400072, India

Aqdas Khan   School of Biotechnology and Bioinformatics, Dr. D. Y., Patil Deemed to be University, CBD Belapur, Navi Mumbai, Maharashtra, 400 614, India

Ahmet Katı   University of Health Sciences Turkey, Biotechnology Department, Uskudar, 34662, Istanbul, Turkey
University of Health Sciences Turkey, Validebag Research Park, Experimental Medicine Research and Application Center, Uskudar, 34662, Istanbul, Turkey

Amit Kumar Mandal   Raiganj University, Centre for Nanotechnology Sciences (CeNS) & Chemical Biology Laboratory, Department of Sericulture, North Dinajpur, West Bengal-733134, India

Arkaprabha Basu   Harvard John A. Paulson School of Engineering and Applied Sciences, Harvard University, Cambridge, Boston, MA 02134, United States

Atul Kumar Singh   Central Research Facility, IIT-Delhi Sonipat Campus, Rajiv Gandhi Education City, Sonipat, Haryana 131029, India

Bharti Bisht   Department of Microbiology, Kasturba Medical College, Manipal Academy of Higher Education, Manipal, Karnataka, 576104, India

Bipasha Bose   Stem Cells and Regenerative Medicine Centre, Yenepoya Research Centre, Yenepoya (Deemed to be University), Mangalore, Karnataka, 575018, India

Beverly Rothermel   Department of Internal Medicine, Division of Cardiology, University of Texas Southwestern Medical Center, 5323 Harry Hines Blvd, Dallas, TX 75390, Texas
Department of Molecular Biology, University of Texas Southwestern Medical Center, 5323 Harry Hines Blvd, Dallas, TX 75390, Texas

Chrianjay Mukhopadhyay   Department of Microbiology, Kasturba Medical College, Manipal Academy of Higher Education, Manipal, Karnataka, 576104, India

Dilara Genc   Kadir Has University, Undergraduate Program of Bioinformatics and Genetics, Fatih, 34230, Istanbul, Turkey

Hanen Salami   Laboratory of Treatment and Valorization of Water Rejects (LTVRH), Water Researches and Technologies Center (CERTE), Borj-Cedria Technopark, University of Carthage, 8020, Soliman, Tunisia

Isha Zafar   Cancer and Translational Research Centre, Dr. D. Y. Patil Biotechnology and Bioinformatics Institute, Dr. D. Y. Patil Vidyapeeth, Pune, Maharashtra, 411 033, India

Jyotirmoi Aich   School of Biotechnology and Bioinformatics, Dr. D. Y., Patil Deemed to be University, CBD Belapur, Navi Mumbai, Maharashtra, 400 614, India

Janvie Manhas   Department of Biochemistry, All India Institute of Medical Sciences, New Delhi, India

Kaplana Mandal   Terasaki Institute for Biomedical Innovation, 11570 Olympic Blvd, Los Angeles-90064, CA, USA

Khushboo Dutta   Department of Integrative Biology, School of Bio Sciences and Technology, Vellore Institute of Technology, Vellore-632014, Tamil Nadu, India

Keshav S. Moharir	Department of Pharmaceutics, Gurunanak College of Pharmacy, Nagpur, Maharashtra, India
Muhammad Nihad	Stem Cells and Regenerative Medicine Centre, Yenepoya Research Centre, Yenepoya (Deemed to be University), Mangalore, Karnataka, 575018, India
Manash K. Paul	Department of Radiation Biology and Toxicology, Manipal School of Life Sciences, Manipal Academy of Higher Education, Manipal, Karnataka, 576104, India Division of Pulmonary and Critical Care Medicine, David Geffen School of Medicine, University of California Los Angeles, Los Angeles, CA, 90095, USA
Malay Chaklader	Department of Internal Medicine, Division of Cardiology, University of Texas Southwestern Medical Center, 5323 Harry Hines Blvd, Dallas, TX 75390, Texas
Pravin D. Potdar	Former Head, Department of Molecular Medicine and Biology, Jaslok Hospital & Research Centre, Mumbai, 400053, Maharashtra, Chairman, Institutional Ethics Committee, Dr. A. P. J. Kalam Educational and Research Centre, Mumbai, Maharashtra, India
Rohit Gundamaraju	Division of Gastroenterology, Department of Medicine, Washington University School of Medicine, St Louis, MO, USA
Shalaka Wahane	Departments of Neurobiology and Neurosurgery, David Geffen School of Medicine, University of California, Los Angeles, Los Angeles, CA90095-1763, USA
Sangeeta Ballav	Cancer and Translational Research Centre, Dr. D. Y. Patil Biotechnology and Bioinformatics Institute, Dr. D. Y. Patil Vidyapeeth, Pune, Maharashtra, 411 033, India
Shubhi Singh	School of Biotechnology and Bioinformatics, Dr. D. Y., Patil Deemed to be University, CBD Belapur, Navi Mumbai, Maharashtra, 400 614, India
Shine Devrajan	School of Biotechnology and Bioinformatics, Dr. D. Y., Patil Deemed to be University, CBD Belapur, Navi Mumbai, Maharashtra, 400 614, India
Soumya Basu	Cancer and Translational Research Centre, Dr. D. Y. Patil Biotechnology and Bioinformatics Institute, Dr. D. Y. Patil Vidyapeeth, Pune, Maharashtra, 411 033, India
Sunita Nayak	Department of Integrative Biology, School of Bio Sciences and Technology, Vellore Institute of Technology, Vellore-632014, Tamil Nadu, India
Suvankar Ghorai	Virology Laboratory, Department of Microbiology, North Dinajpur, West Bengal-733134, India
Sare Nur Kanari ElHefnawi	University of Health Sciences Turkey, Validebag Research Park, Experimental Medicine Research and Application Center, Uskudar, 34662, Istanbul, Turkey
Sevde Altuntas	University of Health Sciences Turkey, Validebag Research Park, Experimental Medicine Research and Application Center, Uskudar, 34662, Istanbul, Turkey University of Health Sciences Turkey, Tissue Engineering Department, Uskudar, 34662, Istanbul, Turkey
Swati Tripathi	Section of Electron Microscopy, National Institute of Physiological Sciences, Okazaki, Japan

Shivaji Kashte   Department of Stem Cell and Regenerative Medicine, Centre for InterdisciplinaryResearch, D. Y. Patil Education Society (Institution Deemed to be University), Kolhapur 416006, India

Shahabaj Mujawar   Department of Stem Cell and Regenerative Medicine, Centre for InterdisciplinaryResearch, D. Y. Patil Education Society (Institution Deemed to be University), Kolhapur 416006, India

Sachin Kadam   Manipal Center for Biotherapeutics Research, Manipal Academy of Higher Education(Institute of Eminence Deemed to be University), Manipal 576104, India

Shraddha Gautam   Advancells Group, A 102, Sector 5, NOIDA, Uttar Pradesh 201301, India

Tareeka Sonawane   Amity Institute of Biotechnology, Amity University, Pune Expressway, Bhatan, Mumbai 410221, India

CHAPTER 1

Bioengineering Organoids for Disease Modeling and Drug Discovery

Kaplana Mandal[1,#], Shalaka Wahane[2,#], Muhammad Nihad[3], Anubhab Mukherjee[4], Bharti Bisht[5], Chrianjay Mukhopadhyay[5], Bipasha Bose[3,*] and Manash K. Paul[6,7,*]

[1] *Terasaki Institute for Biomedical Innovation, 11570 Olympic Blvd, Los Angeles-90064, CA, USA*

[2] *Departments of Neurobiology and Neurosurgery, David Geffen School of Medicine, University of California, Los Angeles, Los Angeles, CA, 90095-1763, USA*

[3] *Stem Cells and Regenerative Medicine Centre, Yenepoya Research Centre, Yenepoya (Deemed to be University), Mangalore, Karnataka, 575018, India*

[4] *Esperer Onco Nutrition Pvt Ltd, 4BA, 4th Floor, B Wing, Gundecha Onclave, Khairani Road, Sakinaka, Andheri East, Mumbai, Maharashtra, 400072, India*

[5] *Department of Microbiology, Kasturba Medical College, Manipal Academy of Higher Education, Manipal, Karnataka, 576104, India*

[6] *Department of Radiation Biology and Toxicology, Manipal School of Life Sciences, Manipal Academy of Higher Education, Manipal, Karnataka, 576104, India*

[7] *Division of Pulmonary and Critical Care Medicine, David Geffen School of Medicine, University of California Los Angeles, Los Angeles, CA, 90095, USA*

Abstract: Organoid technology has been used to model diseases across different organ systems, drug screening, and regenerative medicine. Organoid technology better mimics human physiology and can provide a better alternative to *in vivo* animal models. Recent advances in organoid technology, including developing the novel organoid platform, engineering complex organoids, and introducing pathological aspects, have provided significant progress toward producing miniaturized tissue or organs on a dish. Novel technologies like high-resolution 3D imaging, organ on a chip, 3D printing, gene manipulation, nanotechnology advances, and single-cell sequencing have led to a massive thrust in the organoid technology that can provide a unique insight into the behavior of stem cells, cater to preclinical research and theranostics (therapy plus diagnostics).

*Corresponding authors **Bipasha Bose** and **Manash K. Paul:** Stem Cells and Regenerative Medicine Centre, Yenepoya Research Centre, Yenepoya (Deemed to be University), Mangalore, Karnataka, 575018, India & Department of Radiation Biology and Toxicology, Manipal School of Life Sciences, Manipal Academy of Higher Education, Manipal, Karnataka, 576104, India & Division of Pulmonary and Critical Care Medicine, David Geffen School of Medicine, University of California Los Angeles, Los Angeles, CA, 90095, USA; E-mails: bipasha.bose@yenepoya.edu.in, manash.paul@manipal.edu
Equal authorship

Keywords: Bio-banking, Bioengineering, Cancer, Disease model, Drug screening, Infectious diseases, Organoid, Organ, Regenerative medicine, SARS-CoV2, Theranostics, 3D-printing.

INTRODUCTION

Since the late twentieth century, the preeminent way of modeling the physiology of human organs has been the usage of cell lines and animals. Global biological research has refined our perception of various cellular signal transductions, screening and identifying drug targets, and discovering novel drugs to combat dreadful diseases such as cancer, infectious diseases, and some non-communicable diseases. Their universal operations and application portray the historical dependence of today's biomedical research on these model systems [1]. The impeccable accuracy of the natural selection process and the *sine qua non* of evolutionarily conserved biological mechanisms across species have elicited the primary focus of researchers toward a few model organisms. They are generally rapidly growing species that can spawn a copious number of offspring in a short interval and can be propagated in a laboratory in a cost-effective manner, such as *Saccharomyces cerevisiae* (yeast), *Caenorhabditis elegans* (nematode), *Drosophila melanogaster* (fruit fly), *Danio rerio* (zebrafish) and the common laboratory *Mus musculus* (mouse) and *Rattus norvegicus* (rat). As time passed, a common discovery thread became traceable while scanning the pathophysiology of maladies and various animal models of human diseases. Functional aspects of biological organisms were deciphered by genomic screens in invertebrates, with subsequent anatomization of evolutionarily conserved genetic aspects in mammalian models, which would pave their way to humans. These commonalities in principles and their applications have begotten an extensive mechanistic understanding of various human diseases [2].

Nevertheless, the triumph was dampened by the fact that the biophysical and physiological functions of the model organisms were not consistently replicated in humans owing to the human system's intrinsic complexity. Sometimes, it has been witnessed that extrapolating results from animal models to humans has become a significant bottleneck in drug discovery processes [1 - 3]. Moreover, recent insights into the phenomenon also reveal some biological processes specific to the human system alone (such as brain development, lung development, metabolism, and evaluation of drug efficacy) beyond the scope of modeling in animals. Obviously, the above challenges have prompted many attempts to model functions of human organs, which include stem cell differentiation in two-dimensions (2D) and three-dimensions (3D) with/without matrix, organoid bioengineering, bio-printing, and organ on a chip using microfluidics [4]. To bridge the disconnect between the actuality and models, a fecund technology, organoid bioengineering, has recently substantially impacted global biomedical research. These essentially undifferentiated stem cell/ differentiated progeny/ cancer stem cell-derived 3D culture systems made the re-creation of precise tissue/ organ architecture and physiology possible.

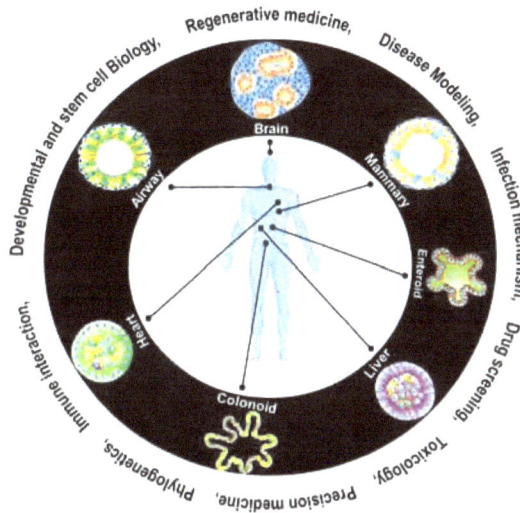

Fig. (1).   Schematic showing the bioengineering of tissue-specific organoids and their use in many biomedical and commercial uses. Figure reproduced from reference 7.

There is no denying that human-specific organoids have the unique potential to produce 3D cultures that closely mimic the human organ of choice and can represent human diseases more accurately than animal-based models [5, 6]. The most common feature of all organoids is attributed to the origin of organoids based on the recapitulation of various 3D-designed protocols of *in vitro* organogenesis from pluripotent or adult stem cells [6]. It is, therefore, limpid and conspicuous that upon analyzing the organoid generation, the scientific minds will extract precious mechanistic information regarding human development, organ regeneration, and disease modeling. 3D culture platforms, especially organoids/ spheroids also have demonstrated their potential and benefited human-specific disease modeling, pharmaceutical drug screening, and molecular medicine-based discovery. The perception of ramifications of its relevance has already triggered extensive application of this model in biomedical research, complementing the already existing ones [2, 4 - 6]. Fig. (**1**) presents a schematic view of tissue-specific organoids and their applications. The discovery and development of organoid technology are still in their juvenescence, compared to the existing cell/ animal-based models, with obstacles to circumvent. In this chapter, besides providing an overview of organoid technology, we shall briefly highlight the major advances in tissue-specific organoids, drug screening, organoid imaging, tissue regeneration, 3D printing organoid, and disease modeling. In tune with the needs of current times, we have also focused on the commercial prospects of organoid technology and biobanking. Finally, we have concluded the chapter dealing with the limitations of the technology and outlining future perspectives, hoping that this emerging platform will intrigue our readers, bringing forth more intense research in the field.

ORGANOID TECHNOLOGY

As mentioned above, various model systems have been used to understand basic biology and model diseases. In order to model cancer, patient-derived xenografts (PDX) and various cancer cell lines are generally used. Each model has its unique advantages but suffers from the limitation that they cannot precisely replicate human diseases [3, 7]. Therefore, *in vitro* human model systems are urgently required for deciphering human biology. The burning need for human model systems is primarily attributed to human-specific phenomena being non-acquiescent for replication in other model animals. As pointed out earlier, as long as the complexity and sophistication are concerned, the human brain is in no way comparable to rodents [3]. There also exist significant differences in metabolism between humans and rodents. For example – ibuprofen and warfarin are toxic to rodents [8], unlike their therapeutic action on humans. Unlike laboratory animals, humans are born, above all, not grown. Ratiocinating the heterogeneity of human genetics and its impact on the initiation and propagation of diseases, understanding the respective drug responses in different genetic populations/individuals must be considered an underlying criterion for personalized medicine. Hence, it is therefore imperative to develop human-specific model systems. Induced pluripotent stem cell (iPSC) and adult stem cell (ASC) technology has recently facilitated the production of human-specific laboratory models [2, 4 - 6]. *in vitro* human-specific models, namely iPSC-derived organoids and ASC-derived organoids, are reflected in the following sections.

Pluripotent Stem Cell (PSC)-derived Organoids

Reprogramming already differentiated mature somatic cells (such as fibroblasts) into iPSCs has become a part of the zeitgeist ever since four specific genes (namely, Myc, Oct3/4, Sox2, and Klf4), together called Yamanaka factors, were introduced in 2006 and the discoverer was rightly awarded the Nobel prize in 2012. Initial methods looked for the iPSCs to be differentiated into one type of cells, such as neurons and cardiomyocytes, in 2D cell culture, until recently, when more complex structures of tissues comprising various cells were modeled in 3D culture [7, 9, 10]. For successful simulation of different stages of a human developmental process, the technology allows a sequential exposure of human iPSCs to a series of differentiation cues, followed by agglomeration of iPSCs resulting in organoid formation [9]. The iPSC technology provides researchers with a limitless supply of patient-specific stem cells and, thereby differentiated tissues. It is worth mentioning here that folks upheld squeamish moral views about using human PSC as the same was being collected from human embryos after they are sacrificed at the blastocyst stage, and the quibbling was resolved once iPSC methodology came to the fore [9, 10]. Generally speaking, the generation of organoids requires three major steps - (i) Activation (or inhibition) of crucial developmental signaling pathways towards establishing a proper regional identity, (ii) Formulating correct media inducing terminal stem cell differentiation, and (iii) Expansion of the cultures in 3D.

Adult Stem Cell-derived Organoids

Human ASC-derived organoids, on the other hand, are simpler alternatives to the iPSC organoid, which comprise majorly cell types residing in the epithelium [4, 11]. Unlike scrupulously reprogrammed iPSCs with subsequent differentiation to the desired organ, these tissue-specific organoids can be generated from biopsies isolated from patient tissue or the organ of interest. ASCs are cultured along with the niche factors that maintain the undifferentiated state of the stem cell allowing their division, resulting in ASC-derived organoids with the epithelial monolayer. These constructs perfectly simulate the 3D architecture and the types of cells of the organ intended to be modeled [11]. The foremost prerequisites for the development of ASC-derived organoids are – (i) Identification and isolation of the appropriate population of ASCs (by enzymatic digestion or mechanical dissociation followed by flow cytometric analysis) and (ii) Delineating their niche requirements [4 - 6, 12]. However, the development of ASC-derived organoids is limited by the availability of the tissue and prior information regarding its cultural condition. Secondly, it is possible that the ASCs have limited self-organizing potential to become a potential organoid, as compared to PSC-derived organoids. However, adult cancer tissue-derived self-organizing organoid is easy owing to the tumorosphere forming potential of cancer cells [13].

Germ-line Specific Organoids

All three germ layers, namely, ectoderm, mesoderm, and endoderm are the basis of the formation of an entire organism. Organoids can be generated either from unspecialized ectoderm, mesoderm, and endoderm, or specific cell types of respective three germ layers to unravel the stages in developmental biology. Combinatorial efforts of cell atlases and organoid technologies have now highlighted the sequential formation of human organ development [14]. The following sections of this chapter will elucidate the organoids derived from all three germ layers starting from the innermost, namely endoderm, next, the middle-mesoderm, and then outermost-ectoderm.

ENDODERM-DERIVED ORGANOIDS

3D *in vitro* mimics of endodermal tissues comprising all adjacent cell types of specific endoderm are referred to as endoderm-derived organoids. Although endodermal organoids are *in vitro* mimics, they can be considered a complete biological system worth exploring. Such an endodermal organoid can be used for *in vitro* organogenesis, disease modeling for research, and translational applications. Moreover, due to their high stemness, pluripotent stem cells can be used for differentiation into the core cell types and the adjacent cell types of the organoid milieu. Alternatively, only the core cell types can be obtained by differentiating PSCs, and other adjacent cell types such as fibroblasts, endothelial cells, myoblasts, *etc.*, can be obtained from direct sources such as commercial sources or else primary cultures. Various endodermal tissues that exist in the human body and have possibilities of obtaining organoids are salivary glands, thyroid, lung, trachea, esophagus, stomach, liver, bile duct, pancreas, small intestine, large intestine, bladder, and prostate. However, the formation of 3D organoids in culture has only been reported in the thyroid, lung, esophagus, stomach, liver, bile duct, pancreas, small intestine, large intestine, and prostate [15].

As organoids are obtained from *in vitro* cultured cells, ethical concerns are less than those in animal models. The assembly of organoids can either be spontaneous self-assembly in low-attachment dishes or mediated by encapsulation inside any biomaterial. The critical point is maintaining the tissue architecture and the vascularization for proper blood supply and tissue drainage. Finally, the functional aspect of the organoid is important, which is attributed to the interplay of various biophysical factors such as elastic modulus, tensile strength, physiological homeostasis, along with anti-fouling properties for better biological sustenance. Hence, the appropriate biomaterial is essential for tissue scaffolding of 3D organoids [16]. Accordingly, this section includes the PSC differentiation into various common endoderm derivatives originating from different parts of the gut, namely the lung from the anterior foregut; gastric/stomach, liver, and pancreas from the posterior foreguts; small and large intestine from the midgut and hindgut respectively. Such endoderm derivatives are then modulated to generate respective functional organoids. We have attempted to cover various points in this section so as to give a complete picture of endoderm organoids. Such points include the protocols/ methods of endoderm spheroid induction, signaling pathways and long-term culture and maturation. The self-organization mechanism of organoids, precisely, coming together of the cells to form the 3D spheroids/ball-like structure, is also being covered. Bioengineering approaches to organoid culture systems and the development of disease models using 3D endodermal spheroids are two such wings under the same topic.

Lung Organoids

The lung is a complex organ comprising about 40 different cell types. As per embryonic development, the lung tissue develops from the anterior foregut [17]. For differentiating hPSCs into lung organoids, signaling pathways need to be modulated to obtain the ventral-anterior foregut spheroids, followed by the development of human lung organoids (HLO). Lung organogenesis involves sequential stages to morphologically give rise to differentiated multiple cell types meant for performing various functions. Cellular and molecular aspects of lung developmental biology are far from complete due to the unavailability of a complete 3D *in vitro* lung organoid comprising multiple cell types. Moreover, Vazquez-Armendariz, & Herold (2021) have comprehensively reviewed various kinds of 3D *in vitro* lung organoids, namely trachospheres; iPSC derived: alveolar, airway organoids; human airway organoids, fetal bud tip derived organoids, bronchoalveolar organoids (BALO); lung bud organoids (LBO); patterned lung organoids (PLO) and human lung organoids (HLO) [18].

For the completeness of such a lung organoid, the HLO needs to be comprised of epithelial as well as mesenchymal compartments similar to a naturally occurring human lung. Various epithelial cell types present in the human lung and HLO are upper airways like epithelium having basal cells and immature ciliated cells. The mesenchymal cell types in HLO are smooth muscle cells and myofibroblasts [19]. Alveolar domain comprising two morphologically distinct epithelial cell types, namely Alveolar epithelial type 1 (AT1) cells, and cuboidal alveolar epithelial type 2 (AT2) cells could also be differentiated from hPSCs forming a part of HLO. AT1 cell types conduct efficient gas exchange, while AT2 cell types secrete surfactants. Moreover, such HLOs exhibited the properties of fetal lung, thereby indicating a premature 3D differentiation, as studied using a global transcriptomic profiling of the cells [19]. Hence, such HLOs can be used for studying the developmental biology of human lung, disease modeling, and also drug testing applications. Lung organoids can also be derived from the actual lung tissues, and not necessarily by differentiating the hPSCs [20]. Multiple approaches to obtaining lung organoids involve whole lung culture, culture of mesenchyme-free epithelia. Culturing of mouse lung has been reported by embedding the mouse' lung in matrigel and incubation of permeable filters [20]. Alternatively, distal epithelial tips of the lung were harvested by removing the mesothelia and mesenchymal cells, followed by removing the severed tips from the airway tubes and culturing under 3D conditions. Interestingly, culturing 3D whole lungs exhibited branching morphogenesis and gene expression with a consistent Sox9 expression. The essential aspect of this protocol involves the consistent expression of Sox9 positive cells, indicating the existence of distal lung epithelia. Such lung organoids with Sox9 expression can also be used for studying lung pathologies of congenital lung diseases, lung cancer, and possible therapeutic aspects [21]. On the other hand, hPSC-derived lung organoids can also be used for modeling the SARS-COV2 infections from the ongoing pandemic in various ethnicities *in vitro* [22] and therapeutic strategies [23].

Well established protocols for differentiating the hPSCs from lung organoids involve the redirection to human or mouse PSCs first into definitive endoderm using Activin A, followed by anterior foregut endoderm (AFE), ventral anterior foregut endoderm (VAFE) and finally the NKX2.1 positive lung progenitors especially the region for the appearance of the trachea and primary lung buds [24]. This protocol published in Nature Protocols by Huang *et al.* is a long 50-day protocol involving four sequential steps. In this protocol, the definitive endoderm is first coaxed into AFE by sequential inhibition of BMP, TGFβ, and Wnt signaling. In the next step, the anterior foregut is then differentiated into ventral anterior foregut by using the cytokines, namely Wnt, BMP, FGF, and RA signaling giving rise to lung and airway progenitor cells. The fourth and final step of differentiation involves obtaining mature lung epithelial cells using the growth factors Wnt, FGF, c-AMP, and glucocorticoid agonism [24]. This protocol directs the PSCs predominantly into Type II functional alveolar epithelial cells. Similarly, Miller *et al.* (2019) published a protocol in which the hPSCs were differentiated first into ventral-anterior foregut spheroids giving rise to human lung organoids and bud-tip progenitor organoids. In this protocol, the human lung organoids had cellular and structural similarities with developing human bronchi/bronchiole, having the lung mesenchyme localized in the periphery and cell types with the expression of alveolar cells [25].

Finally, the 3D single-aspect differentiation of HLO from hPSCs led to the formation of fetal lung like tissues. Hence, maturation of such tissues is being in demand. Dye *et al.* have incorporated biomaterial approach for obtaining mature human lung organoids in 3D. Upon incorporation of biomaterial scaffolds, there is an enhanced transplantation efficiency and improved functionality. Various aspects of biomaterial scaffold include various physio-chemical features such as the type of polymer, degradation, and interconnectivity of polymer pores. HLO loaded on Polyethylene glycol (PEG) hydrogel scaffolds reportedly formed immature lung organoids. In contrast, HLOs loaded onto poly (lactide-c--glycolide) (PLG) scaffolds or polycaprolactone (PCL) resulted in the formation of tubular structures resembling an airway of an adult lung, both structurally and functionally [19]. Most importantly, biomaterials have the ability to modulate the size and structure of organoids, thereby governing transplantation applications.

More recently, an advanced protocol for generating lung and airway epithelial cells from hPSCs has been reported spanning 50-80 days [26]. In this protocol, the lung and airway epithelial cells were obtained by directed differentiation of hPSCs and maturation of NKX2.1 positive lung progenitors on a biomaterial/collagen I matrix without inhibiting GSK3 signaling. This is, therefore, an advanced protocol involving biomaterial matrix and has resulted in the formation of mature cell types, namely AT1 and AT2 cells, airway ciliated and basal cells, and neuroendocrine cells. Matured cells expressing NGFR have also been purified using flow cytometric cell sorting for various downstream applications.

Gastric Organoids

As a rule of thumb, all differentiation protocols from human pluripotent stem cells mimic in-vivo organogenesis. In the same direction, McCraken *et al.* (2014) reported the differentiation of human PSCs into gastric organoids by following *in vivo* stomach organogenesis. In this protocol, the first step is to differentiate PSCs into fundic epithelium. This was obtained by the modulation of the Wnt/β-catenin signaling. Activation of Wnt/β-catenin signaling in human pluripotent stem cells derived from foregut progenitors led to the development of human fundic type gastric organoids (hFGO). Such hFGOs can be used for understanding the developmental biology of lineages derived from the human fundus, with various cell types present, namely functional parietal and chief cells [27]. Such hFGOs can also be used for studying human gastric physiology/pathophysiology and drug testing applications [27]. McCraken *et al.* (2014) had earlier reported the differentiation of human pluripotent stem cells into human antral gastric organoids (hAGOs). The protocol for obtaining hAGOs was the differentiation of hPSCs first into definitive endoderm followed by patterning into posterior foregut and differentiation into antral gastric organoids [28].

Broda *et al.* (2019) explained the 34-day step-wise protocol for the differentiation of PSCs into both the domains of the stomach namely fundic epithelium and antrum [29]. Again, in lines of recapitulating *in vivo* organogenesis, for generating gastric organoids, an endodermal lineage, the hPSCs were subjected to the first step of differentiation; namely, the induction of definitive endoderm till day 06. The first 03 days of DE differentiation involve 2D conditions with Activin A treatment confirmed by the expression of DE markers: Sox17 and FoxA2. Indeed, the first three days of differentiation were for 2D differentiations in the entire multi-step protocol. Subsequently, for the next three days, the 3D differentiation of free-floating posterior foregut spheroids was induced using a combination of growth factors and small molecules such as FGF4, GSK3 antagonist, and Wnt activator CHIR99021 (CHIR), BMP antagonist Noggin and Retinoid Acid (RA). The characterization markers for the posterior foregut spheroids are SOX2 and HNF1β. Indeed, the "posterior foregut organoids/spheroids (PFS)" can be the starting material of various other endoderm-derived organoids such as liver and pancreatic organoids, while the anterior foregut organoids can be excellent starting material for lungs. Moreover, the intestinal organoids can be derived from the midgut and hindgut organoids as per the speciation of the developmental origin. Various signaling pathways are modulated, thus, in the development of gastric organoids such as the inhibition of BMP signaling using Noggin for obtaining first the foregut spheroids. Following the foregut spheroid formation, RA mediated the induction of posterior foregut spheroids [30].

The continued 3D protocol by Broda *et al.* (2019) was ensured by embedding the PFS in matrigel till the end of differentiation. After day 06 of differentiation, the 3D PFS were simultaneously subjected to hAGOs and hFGOs [29]. In fact, the time-dependent expression of subsequent lineages derived from the posterior foregut endoderm depends on the signaling between the definitive endoderm (DE) and septum transversum mesenchyme (SM). Such signaling has recently been deciphered in detail using single-cell transcriptomics carried out using mouse embryos from various development stages [31]. Finally, in the gastric organoids protocol by Broda *et al.* [29], the matrigel-embedded PFS were treated for three days with EGF, Noggin, and Retinoic acid, followed by continued treatment with EGF for another 21 days to obtain hAGOs. The hAGO shows the development after the first three days in 3D culture (*i.e.*,-day 9 from the start of differentiation). However, the differentiation of hFGO was a bit more complex, requiring additional signaling such as Wnt/β-catenin using the small molecule CHIR. Essentially, for hFGO, from day 6 to day 9, the spheroids were treated with CHIR; from day 9 till day 20, the spheroids culture media was treated with CHIR, EGF; from day 20 to day 30, the media was enriched with an addition of FGF10. Finally, from day 30, till the end of differentiation (day 34), the hFGO media was found to contain CHIR, EGF, FGF10 and BMP4 and the MEK inhibitor PD0325901. EGF concentration is being a bit critical which was being tapered at day 30 for both hAGO and hFGO so as to facilitate the development of endocrine cells namely G cells that secrete the hormone gastrin.

The self-organization of gastric organoids, similar to other organoids, is a desirable feature. A great degree of physiological function has indeed been achieved for gastric organoids, namely histamine inducible acidification and production of the hormone gastrin [27]. However, there are drawbacks to this self-organization such as lacking a fully functional aspect. Here in the case of gastric organoids, separate organoids, namely hAGO and hFGO have been obtained. However, the co-existence of both the organoids together in the 3D system in a symbiotic system is highly desirable. Moreover, for the other specialized cell types, the absence of mesenchymal and vascular cells renders such organoids a little better than 2D culture systems but not equivalent to the stomach.

Liver Organoids

Tissue-specific stem cells are the guardians to replenish the diseased or damaged cells that happen due to normal wear and tear. In response to toxic damage, the actively proliferating cells in adult mouse liver were found to express the LGR5 (leucine-rich-repeat-containing G-protein-coupled receptor 5) receptor, which is a target for Wnt [32]. These LGR5+ cells have the potential to differentiate into hepatocytes and cholangiocytes *in vivo*, which potentiates them as stem cells. It was observed that these cells have the potential to expand and organize themselves into 3D cystic structures when cultured using RSPO1 (R-spondin1), a Wnt agonist, which was identical to organoids formed from healthy biliary ducts [32]. Thus, Huch *et al.* established human hepatic organoids from the healthy liver as well as single cells expressing EPCAM [33]. These liver organoid cultures can self-renew for up to a year and are genetically stable [32, 33]. Vyas *et al.* have shown that progenitor cells from the human fetal liver can self-assemble to form 3D liver organoids when cultured in decellularized liver scaffolds [34]. These organoids recapitulated the *in vivo* organogenesis of hepato-biliary structures. Their results prove that cell-ECM interactions are important for the proper differentiation of hepatocytes and cholangioblasts [34]. The importance of liver-derived ECM was again confirmed by the work of Saheli *et al.*, where they transferred it into a hydrogel to improve the liver organoid function [35]. Hu *et al.* described functional organoids from mouse and human hepatocytes that showed similar regenerative responses compared to those in the adult liver following partial hepatectomy [36]. Patient-specific organoids can be created from iPSCs that can be used for personalized drug toxicity and efficacy screening [37]. This approach has the potential to develop a large number of organoids as compared to culturing cells from liver biopsies. To improve organoids' functionality, microfluidic platforms have been described where organoids are grown on a chip with perfusion chambers that facilitate the continuous flow of nutrients and oxygen [38]. In current times, various companies have commercialized excellent clinical grade microfluidics platform. These organoids can be used as a disease model for various liver diseases and they can also be used for transplantations.

Pancreas Organoids

Shifting gears, we now reflect upon the same stem cell marker LGR5 (discussed in the previous section as a liver-stem cell marker), expressing stem cells in the pancreas. Considering the upregulation of Wnt signaling and LGR5 expression in the regenerative response of an adult pancreas, Huch *et al.* defined a culture medium containing the Wnt activator, RSPO1, that favored an unlimited expansion of pancreatic duct fragments [32]. They went further to establish 3D organoid models of human and mouse ductal adenocarcinoma for the first time [39]. As the protocols for differentiating PSCs into pancreatic lineages were getting established, Huang *et al.* devised a protocol to create pancreatic organoids from PSCs [40]. Initially, these organoids were developed as tumor models for drug screening but later the focus shifted to creating functional β islets. By mimicking the sequential signaling events responsible for lineage commitment and the formation of tissues during the organogenesis of the pancreas, as well as providing a biomimetic ECM scaffold, Wang *et al.* generated islet organoids from hESCs [41]. Thus, with the utilization of tissue engineering methodologies, the morphogenesis of organoids improved. Various scaffolds with different pore sizes and ECM components have been developed, which can provide a microenvironment suitable for the spatiotemporal regulation of growth factors [42].

Tao *et al.* used a multi-layered microfluidic device to generate islet organoids that resembled human pancreatic islets in their morphology and cellular complexity [43]. This organ-on-chip platform due to its continuous perfusion system enabled the long-term culture of 3D islet organoids with improved viability and functionality [43]. Since *in vivo* organogenesis is a complex process that involves interaction between multiple tissues that are interconnected at the boundaries, a human hepato-biliary-pancreatic organoid model was developed by Koike *et al.* to study endoderm organogenesis [44]. Recently, more advanced micro-physiological systems, Acry-Chip and Oxy-Chip, were developed that allowed imaging and functional assessment of perfused 3D organoids with great ease [45]. It is evident that a dynamic microenvironment is important for the maintenance of an organoid function. Due to the complexity of the organoids, proper diffusion of nutrients, growth factors, and oxygen was a challenge in the earlier organoid models. With the development of dynamic perfusion systems, it was possible to culture functional organoids for a long term. For the transplantation of pancreatic organoids, a 3D printed tissue trapper was created by Soltanian *et al.* that was implanted in the peritoneal cavity of mice which developed vasculature and remained for 90 days [46]. By combinatorial efforts using various 3D scaffolds and microfluidic platforms, it is possible to generate 3D pancreatic organoids that mimic the *in vivo* organogenesis of pancreas, which can be used for various applications such as drug screening and islet transplantations.

Intestinal Organoids

The continuous coiled tubular structures are defined as, one 20 feet long and the other about a foot long, with small and large intestines being the concluding sections of the digestive tract. Essentially, the embryonic midgut gives rise to the small intestine, and the hindgut gives rise to the large intestine from the developmental biology perspective. Although the small intestine is made of duodenum, jejunum, and ileum, the composition of intestines comprises various cell types. For example, the epithelial mucous secreting goblet cells perform their role by producing mucous that prevents friction caused due to food passing through the intestine [47]. Moreover, the intestines are composed of epithelial cells involved in food absorption, namely goblet cells. Another cell type namely, paneth cells, is involved in the role of maintaining the relationship between host immunity and the gut microbiome. Hence, the purpose of developing intestinal organoids must involve the presence of goblet and paneth cells under 3D structured conditions.

However, the inclusion of the lymphovascular system in the intestinal organoid has significant big implications for its functionality. Sugimoto *et al.* (2021) reported the transplantation of functional organoids in the small intestine (SIC - small intestinalized colon) in which native colonic epithelia has been replaced by ileum-derived organoids. In this work, the authors first transplanted the human ileum-derived organoids into the mouse colon to obtain villus-like structures. Another xenotransplantation experiment was performed in which the SICs were positioned at an ileocaecal junction in rats for constant exposure to the intestinal juice. Moreover, the xenotransplantation at the ileocaecal junction provided the added advantage of developing an intact vasculature, innervation, structured villi, and lacteal (a fat-absorbing lymphatic property of the small intestine). However, if the colonic organoids had been used instead of ileum organoids, there happened to be an intestinal failure and mortality in the rodent model system, thereby indicating the applicability of ileum organoids.

Derivation of intestinal organoids from hPSCs is another common approach recently reported by Onozato *et al.* [48]. Moreover, similar to the challenges faced in differentiation into other endodermal lineage organoids, this protocol also resulted in the generation of immature intestinal organoids with structural deficiencies namely lack of crypt-villus-like structures. Moreover, budding-like organoids bearing epithelial-like tissue and other functional features of mature intestines were formed. Such functional features involved the high expression of drug transporters which could induce the expression of cytochrome P450 3A4 and P-glycoprotein. Moreover, the physical properties of the organoids allowed a trans-epithelial electric resistance of the order of 400 Ω in the polarised intestinal folds. Moreover, such organoids exhibited fibrosis upon treatment with tumor necrosis factor-α and/or transforming growth factor-β, thereby proving to be an appropriate model for intestinal fibrosis, mucosal damage, and drug screening [48]. Another challenge is that, though directed differentiation of hPSCs yields gut-specific cell types, these structures lack crucial functional links across germ layers that are involved in gut development. Uchida *et al.* devised a straightforward approach for creating xenogeneic-free functioning intestine organoids from hPSCs. These organoids showed the presence of classical morphological features and differentiation markers (Fig. **2**). Moreover, in reaction to histamine and anticholinergic drugs, these organoids demonstrated intestinal activity, including peptide absorption and innervated bowel movements. These gut organoids might be cultured and engrafted *in vivo*. Thereby, these gut organoids from hPSCs, may serve to model intestinal diseases and screen therapeutics [49].

MESODERM-DERIVED ORGANOIDS

The middle germ layer of the human body is the mesoderm. Kidney, bone, muscle, heart, and reproductive organs are derived from embryonic mesoderm while precursor tissues of the kidney, epithelial structure and mesenchymal tissues are derived from the intermediate mesoderm [9]. Mature kidney structure is generated by the inducive interaction of these two structures. Organoids are derived using PSCs, which have intensively been used for recreating the human morphogenesis process *in vitro* [50]. Kidney organoids are used as an alternative to or as a part of other emerging *in vitro* model systems, such as the PDX models or kidney-on-a-chip [51]. This section will focus on recent advances in organoid differentiation, animal models, organ-on-a-chip, and genome editing highlighting cutting-edge technologies for medical applications for kidney and bone organoids. For renal disease in particular, we will discuss the major advances that aim to provide additional insights into pathophysiological mechanisms, in addition to technical barriers and key perspectives in the field [52].

Renal Organoids

Kidney disease is one of the leading problems in adults. Almost 15% of adults in the USA are affected with chronic kidney disease (CKD) that often leads to interstitial fibrosis, anemia, hyperphosphatemia and cardiovascular disease, and hypertension. Kidney disease is marked by a slow and irreversible loss of nephrons which are a functional unit of the kidney [51]. Although the kidneys have self-regeneration capacity after injury, a process known as nephrogenesis, the loss of nephrons in CKD is rather permanent. Currently, there are no available efficient therapies to treat and reverse kidney dysfunction. Patients with advanced kidney diseases are left with options of hemodialysis or kidney transplantation only. Dialysis reduces the quality of life, significantly increases mortality and morbidity, and has side effects such as hypotension, high infection rate, and loss of erythropoietin [53]. Likewise, there are problems with kidney transplantation such as shortage of organ donors and risk of infection or tumor formation post-transplantation. Hence, renal regeneration using organoids has become a topic of emerging interest. Renal organoids made from whole kidney cells exhibit similar characteristics as the kidney. For this reason, these organoids are gaining interest as a high-throughput, personalized tool for lead identification and nephro-toxicology assessment as part of the drug development process. Human renal cell therapy is still in the developmental stage, but it holds great potential as regenerative medicine continues to advance.

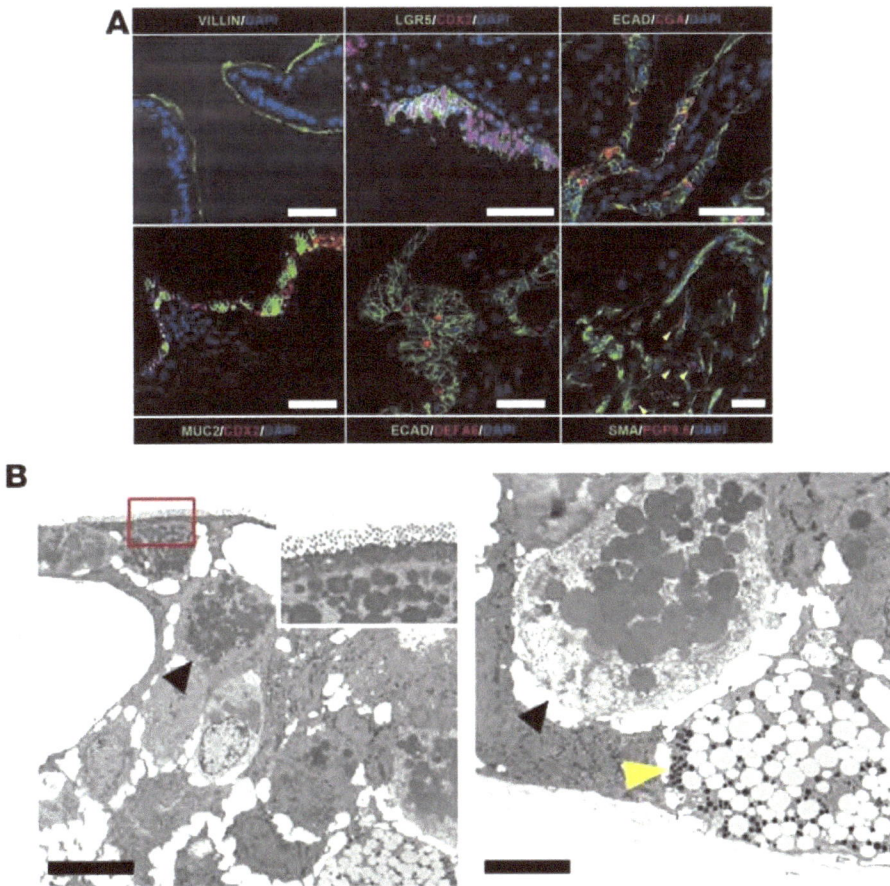

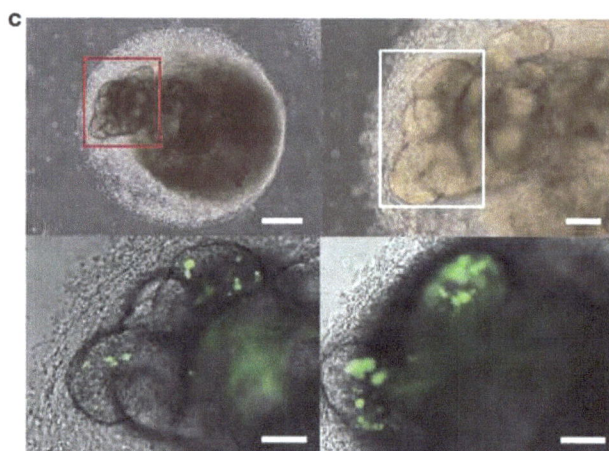

Fig. (2). Characterization of intestinal organoids and identification of LGR5-EGFP+ cells during intestinal organogenesis. (**A**) Day 50–60 organoids derived from human embryonic stem cells (hESCs) (SEES1 cells) immunostained with intestinal differentiation markers: villin, leucine-rich repeat-containing G protein-coupled receptor 5 (LGR5), CDX2, E-cadherin (ECAD), chromogranin A (CGA), mucin-2 (MUC2), defensin -6, Paneth cell–specific (DEFA6) (PGP9.5). DAPI was used to counterstain cell nuclei. Yellow arrows show neuronal cells inside the myenteric region that are positive for PGP9.5 and -SMA. 50 m (VILLIN, LGR5, SMA); 100 m (scale bar) (ECAD and MUC2). (**B**) An enterocyte with a brush-like border (left), Paneth cells with secretory granules (black arrowhead), and goblet cells with mucin granules (yellow arrowhead) (right). Scale bar: 10 µm (left); 5 µm (right). (**C**) SEES1-derived organoids produced EGFP (green) under the LGR5 promoter, showing that they were LGR5-positive+ gut organoids. Architecture resembling a gastrointestinal tube in a day 34 organoid (red square; magnified picture, white square) (top row). By fluorescence microscopy, a modest number of EGFP-positive+ cells were discovered on day 34 organoid (white square) (bottom left). At day 41, the number of EGFP+ cells increased (bottom right). Scale bar: 300 m (upper left) and 100 m (bottom right) (top right and bottom row). Figure from reference [49].

The human kidney has an extremely complex structure and function with more than 25 cell types involved. Typically, renal organoids include all cell types to recapitulate the *in vivo* micro-environment and therefore enable more effective drug screening, improved understanding of disease etiology, and personalized therapies [54]. In addition, iPSC-derived organoids could be used for renal cell therapy where they may improve the inherent ability of renal cells to regenerate and repair. In this regard, recent efforts have focused on the generation of kidney tissues for therapies and understanding kidney development. Renal tissues can be generated from PSCs that are differentiated into complex organoid structures composed of multiple cell types, including nephron epithelial cells. The nephrons are a multicellular structure including glomeruli, which filter the blood plasma into multicomponent tubular water to produce urine [54]. Nephron progenitor cells are derived from hPSCs that can give rise to three germ layers of embryo making them a popular choice of cell source for organoid development [55]. The first attempt to create kidney organoids was made in 2014 by Taguchi *et al.,* who generated metanephric mesenchyme (MM), which was induced to generate nephrons [56]. Similarly, Takasato *et al.* developed an independent protocol to efficiently differentiate functional kidney organoids [57]. However, in order to make a functional renal organoid, multiple cell types, namely epithelial (for renal tubules and corpuscles), endothelial cells, and neurons, must be incorporated. All such cell types can be directed either in a self-assembly process or else in a sequentially *in vitro* directed process for vascularization and neurogenesis [58].

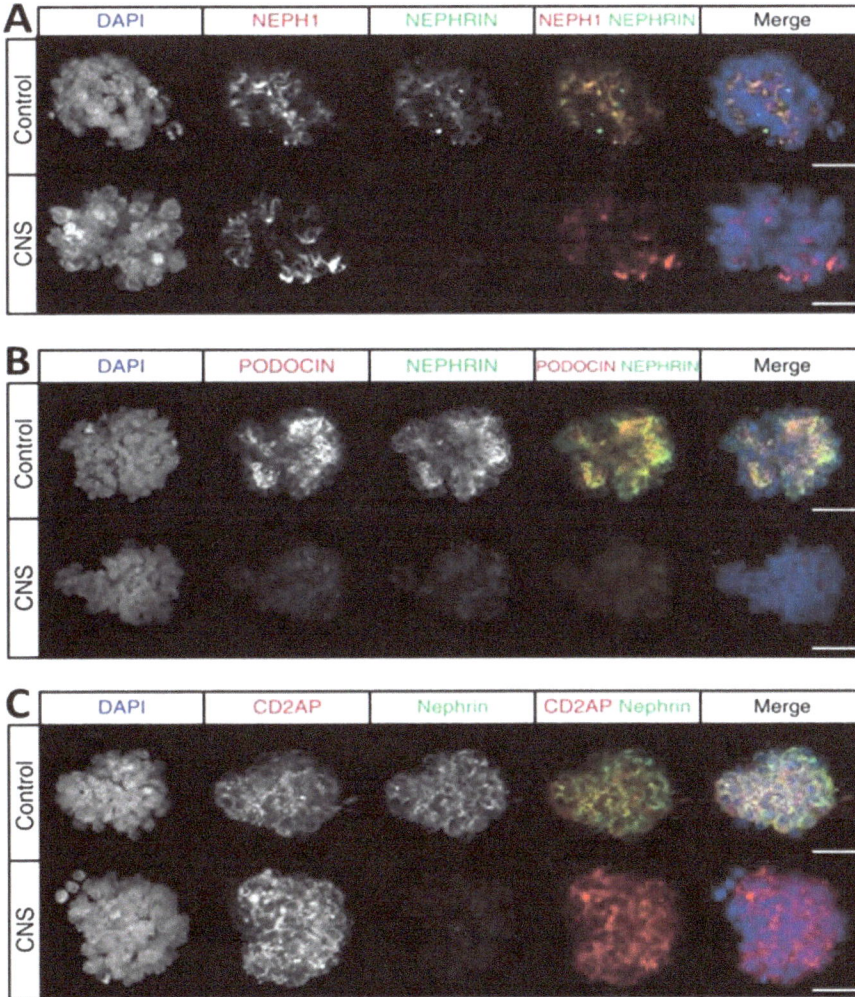

Fig. (3). Organoid glomeruli model of congenital nephrotic disease *in vitro*. Organoid glomeruli from patient lines with congenital nephrotic syndrome will be characterized. B-D. Immunostaining of OrgGloms extracted from control organoids and congenital nephrotic syndrome patient organoids reveals decreased NEPHRIN and PODOCIN protein levels in organoids produced from patient-iPSC, as shown by >3 biological replicates. 10 µm scale bars. This Figure is a part of Fig. (**6**), and reproduced from reference [54].

Induced-PSCs-derived kidney cells from patients have been increasingly used to generate organoids for kidney disease modeling, one of the critical technologies that enable personalized therapy [59]. In addition to primary cells and established kidney cell lines, these iPSCs recently have been integrated into a microfluidic culture system called the "organ-on-a-chip" platform to simulate the microenvironment of the kidney, such as the glomerulus [54]. Hale *et al.* used iPSC-derived kidney organoids and reported enhanced podocyte-specific gene expression, polarized protein localization, and glomerular basement membrane matrisome in comparison to 2D cultures (Fig. **3**). Organoid-derived glomeruli maintain marker expression over 96 hours, allowing for toxicity evaluation studies [54]. A congenital nephrotic syndrome patient with NPHS1 mutations had reduced NEPHRIN and PODOCIN protein levels in 3D organoid glomeruli. Organoid glomeruli generated from patient iPSCs enable *in vitro* modeling of human podocytopathies and screening for podocyte toxicity.

One of the significant challenges of traditional *in vitro* models lies in recapitulating renal diseases with genetic biases such as autosomal dominant polycystic kidney diseases associated with PKD1 and PKD2 genetic mutations, which lead to end-stage renal diseases for 5% and interstitial fibrosis resulting in chronic kidney diseases for 50% of patients with such a condition [60]. As a solution to this problem, 3D bioprinting is used to develop 3D constructs for kidney-on-a-chip technology using patient-derived kidney organoids [61]. A recent study used 3D bioprinting technology to create proximal renal tubules embedded in an extracellular matrix (ECM) on a microfluidic chip with perfusion to apply physiologically relevant shear stress by blood flow. This organoid-based organ-on-a-chip system with tubules helps in the formation of kidney tissue-like epithelium with more *in vivo*-like phenotype and thus facilitates in reproducing organ-scale drug delivery or screening [62]. Moreover, in another study, the authors have shown the advancement of kidney-on-a-chip by integrating the renal epithelium layer on top of the interstitial layer with a basement membrane in between [54]. However, despite all the improvements, kidney organoids still lack the structural complexity of the organ and its functionality, such as an excretion system in the bioprinted constructs. Another application of kidney organoids is to study the toxicological aspects of the drugs to be used. The major importance is that the drugs have been shown to affect specific cell types among all, when used after nephrotoxic injuries. For example, cisplatin has shown toxicity for tubular cells in organoids [59]. Incorporating cellular interactions that occur during organogenesis and also including multiple germ layers to generate innervated and vascularized organoids would be next generation tissue engineering and crucial regenerative medicine.

Bone Organoids

Bone and cartilage are derived from embryonic mesoderm. Osteogenesis or bone formation is a complex process involving cellular differentiation and the generation of a mineralized organic matrix. The bone provides stability and protection for the inner organs consisting of many cell types, including osteocytes and ECM of 30% organic and 60% inorganic crystalline phases. The inorganic part includes calcium, phosphorous, and hydroxyapatite (HA), which provides mechanical stability and rigidity to the bones, whereas the organic component is responsible for elasticity [63]. Under the mechanical influence, the bones are subjected to changes in local cellular behaviors [64]. One of the major applications of bone tissue engineering is the repair of bone defects after surgery and inflammation or excessive loss of bone following trauma or tumor resection. Moreover, organoid-based *in vitro* models allow superior control of experimental conditions such as the use of biomaterials or material processing [65]. Bone graft transplants, implants of different biomaterials, or bone transport methods are the most common treatment approaches although they remain challenging today. The first bone implantation was performed in 1998 using a porous ceramic structure [66]. Bone regeneration using autologous bone grafts is the standard method, but it requires surgery. Alternatively, stem-cell-based bone tissue engineering is used. Such stem-cell derived cells have osteo-inductive properties that are important for bone repair. Although many advanced approaches are suggested to construct large tissue constructs, creating bone organoids with a multicomponent tissue architecture still remains a complex process.

Moreover, bone is a heterogeneous, multimodular tissue with complex geometric features. Hence, proper bone organoids are yet to be created. Kale *et al.* outlined a fascinating approachable method to shape spheroids of bone precursor cells [67]. Bone cell spheroids contain a heterogeneous population of various bone progenitor cells, including osteoblasts capable of creating microscopic crystalline bone named "microspicules". Often, this calls for the 3D aggregation of cells as well as the elimination of serum and the TGF-β1 requirement. Since these structures lack different cell types, therefore, further research is needed to establish these spheroids into bona fide bone organoids [68]. In order to study osteogenesis in physiological and pathological conditions, as well as under the influence of environmental factors, the recent study by Akiva *et al.* reported a complete 3D living *in vitro* bone model [69]. The researcher used human bone marrow stromal cells that were differentiated *in vitro* into a functional 3D self-organizing co-culture of osteoblasts and osteocytes, resulting in an organoid representing early-stage bone (woven bone) development. The woven bone organoids showed osteocytes embedded within the collagen matrix, which is produced by osteoblasts and mineralized under biological control. The embedded osteocytes in the organoid, just like *in vivo* osteocytes, showed network formation and communication by sclerostin secretion.

In bone tissue engineering, scaffold-based approaches are commonly used to mimic the structure and function of the ECM by changing HA or tricalcium phosphate and its composition to create different stiffness or porosity of the ECM in order to guide bone tissue formation. Hydrogel-based methods are also one of the greatly appreciated approaches due to their potential application and ability to provide flexibility and permeability to the constructs for water, nutrients, and/or drugs [70]. To fabricate bone tissue hydrogel-based scaffolds, naturally derived hydrogels such as fibrin, gelatin, hyaluronic acid, fibroin, alginate, and collagen are used. Another option is synthetic polymer-based scaffolds, which are typically made of PEG or peptides using various techniques such as electrospinning, self-assembly, or salt-particle leaching [71]. These biomaterials are also used in a number of organ-on-a-chip platforms. Brady *et al.* showed that the mechanical stimulation of osteocytes (MLO-Y4) cells with fluid shear stress using a rocking platform, which oscillated at a frequency of 0.5 Hz and with a tilt angle of 7°, secreted paracrine factors, enhanced MSC migration, proliferation and osteogenesis, and furthermore significantly increased osteoblast (MC3T3) migration and proliferation [72]. Park *et al.* proposed a mechanically stable, semitransparent, trabecular bone organoid model with controlled thickness using an un-mineralized bone ECM and demineralized critical bone, named demineralized bone paper (DBP). They were able to recreate bone remodeling cycle by co-culturing primary murine osteoblasts and BMMS with chemical stimulation. Such a co-culture demonstrated the spatiotemporal profile of regulatory molecules as concentrated paracrine signaling that is important for trabecular bone cavities [65].

Natural fracture healing begins at the cartilage intermediary called a "soft callus", which then transforms into the bone in a mechanism that mimics developmental events. Cells originating from the periosteum, which include potent skeletal stem cells, are the key contributors to the soft callus. Gabriella *et al.*, 2020 used human periosteum-derived cells to produce micro spheroids that can be differentiated into callus organoids [73]. In addition, *in vitro*, callus organoids spontaneously bioassemble into massive, engineered tissues capable of healing critical-sized long bone defects in mice. Another study showed the advancement of organoids that can differentiate into callus organoids using human-periosteum-derived cells (hPDCs) and exhibit similar morphological properties as the native tibia, which could be used as bio-ink for tissue manufacturing. Moreover, they showed bio-fabrication of cartilage intermediate tissues using microspheroids that are scalable and have the capacity to form ectopic bone micro-organs and to heal critical-sized long bone defects *in vivo*. They also correlated it with gene patterns that are found in developing and healing bones [39]. Such regenerated bone exhibited morphological properties that are close to those of native tibia. These callus organoids are reported as a living "bio-ink". Latest developments in 3D bone organoid culture have taken advantage of the thin demineralized bone slices by utilizing demineralized cortical bones called DBP. This DBP is used to develop trabecular bone organoids that mimic the critical and essential complexity of trabecular bone cavities. To replicate the bone remodeling method, researchers cocultured primary murine osteoblasts and bone marrow mononuclear cells (BMMs) and used chemical stimulants on DBP discs [65].

ECTODERM-DERIVED ORGANOIDS

After an exhaustive coverage of the innermost and middle germ-layer-derived organoids, the current section of this review encompasses the outermost layer-ectoderm-derived organoids. Neuro-epithelium/brain, ocular tissue, and mammary gland arise from the ectoderm. With regard to the applicability of ectoderm-derived organoids, mammary gland-derived organoids, and brain organoids might find their applications in studying developmental biology, drug testing, and transplantation. Somatic and stem cell organoids will indeed have their applications in studying developmental biology and transplantations. However, cancer organoids will be more suitable for drug testing applications.

Mammary Gland-derived Organoid

The mammary glands (MG) are epithelial tubes whose primary function is to produce and secrete milk. MGs are highly dynamic and undergo dramatic changes in architecture during the formation of a highly elaborate ductal tree that fills the entire fat pad, with lateral branching during the adult stage. Mammary organoid (MO) models were developed to study the morphogenesis of MG or disease development such as cancer [74]. Qu *et al.* proposed a method to derive mammary cells by differentiating human iPCS [75]. Another study showed branching morphogenesis using primary mammary epithelial organoids. They showed that the branching mechanism is stimulated by stromal fibroblasts, epidermal and fibroblast growth factors that procude matrix metalloproteinases, and recombinant stromelysin 1/MMP3 [76]. Koledova *et al.* showed that sprouty 1 (SPRY1) modulates the microenvironment to enable the proper mammary branching morphogenesis. A loss of SPRY1 results in enhanced branching in the mammary. Branching morphogenesis occurs due to negative regulation of epidermal growth factor receptor (EGFR) signaling in the mammary stroma [77]. A recent study showed that the mammary glands contain a Wnt-responsive cell population that is enriched for stem cells [78]. Wnt signaling has also been implicated in different stages of mammary development as well as in mammary oncogenesis [79, 80].

These engineered MOs are intensively used to study the mechanism of breast cancer development or treatment and breast cell transformation. Rosenbluth *et al.* proposed an organoid culture protocol that preserves the complex stem/progenitor and differentiated cell types *via* long-term propagation of normal human mammary tissues [81]. Breast cancer organoids capture the diversity, pathological, and clinical heterogeneity of the disease. These breast cancer organoids represent the hormone receptor status or HER2 expression level of the parent tumor. In addition, these organoids are capable of comprehensively capturing the genomic heterogeneity and therefore may improve *in vitro* drug screening or the development of personalized medicine. Recently, the role of matrix stiffness, organization, and composition in breast cancer cell invasion was demonstrated *via* a 3D scaffold with tunable mechanical properties such as stiffness. The scaffold was made of methacrylated gelatin/collagen I to recreate physiologically relevant stiffness while maintaining the matrix density constant. They generated soft (2 kPa) and stiff (12 kPa) breast tumor spheroids using MDA-MB-231 cells and the comparisons revealed that cell invasion is delayed in the stiff microenvironment [82].

Brain Organoids

The human brain's complex architecture and function enable us to perform a wide range of higher cognitive functions. Severe neurological and psychiatric disorders may result in the event of abnormalities in the brain's structure or function. Current understanding of research, has shown evidence that neurological and psychiatric disorders have their roots in neurodevelopment [83, 84]. Despite this knowledge, it is quite challenging to understand the neurodevelopmental causes and mechanisms of brain disorders, and a major hindrance to these studies is the complexity of human brain regions and their physical inaccessibility without causing secondary damage. Thus, to study phenotypes associated with human brain development, function, and disease, alternative experimental systems are required that are accessible, ethically justified, and replicate the human context. Human/induced pluripotent stem cell (hPSC / iPSC)-derived brain organoids help recapitulate features of early human neurodevelopment *in vitro*, including the generation, proliferation, and differentiation of central nervous system (CNS) progenitor cells into major CNS cell types. The functioning of such an organoid requires the self-assembly and interaction of adult neurons and glia and their complex interactions in the developing brain in 3D. Recently, a number of brain organoid protocols from multiple labs have been developed. They have been able to recapitulate aspects of embryonic and fetal brain development in a reproducible and predictable manner. These different organoid technologies provide distinct bioassays to unravel novel, disease-associated phenotypes and mechanisms. In this part, we summarize how the diverse brain organoid methods can be utilized to enhance our understanding of brain disorders.

Brain organoids are a powerful tool to model CNS disorders since they are capable of reproducing specific murine or human brain structures, and cellular lineages and have been used to simulate different human brain regions *in vitro*, including the hippocampus [85], midbrain [86, 87], pituitary gland [88], hypothalamus [89], and cerebellum [90]. This makes brain organoids an excellent model for investigating brain development and the mechanisms of related diseases. A range of newly established protocols mention ways to successfully cultivate brain and spinal cord organoids *in vitro* and are capable of keeping them in culture for 6 months [91]. Slice cultures of these brain organoids could be maintained over extremely long periods, up to and beyond one year, after which they were still successfully used for endpoint analyses like patch-clamp studies. Under their reported growth conditions, neurons within the organoid showed enhanced survival (~66% decrease in the presence of TUNEL-positive cells), a striking degree of morphological maturity, which translated to functional maturity with the ability to form neural networks and even able to drive muscle contractions [92].

History of Brain Organoid Research

Organoid technology started from a basic understanding of re-aggregation of sponge cells, that highlighted the role of self-organization in organ formation [93 - 96]. Embryonic and induced pluripotent stem cells, both have come under the same umbrella of PSCs, which have been established subsequently [97 - 99]. Using these stem cell lines, a number of different types of organoids were developed, ranging from the intestine to the hippocampus [85, 100 - 103]. Organoid models rapidly gained popularity as tools to study a number of human aging, cellular connectivity-related, and developmental disorders [104 - 106]. Early organoid culture protocols largely depend on intrinsic signaling and self-assembly of stem cells. However, with region-specific differentiation factors and more elegant and elaborate culture systems, the growth of other regions of the brain can be induced *in vitro* as region-specific organoids. These brain region-specific organoids have been shown to recapitulate the molecular, cellular, and structural features of specific areas of the human brain. Fig. (**4**) depicts a schematic of the timeline for the development of brain organoid technology.

Modelling Neural Networks In Vitro

Intricate as they are, with the possibilities of drug research, even brain and spinal cord organoids cannot be studied in isolation owing to the network they form and control with the rest of the body. The Paşca lab and researchers elsewhere have linked organoids to assembloids. One such assembloid involves an organoid enriched for excitatory and another enriched for inhibitory neurons [107]. The model captures aspects of axonal pathfinding and also migration, such as how some populations of neurons migrate from the ventral to the dorsal forebrain.

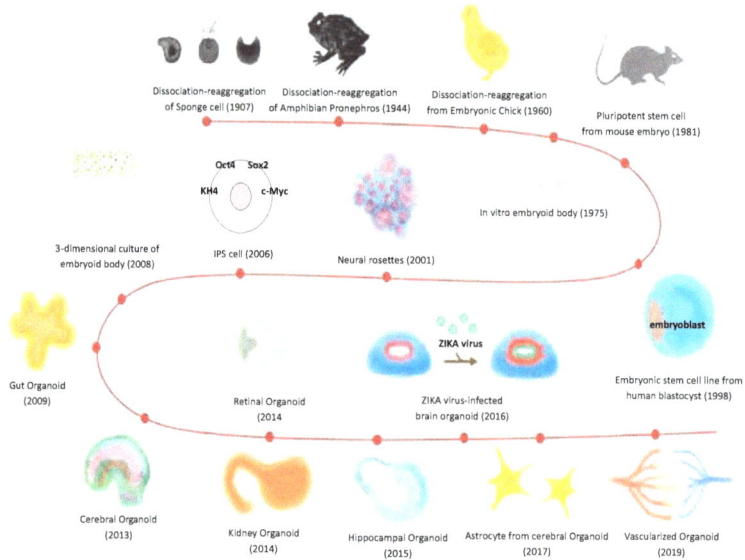

Fig. (4). Schematic showing the historical time points of brain organoid technology. Organoid technology originated from studies on the re-aggregation of sponge cells, indicating that self-organization plays a crucial part in organ creation. Establishment of PSCs, including embryonic stem cells and induced pluripotent cells. With these stem cell lines, numerous organoids, ranging from the gut to the hippocampus, are created. Human brain problems, such as ZIKV infection, have been extensively studied using brain organoid models. Existing limitations of organoids, such as oxygen depletion during long-term culture, are being addressed through bioengineering technology. The figure is reproduced from Reference [113].

Applications of CNS Organoid Models

Owing to the fact that brain organoids go through developmental steps very similar to those of a human fetal brain, they are suitable to model neurodevelopmental disorders with actual pathogeny. In the more current state of affairs, bioengineering technologies are being developed to address the present limitations of organoids such as oxygen deprivation during long-term culture [108]. Amongst all the brain disorders, traumatic brain injury (TBI) remains a prominent public health concern, with about 2.87 million TBI-related emergency department (ED) visits, hospitalizations, and deaths in 2014 in the United States alone. TBI events also include over 837,000 children [109]. This number is staggering on a global scale, with over 60 million new TBI cases occurring worldwide each year [110]. Apart from the physical trauma that could be mild (concussions) to severe, there is a range of neurophysiological and psychological defects manifested by TBI, ranging from coma, epileptic seizures, paralysis, cognitive and memory defects, to difficulty in communication, as well as depression. Repeated or severe traumatic brain injuries might increase the risk of degenerative brain diseases like Alzheimer's or Parkinson's disease. Depending on the region affected, spinal cord injury with mild to severe trauma may lead to conditions like paralysis, loss of independent breathing, and motor deficits.

From initial studies focused on easier manipulations to current strategies involving targeted deletions and including rare cells, brain organoid research has made immense progress in the last decade [111]. Using high-intensity focused ultrasound on inducing mechanical injury in 3D human PSC-derived cortical organoids *in vitro*, Lai and colleagues reported that the injured organoids recapitulated key hallmarks of traumatic brain injury, phosphorylation of Tau and TDP-43, neurodegeneration, and transcriptional programs indicative of energy deficits [112, 113].

Prenatal and adult oxygen availability is essential for brain function, and models of hypoxia have been developed using brain organoids. During hypoxia, the CNS adopts complex mechanisms involving activating the Hif1a pathway that allows it to survive. Low oxygen tolerance is thought to stimulate brain plasticity *via* a combination of energy conservation and improved homeostatic control of subsequent hypoxic damage. In a model of transient hypoxic injury, reoxygenation restored neuronal cell proliferation but not neuronal maturation [114].

In premature birth [before the post-conception week (PCW)-28], infants are at high risk of developing hypoxic episodes because of lung immaturity, hypotension, and lack of cerebral-flow regulation. Thus, creating human-specific models is key to understanding and preventing PCW-associated risks. In an effort to model this, Paşca *et al.* reported a decreased number of Tbr2-positive cells in human cortical spheroids exposed to low oxygen, identified specific defects in intermediate progenitors, a cortical cell type associated with the expansion of the human cerebral cortex, and demonstrated that these are related to the unfolded protein response and changes [115]. Others have shown that transient hypoxia resulted in immediate and prolonged apoptosis in cerebral organoids, with outer radial glia, a progenitor population more prominent in primates, and differentiating neuroblasts/immature neurons suffering from larger losses. At the same time, neural stem cells in the ventricular zone displayed relative resilience to HI and exhibited a shift of cleavage plane angle favoring symmetric division, thereby providing a mechanism to replenish the stem cell pool [116]. Human iPSC-derived spinal organoids have been instrumental in helping to understand molecular mechanisms of cell-type diversification [85, 117] in health and underlying disease mechanisms like spinal motor atrophy [117], leading to drug discovery and therapeutic interventions. More complex works have shown a functional integration of lab-grown cerebral organoids transplanted back into rat brains after an injury that showed improved neurological motor function [118]. The brain organoid injury model has great potential in unravelling the intricate molecular mechanisms of TBI and drug testing for therapeutics. Thus, the use of organoids in human disease modeling has received a significant boost and may aid in disease management and the development of precision therapy.

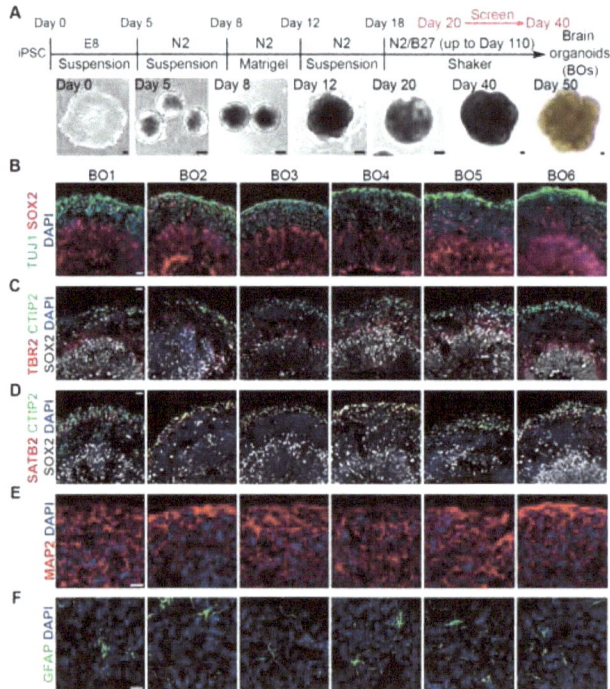

Fig. (5).   Generation and characterization of brain organoids generated from hiPSCs. A) Schematic representation of the brain organoid technique and typical phase pictures at various developmental stages. Scale bar, 100 μm. B) Immunostaining of neural progenitor cell marker SOX2 and neuronal marker TUJ1 in day 50–60 brain organoids. C) Immunostaining to probe SOX2, TBR2, and CTIP2 on day 50–60 brain organoids. D) Images of SOX2, CTIP2, and the cortical top layer marker SATB2 immunostaining in day 50–60 brain organoids. E) Immunostaining of the neuronal cell marker MAP2 in 90-day-old brain organoids. F) Immunostaining of the GFAP astrocyte marker in day 110 brain organoids. Abbreviation: BO: brain organoid. B–F) Scale bar, 20 μm. The figure is reproduced from reference [119].

Alzheimer's disease (AD) is progressive, incurable, and lacking effective drugs. Understanding the complex molecular pathophysiology has been challenged due to the lack of appropriate human-specific models, leading to the high failure rate of therapeutic discovery. Chen *et al.* used hiPSCs-derived 3D brain organoids to replicate sporadic AD and simulate blood-brain barrier (BBB) leakage. Fig. (**5**) shows the schematic of the developmental timeline of the hiPSC-derived brain organoids and the expression of classical markers. They showed that the brain organoids exposed to serum may replicate AD-like pathology, such as amyloid beta clusters, increased phosphorylated tau protein, synapse loss, and a disorganized neural network (Fig. **5**) [119]. Single-cell transcriptomics of the brain organoids reveals that serum inhibits synaptic activity in neurons and astrocytes and causes an immunological response in astrocytes. Recently, Smits *et al.* in an effort to model Parkinson's disease (PD), generated midbrain floor plate neural progenitor cells (mfNPCs)-derived organoids from normal and PD patients carrying the LRRK2-G2019S mutation. They reported that the mfNPC-derived organoids resembled the PD pathophysiology [120]. Thus, organoid-based models can replicate CNS-disease phenotype and may be used for future therapeutic research.

DISEASE MODELING USING ORGANOIDS

The concept and subsequent execution of organoid models emerged due to the quest to understand organogenesis *in vitro*. Moreover, to understand the biology, self-assembling organoids from cells of ectodermal, mesodermal, and endodermal lineage have played a great role. Hence, such organoids have been successful in modeling various non-communicable metabolic disorders for understanding basic biology. Finally, cancer research is incomplete without understanding the dynamics of self-assembling metastatic cancer stem cells under 3D conditions [121]. Moreover, the co-culturing of cancer stem cells (CSCs), along with the mesenchymal stem cells (MSCs) that are present in the CSC milieu to form co-culture spheroids, has been highly successful in predicting cancer metastasis [122]. For example, multicellular 3D spheroids have been generated from human adipose-derived mesenchymal stem cells and triple-negative breast cancer (MDA-MB-231) to mimic the tumor microenvironment *in vitro* [122]. Secondly, with the advent of the current pandemic, a greater emphasis has been given to COVID-19 research. 3D lung cells/stem cell-derived spheroids have great potential to address such questions and design therapeutics [123]. Hence, this section focuses on modeling diseases such as cancer, COVID-19, and some genetic disorders/diseases using organoids.

Organoid and Cancer Disease Modeling

It became cognizable after successfully growing organoids from human ASCs that organoids can efficiently be grown from patient-derived tumors. Conventionally, to study human cancer, cancer cell lines, syngeneic or xenograft mice models have been used as major experimental platforms. As ASCs-derived organoids came to the fore, transformed tumor cells were also cultured *in vitro* as 'tumoroids' or 'canceroids' or tumorospheres. They showed substantial promise in modeling cancer and further translational research [124, 125]. Up till now, few research groups across the globe have successfully designed and developed organoids from primary biopsy samples, like colon cancer [126, 127], oesophageal cancer [128], pancreatic cancer [39, 40], gastric cancer [129], liver cancer [130], endometrium cancer [131] and breast cancer [132] tissues, also from metastatic colon cancer [133], prostate cancer [134] and breast cancer [135]. Engel *et al.* generated patient-derived colorectal cancer organoids, demonstrating tumour-to-tumour variation in the organoids, and utilized them to evaluate chemotherapeutic (5-FU) treatment susceptibility and drug resistance (Fig. **6**) [136]. Much to the intrigue of the researchers, the above-mentioned reports demonstrated that patient tumor-derived organoids match in phenotype and genotype with the tumor epithelium they had taken their origin from [137]. Importantly, it has been observed that tumoroids [128] fail to grow as fast as their normal counterparts ostensibly due to an enhanced apoptosis as well as inhibition of mitosis [138, 139]. Moreover, under 3D stationary culture platforms, possibly it becomes difficult to replenish the cells located inside the core of the tumoid/tumorosphere. Another bottleneck is the tumor-adjacent normal tissue in tumor biopsies that can generate healthy organoids faster, and need to be removed for tumoroid growth. Scientists surmised that either tumoroids would be cultured by using pure tumor materials or to grow under specific conditions. For instance, plenty of CRCs have shown activating mutations in WNT signaling pathways [68]. Pure tumoroids can be derived by culturing them in a medium devoid of WNT and R-spondins [128]. Likewise, culture medium lacking EGF would be chosen for tumoroids when tumors dwell mutations in EGFR [127, 140].

It turns out that drug responses in patients and their matched cancer organoids are very similar. The drug showing no anticancer activity in organoids remained docile in matched patients, and the one showing antitumor efficacy in tumoroid culture also had its effect on almost 90% of the patients [137]. Other reports having larger cohorts also agree with this observation [141 - 143]. However, more scrupulous investigations are required for tumor organoids to be accepted as *in vitro* replicas of the cancer patient. This efficient *in vitro* screening might pave the way towards the prediction of personalized cancer therapy in the near future. Recently, a vast range of patient-derived tumors, as well as healthy organoids, have been produced and biobanked. These bio-banked organoids can be used for predicting patient-specific drug response [126, 144]. In addition to various correlations that are established between mutation and drug response, numerous compounds were found to have differential cytotoxic effects in the tumoroids without a probable genetic marker [126]. The proteomic analysis also displayed a significant difference in protein profiles between tumoroids and matching healthy organoids. Notably, tumor organoids originating from different individuals showed personalized protein profiles [145]. Together, these observations point to the role of proteomic analysis in personalized treatment. Even so, proteins from the primary tumor should be compared head to head with those from the tumor organoids in order to ascertain the conservation of expression profiles after *in vitro* expansion.

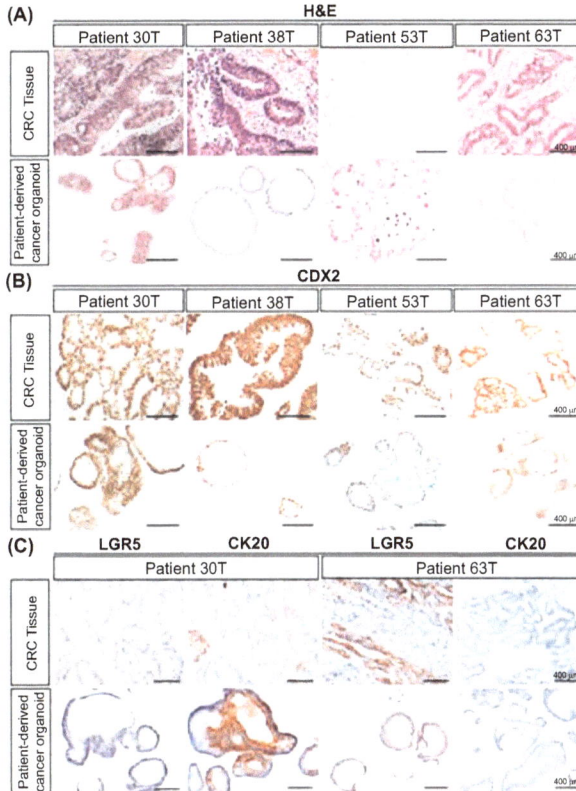

Fig. (6). Cancer organoids generated from patients replicate the histological features of their source tumors and the variability of stem cell markers amongst tumors. Haematoxylin and eosin (H&E) staining of tissue sections from primary colorectal adenocarcinoma and cancer organoids generated from the same tumor. Immunohistochemical identification of CDX2 (adenocarcinomas of intestinal origin marker) in primary colorectal adenocarcinoma vs patient-derived colorectal cancer organoids (PDCOs). Detection of LGR5 (CBC stem cell marker) and CK20 (intestinal epithelial marker) by immunohistochemistry in colorectal adenocarcinomas and PDCOs. This figure is taken from reference [136].

A key benefit of exploring organoids (originating from both tumor and healthy tissue) for drug discovery is that drugs having the potential to specifically target cancer cells can be screened, keeping normal cells intact. When translated, this methodology can reduce toxicity in patients. As is evident from many clinical trials, hepatotoxicity is a significant cause of drug withdrawal or failure. In the recent past, hepatic organoid constructs could actually provide a valuable platform for *in vitro* screening of liver toxicity of the drug molecules [146]. Interestingly, cytochrome P450 enzymes, which mostly mediate drug-induced liver toxicity, are expressed in liver organoids very close to the physiological levels [146, 147]. Moreover, to overcome drug resistance, exploiting the body's immune system towards eradicating malignant cells showed enormous promise. It prompted activation and growth of cytotoxic immune cells *in vitro* for use in patients. Thus, many reports were found in recent times describing a co-culture of hematopoietic cells along with organoids. For example, Finnberg and colleagues co-cultured CD45+ lymphocytes with air-liquid interface human tumoroids and they continued to live for 8 days, whether haematopoietic cells died very fast [148]. In another study, the authors co-cultured organoids from primary human mammary ductal epithelium expanded in the presence of zoledronic acid along with Vδ2+ T lymphocytes, which effectively killed TNBC cells [149].

Organoids have also been explored to establish the link between an infectious agent and tumorigenesis to shed some light on how a pathogenic entity is contributing towards malignant mutagenesis. Co-culturing organoids with pathogens (such as stomach organoids to study the relation between helicobacter pylori and gastric cancer) [129] would actually reveal the causal link present, if any. A clear understanding of the perpetual accumulation of mutation in disease-causing genes that finally engender the onset of cancer is required and can be deciphered by the prolonged genetically stable healthy organoid cultures [150]. Furthermore, organoids can be utilized to observe the initiation and propagation of the disease in different organs. Using CRISPR-Cas9 genome editing technology, combinations of CRC mutations were introduced in healthy human intestinal organoids to produce CRC progression constructs [138, 151]. It was demonstrated that those organoids, having activating mutations in KRAS and inactivating mutations in SMAD4, APC, and TP53, grew irrespective of niche factors for intestinal stem cells such as WNT, EGF and R-spondin [138, 151]. We can easily delineate that organoid technology will find enormous implications in future cancer research with this promising spectrum.

Organoid and Infectious Diseases Discussed in the Context of COVID-19

The COVID-19 pandemic caused by severe acute respiratory syndrome coronavirus 2 (SARS-CoV-2) infection exhibits respiratory illness, with clinical manifestations largely resembling that of earlier SARS [152]. However, patients who survived the infection show increasing evidence of neurological symptoms, including headache, anosmia, ageusia, confusion, seizure, and encephalopathy, suggesting detrimental effects of SARS-CoV2 on the CNS [153]. New studies from Düsseldorf University show that isolates of SARS-CoV-2 infect human brain organoids within two days of exposure, predominantly target neurons, and display altered distribution of Tau protein from axons to soma, hyperphosphorylation, leading to neuronal death, thus highlighting its neurotoxic effect [154]. ACE2, the entry receptor of SARS-CoV-2, is detected and concentrated in several locations of the human brain, including the substantia nigra, middle temporal gyrus, and posterior cingulate cortex [155]. SARS-CoV-2 infected 35-day-old brain organoids were generated using iPSC-derived human neural progenitor cells (hNPCs). SARS-CoV-2 robustly infected organoids in peripheral and deep layers, specifically in Tuj1-positive neuronal and Nestin-positive NPCs, and the culture supernatant was burdened with actively released progeny virus particles, providing insight into the pathognomonic symptoms and other neurological manifestations of COVID-19 [156]. Fig. (7) depicts the use of an organoid platform to explore several aspects of COVID-19 pathology.

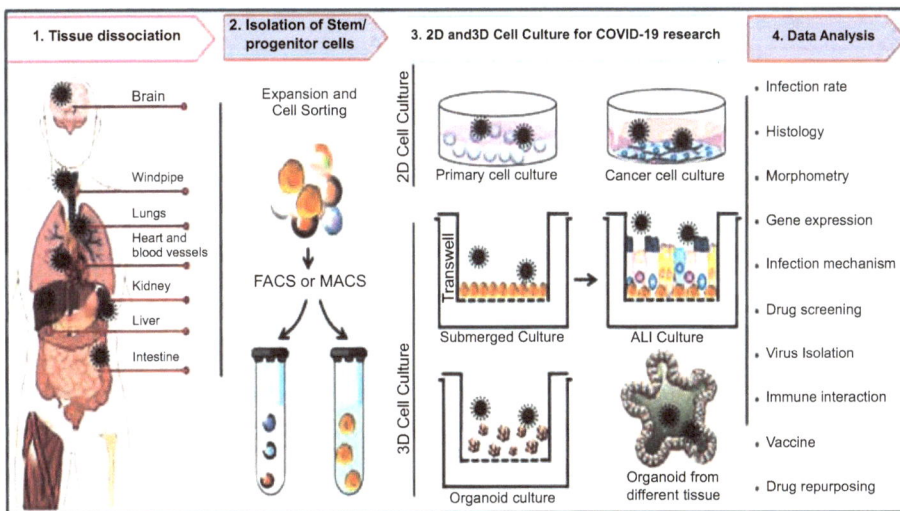

Fig. (7). Schematic diagram depicting the configuration of the COVID-19 research platform employing 3D organoids. Tissue separation from different organs is the initial stage in the process of employing 2D and 3D models for *in vitro* COVID-19 research. Fluorescence-activated cell sorting (FACS) and magnetic-activated cell sorting (MACS)-isolated stem/progenitor cells are cultured in 3D employing extracellular matrix mimetics and niche-specific culture media. Stem/progenitor cells generated from diverse tissues arrange themselves into tissue-specific organoids. SARS-CoV-2 infects organoid cells grown in three dimensions. As shown, several elements of post-infection research may be performed. This figure is taken from reference [158].

Lung and colon organoids have successfully been used to identify the SARS-CoV2 inhibitors, including imatinib, mycophenolic acid, and quinacrine dihydrochloride. Treatment using physiological levels of these drugs significantly inhibited SARS-CoV2 infection of both hPSC-derived lung and colon organoids [157, 158]. Small intestinal enteroids helped demonstrate infection of the virus with the intestine as a potential site of SARS-CoV2 replication and a possible contributor to local and systemic illness [159, 160]. Since SARS-CoV2 has high homology with SARS-related coronavirus identified in horseshoe bats, Zhou and colleagues generated intestinal organoids from the horseshoe bat species *Rhinolophus sinicus* using human intestinal organoid culture, which showed that they were readily infectable [161]. This study is of particular interest as several related coronaviruses have been identified in fecal samples or anal swabs of Chinese horseshoe bats but cannot be cultivated. The new bat organoids can help cultivate and study these viruses to help understand infectivity and gear us for any future pandemics of a similar kind.

Genetic Diseases/disorders

Organoids are well-suited to model genetic diseases and decipher their respective interventions. The range of applications of organoids for various diseases and those specific to brain disorders is very well summarized [162, 163] [164, 165]. One recent example for modeling a brain-related genetic disease namely the rare pediatric disease, Leigh syndrome (LS) accounts for the mitochondrial dysfunction affecting the neuronal system [166]. LS is a rather incurable disease, and there are no laboratory model systems to understand the disease mechanisms. Hence, neural brain organoids have been generated from the iPSC obtained from the patients and subjected to subsequent CRISPR/Cas9 editing of various mutations present in the LS. SURF, the complex IV assembly gene is mutated in LS. Multi-OMICS and single-cell RNA sequencing of such brain organoids obtained from the mutated neurons revealed the defects occurring at the neuro progenitor (NPC) stage, retaining proliferation and glycolysis but failing to form mature neurons. The same defect was further ameliorated in the brain organoids by augmenting the SURF-1 gene expression and inducing the Peroxisome proliferator-activated receptor gamma coactivator 1-alpha (PGC-1α) using Bezafibrate (lipid-lowering agent) treatment, thereby inducing metabolic reprogramming of NPCs and restoration of neurogenesis. Thus, the mechanism and possible interventional strategies to manage the rare mitochondrial disease LS could be deciphered using this organoid model.

ORGANOID BIOENGINEERING AND DRUG SCREENING

Significant attrition in the drug development process occurs in Phase II and III due to statistical reproducibility, efficacy, and safety problems. Though significant advances in research have occurred but, poor research hypothesis, inadequate data reproducibility, absence of human-specific models, and preclinical challenges lead to a significant block in the translation of preclinical candidates to the clinic, leading to significant unmet medical requirements plaguing humanity [167]. Several studies have highlighted the differences between human and animal development, highlighting the latter's inability to predict therapeutic effectiveness in many instances [168]. Hence, human-specific models that can serve as alternatives to animal model is of great importance. Human-specific models not only can increase the success rate of drug screening and can help discover personalized medicine but also can face substantial legal, ethical, and financial hurdles. Scientists have developed several preclinical cellular and animal models in an effort to reciprocate human disease accurately. These models include 2D cell cultures, PDX, genetically modified mouse models (GEMM), humanized animal models, transgenic animal models, engineering large animal species, and 3D organoids [167] [168, 169]. However, 2D cell culture is still the method of choice for biomedical research but is now giving way to 3D organoids and 3D cultures, owing to the convincing proof regarding the enhanced experimental dimensions and granularity that can lead to valuable insights [170]. When 3D cell culture studies are performed, the cell environment may be manipulated to replicate that of a living cell, providing more precise evidence regarding cell-to-cell interactions, tumor characteristics, drug development, metabolic profiling, stem cell science, and other forms of diseases. Human 3D organoids can be manipulated to replicate much of the complexity of a human tissue system or an organ, or to selectively express selected cell types. 3D-cancer organoids and tumoroids (cultivated tumor spheres/tumorospheres) can be generated from a wide variety of cancers and provides excellent opportunity to study potential drug screens and discovery [171].

Cancer organoid model appears to be the preferred method to model physiological disease states. Patient-derived tumor organoids (PDTO) lines were created from various liver tumor areas and their histologic, genetic, and transcriptomic profiles were found identical to those of primary human tumors. Idarubicin and plicamycin were found to be effective amongst 129 drugs tested in a high throughput screening (HTS) experiment on 27 PDOs [172]. Organoids were derived from colorectal carcinoma (CRC) patients provided tumor and natural organoids. Tumor organoids mimic the amount of somatic copies and mutation spectra observed in CRC and used for high-throughput medication testing. Kondo *et al.* used HTS to screen 2427 drugs using 30 cancer-tissue originated spheroids (CTOS) lines. Two pairs of drugs (bortezomib and carfilzomib; docetaxel and cabazitaxel) inhibiting the same pathway were found effective [173]. Sachs *et al.* surgically resected breast cancer (BC) specimens and created a living biobank of BC-organoids and performed high-throughput drug sensitivity screening. The *in vitro* drug sensitivities tend to have physiological relevance, as shown by proof-of-concept xeno-transplantation assays and derived BC organoid line recapitulation of the tamoxifen reaction of metastatic BC patients [132]. More recently, hepatocellular carcinoma PDTOs showed a different dose-dependent reaction to sorafenib therapy, implying that PDTO models may be used to predict patient-specific drug sensitivities to the intended drugs [174]. Wetering *et al.* used a biobank of CRC PDTO organoid and showed that nutlin-3a (MDM2 inhibitor) resistance is associated with TP53 mutants [126]. Similar data was also reported in biliary tract carcinoma-derived organoids [175]. The organoid model system has the ability to enhance drug discovery by better distinguishing which medicines are successful for which indications at an early stage, allowing the development of customized and personalized treatments. Though organoid technology has several advantages but the technology is at its infancy and have some limitations. Bioengineering advances can help in the application of organoids in finding treatment for a myriad of diseases.

ORGANOID IMAGING

Organoid technology has opened up a new experimental window for studying cellular processes that control organ growth, function, and disease. Imaging the organoids is a crucial technique that provides us with a glimpse of the kinetics and dynamics of cellular and sub-cellular behavior. Fluorescence microscopy has been the key to revealing cellular details and phenotyping. Tissue non-destructive optical sectioning techniques such as confocal, multi-photon laser scanning microscopy, and light-sheet fluorescence microscopy are used widely to elucidate organoids architecture. Paul *et al.*, used confocal immunofluorescence microscopy to elucidate symmetrical versus asymmetrical cell division in airway stem cell organoids using γ-tubulin immunostaining for centromeres and quantitated the angle and plane of polarity [17]. Wilkinson *et al.* generated multicellular lung organoids and compared the cellular architecture with adult human distal lung using confocal microscopy [176, 177] (Fig. **8**). Tissue clearing is a promising approach that allows deep imaging and has evolved into a versatile method for three-dimensional imaging and quantification of the organoid. FUnGI optical clearing of organoids, followed by 3D confocal imaging and volumetric image analysis of the complex 3D architecture, was generated by Ineveld *et al.* [178]. Rakotoson *et al.* demonstrated a new single-objective planar-illumination two-photon microscope for rapid 3D imaging of brain organoids [179]. Live cell imaging of organoids to study cell division was performed with simultaneous co-expression of H2B-mNeonGreen and either Centrin-mCardinal or TagRFP-CAAX [180]. Phototoxicity and low signal-to-noise ratios are common imaging issues in the analysis of live organoids, particularly when imaging deep within the sample. Microscopy techniques, such as fluorescence lifetime imaging microscopy (FLIM) or Lightsheet microscopy, have recently become popular for studying live organoids. Okkelman *et al.* used FLIM for NAD(P)H and phosphorescence-based PLIM for real-time oxygenation measurement to study the redox metabolism of Lgr5-GFP intestinal organoids [181]. Imaging serial sections of organoids using transmission electron microscopy (TEM), can be time-consuming but can provide granular data of subcellular components. PDTO breast cancer organoids were sectioned (70–80 nm) and imaged using a TEM, revealing the ultrastructural characteristics [182]. Serial block-face scanning electron microscopy (SEM) and image processing can aid in the 3D visualization of organoids.

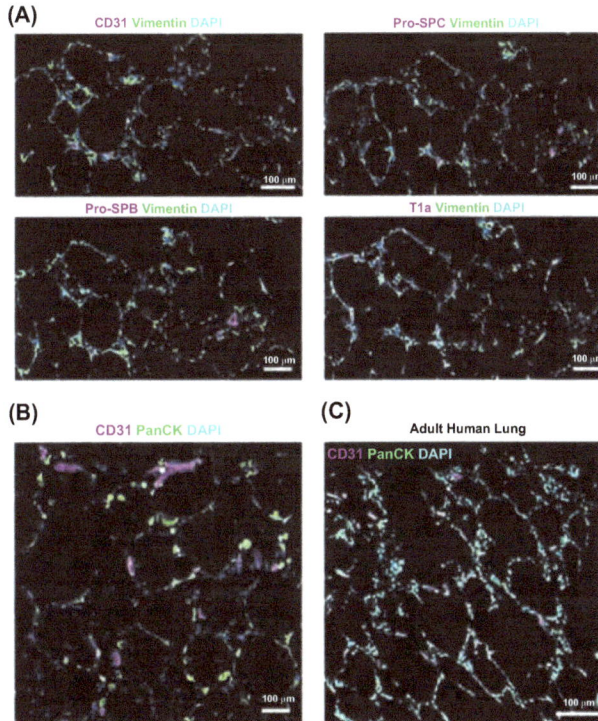

Fig. (8). Development of a 3D bioengineering lung organoids. Comparative immunostaining of 3D, multicellular organoids and adult human distal lung tissue. (A) A confocal image showing cross section of 3D multicellular lung organoids immunofluorescently stained with antibodies against CD31 (HUVECs), vimentin (FLFs), and proSPB and proSPC (type II alveolar epithelial cells) and T1a (type I alveolar epithelial cells; scale bar = 100 m). (B) A confocal image displaying multicellular 3D lung organoids immunostained for CD31 (HUVECs) and PanCK (SAECs). Furthermore, FLPs were also seeded. (C) A confocal micrograph showing a cross-section of a normal adult human lung, immunofluorescence stained for CD31 (HUVECs) and PanCK (SAECs; scale bar equals 100 micrometers). Abbreviations: FLFs: fetal lung fibroblasts; HUVECs: human umbilical vein endothelial cells; SAECs: small airway epithelial cells; SPB: surfactant protein B; SPC: surfactant protein C. Figure used from reference [67].

ORGANOID AND TISSUE REGENERATION

Stem cells possess the ability of limitless division and the potential for tissue regeneration, maintenance and functional activation. Moreover, many disorders can be triggered by anomalies in stem cells or in the way other cells interact with them. Thus, stem cell-derived organ-specific organoids are being used by researchers for tissue regeneration [183]. Breault and Zhou created a system to produce insulin in intestinal epithelial cells and checked it on mouse antral stomach organoids. These cells originate from the same area of the embryo and have developmental similarities to the pancreatic beta cells. They were also able to integrate a new kind of insulin-producing matrix into a diabetic mouse, and found that it controlled the blood sugar levels [184]. Despite the current progress in treating inflammatory bowel disease, a small group of patients appear refractory to traditional medications. It has been found that "mucosal curing" is critical for the improvement of the difficult-to -treat patients. Okamoto *et al.* demonstrated that intestinal stem cell (ISC)-derived organoid-based transplantation onto the colon ulcers can help reconstruct the crypt-villus architecture [185]. The ISC-based endoscopic transplantation therapy may provide significant long-term improvements for people with refractory ulcerative disease.

Dong *et al.* generated human cerebral organoids and grafted them in the medial prefrontal cortex of mice. The organoids survived and grew extensions over a period of one month. Moreover, the cerebral organoids transplanted mice displayed an enhanced freezing reaction to conditioned auditory stimuli [186]. This study demonstrated that subcortical neural networks may be developed using organ micrografting and could have vital therapeutic benefits for neurodegenerative disorders in the future. Another approach was attempted by Eiraku *et al.*, who used a 3D culture to create epithelial vesicles mimicking an embryonic eye's optic cup using mouse embryonic stem cells. Such technologies may be of key importance for retinal regeneration in the future [187]. Hohwieler *et al.* had developed a protocol to generate pancreatic organoid using human PSCs. These organoids develop natural pancreatic duct and acinar tissue upon transplantation into immunodeficient mice. They further screened many CFTR constructs in cystic fibrosis (CF) pancreatic organoids and developed an mRNA-based gene therapy in CF organoids [188]. There is a dearth in transplantation experiments partly due to the organoid bioengineering technology, which is still in its infancy [188]. It is evident that organoids have enormous potential for regenerative medicine applications, but they suffer from several drawbacks. These include methodology-related issues like repeatability, precision, scalability, quality compliance, clonal heterogeneity, pluripotency/multipotency, and genetic stability. Factors like methodologies to investigate the functional activity, preclinical validation, and safety issues are also lacking. Overcoming these bottlenecks can lead to organoid-based innovative methods for the replacement of badly injured tissues and organs.

3D PRINTING AND ORGANOIDS

Charles Hull, in 1984 created 3D prints by rapid plastic component prototyping, a process called Stereolithography (SLA) or 3D printing. He further patented the

SLA 3D printer and co-founded 3D systems in 1986. 3D printing or additive manufacturing relies on layer-by-layer repetitive deposition of print material to create a physical model based on a computerized digital 3D model. The use of 3D printing has now progressed beyond the use of plain resin and plastic geometries to the printing of biological cell scaffolds and bioprinting [189]. Bioprinting is the refined use of 3D printing methodologies to fabricate biological components by using different cell types, growth factors, and biomaterials that replicate the properties of natural tissue to the limit. The area has opened up enormous possibilities to design and fabricate custom-built complex shapes for a wide range of medical applications, including precision medicine, tissue transplantation, disease simulation, and drug screening [63]. Droplet-based, extrusion-based, and laser-assisted bioprinting are the three-primary bio-printing technologies. A detailed comparison of the technologies can be found in other published reviews [63, 190, 191].

A fusion of organoid technology and 3D printing and bioprinting methodologies has opened up numerous possibilities for medical research. Brain organoids are currently known to be a successful *in vitro* research model for the human brain, especially for drug screening applications. Rothenbucher *et al.* cultured brain organoids on 3D printed polycaprolactone scaffold to engineered flat morphology brain organoids (efBOs). The efBOs exhibited good diffusion characteristics and solved the buildup of necrotic tissue center [192]. Reid *et al.* described the generation of chimeric organoids by co-printing normal mammary epithelial cells with tumor cells, thus creating a platform to study tumor induction and microenvironmental control [193]. Skylar-Scott *et al.* have reported a biomanufacturing method for introducing interconnected perfusable vascular channels into biomatrices containing numerous organ building blocks (OBBs) by using embedded multi-material 3D bioprinting. This methodology is referred to as "Sacrificial writing into functional tissue (SWIFT)" and may help in creating bio-specific tissue models of embedded vascular channels [194]. Brassard *et al.* presented an approach to bioprint stem cells directly into biomatrices which then undergo self-organization and formation of macro-scale organoids [195]. Cetnar *et al.* used 3D bioprinting and perfusion bioreactor technology to produce models of the evolving human heart with anatomically precise geometries and study the natural developmental processes and microenvironment interactions [196]. In order to get bioprinting into the clinic, especially for complex organ replacement, researchers still have to solve some tricky problems, but a number of recent developments and the availability of relatively cheap, user-friendly bioprinters have made the technique more accessible than ever. This chapter will examine organoids, commercialization, biobank, and future initiatives.

COMMERCIAL PROSPECTS AND BIOBANK

According to market trends, the need for organoid technology will increase in the next years, and 20 companies are currently into biobanking, manufacturing, commercialization, robotics for organoid synthesis, organoids on a chip, *etc.* Prioritization is given to the heart, brain, gut, and renal organoids [7]. Current organoid technology startups include SYSTEM1 BIOSCIENCES, XILIS, 3DYNAMICS, CELLESCE, PATH BIOANALYTICS, KNOWN MEDICINE, DYNOMICS, and CYPRE. A list of important organoid technology companies is provided in Table **1**. The financial parts include venture capital, partnerships, and direct involvement with sample generators. Although the technology of creating organ-specific organoids has become largely accessible through the years, with more and more companies providing peripherals that allow easy cultivation in basic laboratory settings, the process still remains labor- and cost-intensive [197]. Endpoint manipulations also require sophisticated tools and instrumentation that hampers technology usage for laboratories that cannot fund the associated costs. A significant drawback is the lack of organoid-specific single-cell transcriptome/ epigenome sequencing and spatial profiling data. The Human Cell Atlas (HCA) has initiated a "Biological network" pilot project (https://www.humancellatlas. org/euh2020/) [164]. HCA (https://hca-organoid.eu) is a global collaborative effort to enhance single-cell research and therapy. This HCA pilot research characterizes single cells in organoids and in other *in vitro* settings. It helps compile, manage, distribute, exploit, and connect massive human organoid datasets, a step forward to help construct organoids-based human disease models. The demand for commercial organoids has increased progressively in recent years [198]. Automated biobanks with closed systems under normal BSL-2 conditions or open systems under a c-GMP cleanroom facility are the most appropriate for cell therapy commercialization.

Table 1. The present market leaders in the commercialization of organoids, their locations across the globe, and the types of applications they provide are shown. Table used from reference [7].

ORGANOID TECHNOLOGY COMPANIES		
Company Name	Country	Application Area
XILIS INC.	Durham, North Carolina, USA	Micro-organospheres
CELLESCE	Wales, UK	Patient-derived organoids (PDO)
SYSTEM 1 BIOSCIENCES	San Francisco, CA, USA	Brain organoids
3DNAMICS	Baltimore, Maryland, USA	Brain and liver organoids
PATH BIOANALYTICS	North Carolina, USA	PDO for precision medicine
KNOWN MEDICINE	Salt Lake City, Utah, USA	Patient-specific organoids for cancer drug development

(Table 1) cont.....

ORGANOID TECHNOLOGY COMPANIES		
Company Name	**Country**	**Application Area**
CYPRE	San Francisco, CA, USA	3D tumor model
DYNOMICS INC.	San Francisco, CA, USA	Human cardiac organoid
CHARLES RIVER	Wilmington, MA, USA	Tumor organoid
CROWNBio	San Diego, CA, USA	Tumor organoid

Biobanks are currently existing for human pluripotent stem cells in Europe and the United States. One such example of a Biobank is The European Bank for iPSCs-EBiSC, initiated in 2014 with the participation of major iPSC laboratories/ scientists with generous funding support from the European Union and EFPIA companies (www.imi.europa.eu) [199]. This biobank is primarily a distribution of research-grade iPSCs, culturing protocol, and related information across the globe, and not for commercial purposes. The iPSC lines offered with EBiSCs (non-profit) are exclusively for research, and the fee is nominal, meant to finance the repository's creation, classification, and upkeep. The iPSCs available through EBiSCs involve more than 35 disease areas and are publicized in the EBiSC catalog (cells.ebisc.org/). EBiSC not only does distribute iPSCs but also makes the comprehensive dataset available to the researchers prior to purchase. Such dataset involves the details regarding the source of the tissue sample, reprogramming method, characterization, and QC screening data as per ISO9001:2015. Additional data about the iPSC lines involve whole-genome sequencing, which is also available for specific cell lines and may be obtained from the EBiSC website (https://ebisc.org/).

The American counterpart of such a biobank is WiCell (https://www.wicell.org/). WiCell has existed before the EBiSC and is a reputed biobank for PSCs. WiCell has been meeting the needs of stem cell community by supplying research and c-GMP (Cell therapy grade) PSCs. WiCell produces ESC, iPSCs, GMP PSCs, PSC-based disease models and also a range of testing facilities. WiCell has delivered cell therapy grade iPSC cell lines for translational research and manufacturing iPSC-based cell therapy products. Biobanks of PSCs that can deliver cell therapy products/ organoids globally are needed. Nearby nations on each continent might create a stem cell network/ consortium to build biobanks. As the USA and Europe have been successful in creating PSC biobanks, other continents will require considerable financing and tireless efforts from the local stem cell community.

CONCLUSION

As organoid technology is going to be the key promising technology for *in vitro* research, cell therapy, and drug testing, further refinements are required to scale up the technology. Such refinements involve evolving the protocols for generating organoids, ease of obtaining the starting material, generous funding, infrastructure, trained personnel, and cryopreservation in biobanks. Continued improvement of existing organoid protocols is required to generate standardized methods that recapitulate *in vivo*-like spatial diversity and complexity. Moreover, every organ's local immune component must be incorporated with various tissue-specific organoids for having *in vivo* like readouts using co-culturing of immune cells in the skin [200]. Inclusion of biophysical aspects is needed while bioengineering specific organoid, like mechanical and immunological regulation in wound healing and skin regeneration [200], in the intestine [201], and also in tumor organoid models [202]. Likewise, the neuroimmune component is also a mandatory requirement in brain organoids distinct from the peripheral immune system and required to study brain pathogens. Although microglia may differentiate naturally in cerebral organoids and mount an immune response [67, 203], the technology is still nascent. A co-culture of organoids with other cell types (including pre-infected microglia) may provide new insights into pathogen infections in the brain and the accompanying immune responses. While cerebral organoids are physiologically relevant models of the human/rodent brain, they are currently limited in their applicability. An absence of vascularization leads to progressive necrosis in the organoid core during long-term culturing (months). Despite new technologies for incorporating vasculature are coming to the fore [204, 205], no standardized protocol has been universally adopted. Another challenging area for brain organoid research is the BBB. Recapitulating viral infections or testing drug candidates would heavily rely on the agents' capability to cross the BBB, making it imperative as a feature.

Future brain organoid research needs to aim at later stages of neurodevelopment, linked with neuronal activity and connections, to unravel further disease-associated phenotypes. Since many decades, ceaseless attempts have been made to discover remedies for CNS-trauma associated neurological and cognitive deficits. Challenges still prevail in understanding the complicated neural networks that cause the deficits. Effective brain and spinal cord models to study these conditions are thus necessitated for better defining its pathophysiology and testing therapeutics. Current technologies disallow generating both ventral and dorsal structures within a single spinal cord organoid. Moving forward, microfluidic devices that can maintain distinct gradients of morphogens over time can overcome this problem [206].

All commonly used model systems for fundamental or translation research have pros and cons. Organoid-based model systems are a promising technology to bridge the gap between bench and bedside compared to animal models that often fail to translate the findings to humans. 3D organoids have emerged as a powerful tool for investigating epithelial cell polarization and carcinogenesis [207]. Organoids are generally made of cells from cell lines or patient-derived samples, making it superior to traditional 2D models in recapitulating the physiology of the organs. For example, currently there is no effective treatment to cure severe terminal kidney dysfunction, partially due to the limited ability of the current *in vitro* models to predict human responses. Other than dialysis, transplantation is the only option, but finding a matched donor is difficult which is exacerbated by the donor shortage. The usage of renal organoids for regenerative therapies is a very plausible solution to the problem. However, as per current progress of the organoid research, there will likely be years before we see such therapies get to a translational stage. Furthermore, organoids do not yet fully recapitulate the structural complexity and function of the corresponding organ. In case of trabecular bone remodeling, although the organoids possess relevant osteogenic cell types and demonstrate functional bone remodeling activities, they lack interactions with proper growth factors as observed *in vivo*. Moreover, the quantities and maturation stages of cells in the organoids are different from those in the native microenvironment, and organoid-based constructs typically do not have osteocytes, which make up 95% of bone cells. Recently, Iordachescu *et al.* developed a trabecular bone humanized model to understand the effect of microgravity and degeneration of the cells [208].

A major limitation of organoid models is a lack of vascularization, which leads to hypoxia and poor supply of nutrients although there have been substantial efforts made to address this technical bottleneck, for example, *via* a perfusion bioreactor. Organoid-based models can, however, be used as a solution to one of the major challenges in pre-clinical studies for cancer drug development, which is the lack of physiological relevance by the traditional preclinical models. Organoids provide a means to develop personalized *in vitro* models that can better recapitulate the patient's tumor microenvironment while retaining the intra-tumor heterogeneity. Such organoid models are used for high-throughput drug screening and optimization of personalized therapies. However, most of the current disease models using organoids do not account for the immune system, which plays a crucial role in mediating disease progression. In a recent study, the authors developed a breast cancer-on-a-chip platform to show immune cell infiltration and the effectiveness of immunotherapies by using immune checkpoint inhibitors to block programmed death-1 [209]. The concept of such models may be used to model other diseases such as autoimmune diseases, other cancers, and infections and provide additional insights into the bioengineering of immune organs for implantation. Therefore, future incorporation of the immune component in platforms such as organ-on-a-chip will be critical and merits further investigation.

Another major issue is the regulatory and ethical issues surrounding the use of organoids, especially those-derived from human samples. Regulatory issues regarding the moral and legal status of organoids include ownership, consent, intellectual property rights, safety, and commercialization. Also, medical practitioners utilizing organoids should consider the ethical and moral ramifications. Distributing organoids globally will raise patenting problems. Public-private partnerships may lead to data sharing, entitlement sharing, *etc.* Organoid privacy and patenting restrictions are challenging. Even iPSC-based organoids raise authorization, commercialization, ownership, IP rights, safety, and marketing difficulties [130]. Big data, genomics, biobanking, and biotechnology's globalization make creating ethical norms difficult. For human application, the animal extracellular matrix Matrigel raises safety concerns. The argument surrounding the trade or donation of human tissue as a commodity continues. Few rules should be established and acknowledged by the worldwide intellect to address these issue [7, 67]. Strong global rules should be framed to address intellectual property (IP) disputes about the creation and use of organoids from human tissue and in order to preserve the dignity and rights of the donor [7, 67, 21]. When these constraints are addressed, commercial applications for drug discovery, disease modeling, and research and development may occur more quickly.

ABBREVIATIONS

2D	Two-dimension
3D	Three-dimension
ACE2	Angiotensin-converting enzyme 2
ASC	Adult stem cell
AT1	Alveolar epithelial type 1 cell
AT2	Alveolar epithelial type 2 cell
AFE	Anterior foregut endoderm
BALO	Bronchoalveolar organoid
BMM	Bone marrow mononuclear cell
BBB	Blood-brain barrier
BC	Breast cancer
CSC	Cancer stem cell
CRC	Colorectal carcinoma
CTOS	Cancer-tissue originated spheroid
CHIR	CHIR99021
CKD	Chronic kidney disease
CNS	Central nervous system

CF	Cystic fibrosis
DE	Definitive endoderm
DBP	Demineralized bone paper
ECM	Extracellular matrix
EGFR	Epidermal growth factor receptor
efBO	Engineered flat morphology brain organoid
FLIM	Fluorescence lifetime imaging microscopy
GEMM	Genetically modified mouse model
hPSC	Human pluripotent stem cell
HLO	Human lung organoid
hFGO	Human fundic type gastric organoid
hAGO	Human antral gastric organoid
HA	Hydroxyapatite
hPDC	Human-periosteum-derived cell
hNPC	Human neural progenitor cell
HTS	High throughput screening
HCA	Human Cell Atlas
iPSC	Induced pluripotent stem cell
IP	Intellectual property
LBO	Lung bud organoid
LGR5	Leucine-rich-repeat-containing G-protein-coupled receptor 5
LS	Leigh syndrome
MM	Metanephric mesenchyme
MG	Mammary glands
MSC	Mesenchymal stem cell
MO	Mammary organoid
mfNPC	Midbrain floor plate neural progenitor cells
NPC	Neuroprogenitor cell
OBB	Organ building blocks
PD	Parkinson's disease
PGC-1α	Peroxisome proliferator-activated receptor gamma coactivator 1-alpha
PDTO	Patient-derived tumor organoid
PDX	Patient-derived xenografts
PSC	Pluripotent stem cell
PLO	Patterned lung organoid

PFS	Posterior foregut organoid/spheroid
PEG	Polyethylene glycol
PLG	Poly (lactide-co-glycolide)
PCL	Polycaprolacton
PCW	Post-conception week
RA	Retinoid acid
RSPO1	R-spondin1
SIC	Small intestinalized colon
SM	Septum transversum mesenchyme
SARS-CoV-2	Severe acute respiratory syndrome coronavirus 2
SEM	Scanning electron microscopy
SLA	Stereolithography
SWIFT	Sacrificial writing into functional tissue
TEM	Transmission electron microscopy
TBI	Traumatic brain injury
VAFE	Ventral anterior foregut endoderm

ACKNOWLEDGEMENTS

MKP acknowledges Professors, S. Dubinett, B. Gomperts, V. Hartenstein from UCLA for providing constant support and mentoring.

REFERENCES

[1]   Mahapatra C, Lee R, Paul MK. Emerging role and promise of nanomaterials in organoid research. Drug Discov Today 2022; 27(3): 890-9.
[http://dx.doi.org/10.1016/j.drudis.2021.11.007] [PMID: 34774765]

[2]   Glanz VY, Orekhov AN, Deykin AV. Human Disease Modelling Techniques: Current Progress. Curr Mol Med 2019; 18(10): 655-60.
[http://dx.doi.org/10.2174/1566524019666190206204357] [PMID: 30727892]

[3]   Ferreira GS, Veening-Griffioen DH, Boon WPC, Moors EHM, van Meer PJK. Levelling the Translational Gap for Animal to Human Efficacy Data. Animals (Basel) 2020; 10(7): 1199.
[http://dx.doi.org/10.3390/ani10071199] [PMID: 32679706]

[4]   Kim J, Koo BK, Knoblich JA. Human organoids: model systems for human biology and medicine. Nat Rev Mol Cell Biol 2020; 21(10): 571-84.
[http://dx.doi.org/10.1038/s41580-020-0259-3] [PMID: 32636524]

[5]   Li M, Izpisua Belmonte JC. Organoids — Preclinical Models of Human Disease. N Engl J Med 2019; 380(6): 569-79.
[http://dx.doi.org/10.1056/NEJMra1806175] [PMID: 30726695]

[6]   Lancaster MA, Huch M. Disease modelling in human organoids. Dis Model Mech 2019; 12(7)dmm039347
[http://dx.doi.org/10.1242/dmm.039347] [PMID: 31383635]

[7] Mukherjee A, Sinha A, Maibam M, Bisht B. K Paul M. Organoids and Commercialization. Organoids 2022. [Working Title]

[8] Demetrius L. Of mice and men. EMBO Rep 2005; 6(S1) (Suppl. 1): S39-44.
[http://dx.doi.org/10.1038/sj.embor.7400422] [PMID: 15995660]

[9] McCauley HA, Wells JM. Pluripotent stem cell-derived organoids: using principles of developmental biology to grow human tissues in a dish. Development 2017; 144(6): 958-62.
[http://dx.doi.org/10.1242/dev.140731] [PMID: 28292841]

[10] Vossaert L, Scheerlinck E, Deforce D. Embryonic Stem Cells: Keeping Track of the Pluripotent Status. Pluripotent Stem Cells - From the Bench to the Clinic 2016.

[11] Tran F, Klein C, Arlt A, *et al.* Stem Cells and Organoid Technology in Precision Medicine in Inflammation: Are We There Yet? Front Immunol 2020; 11573562
[http://dx.doi.org/10.3389/fimmu.2020.573562] [PMID: 33408713]

[12] Corrò C, Novellasdemunt L, Li VSW. A brief history of organoids. Am J Physiol Cell Physiol 2020; 319(1): C151-65.
[http://dx.doi.org/10.1152/ajpcell.00120.2020] [PMID: 32459504]

[13] Ogundipe VML, Groen AH, Hosper N, *et al.* Generation and Differentiation of Adult Tissue-Derived Human Thyroid Organoids. Stem Cell Reports 2021; 16(4): 913-25.
[http://dx.doi.org/10.1016/j.stemcr.2021.02.011] [PMID: 33711265]

[14] Yu Q, Kilik U, Holloway EM, *et al.* Charting human development using a multi-endodermal organ atlas and organoid models. Cell 2021; 184(12): 3281-3298.e22.
[http://dx.doi.org/10.1016/j.cell.2021.04.028] [PMID: 34019796]

[15] Lewis A, Keshara R, Kim YH, Grapin-Botton A. Self-organization of organoids from endoderm-derived cells. J Mol Med (Berl) 2021; 99(4): 449-62.
[http://dx.doi.org/10.1007/s00109-020-02010-w] [PMID: 33221939]

[16] He J, Zhang X, Xia X, *et al.* Organoid technology for tissue engineering. J Mol Cell Biol 2020; 12(8): 569-79.
[http://dx.doi.org/10.1093/jmcb/mjaa012] [PMID: 32249317]

[17] Paul MK, Bisht B, Darmawan DO, *et al.* Dynamic changes in intracellular ROS levels regulate airway basal stem cell homeostasis through Nrf2-dependent Notch signaling. Cell Stem Cell 2014; 15(2): 199-214.
[http://dx.doi.org/10.1016/j.stem.2014.05.009] [PMID: 24953182]

[18] Vazquez-Armendariz AI, Herold S. From Clones to Buds and Branches: The Use of Lung Organoids to Model Branching Morphogenesis *Ex Vivo*. Front Cell Dev Biol 2021; 9631579
[http://dx.doi.org/10.3389/fcell.2021.631579] [PMID: 33748115]

[19] Dye BR, Hill DR, Ferguson MAH, *et al. in vitro* generation of human pluripotent stem cell derived lung organoids. eLife 2015; 4e05098
[http://dx.doi.org/10.7554/eLife.05098] [PMID: 25803487]

[20] Shibuya S, Allen-Hyttinen J, De Coppi P, Michielin F. *in vitro* models of fetal lung development to enhance research into congenital lung diseases. Pediatr Surg Int 2021; 37(5): 561-8.
[http://dx.doi.org/10.1007/s00383-021-04864-8] [PMID: 33787982]

[21] Huang JQ, Wei FK, Xu XL, *et al.* SOX9 drives the epithelial–mesenchymal transition in non-smal--cell lung cancer through the Wnt/β-catenin pathway. J Transl Med 2019; 17(1): 143.
[http://dx.doi.org/10.1186/s12967-019-1895-2] [PMID: 31060551]

[22] Bose B. Induced Pluripotent Stem Cells (iPSCs) Derived 3D Human Lung Organoids from Different Ethnicities to Understand the SARS-CoV2 Severity/Infectivity Percentage. Stem Cell Rev Rep 2021; 17(1): 293-5.
[http://dx.doi.org/10.1007/s12015-020-09989-2] [PMID: 32500482]

[23]   Bose B, Kapoor S, Nihad M. Induced Pluripotent Stem Cell Derived Human Lung Organoids to Map and Treat the SARS-CoV2 Infections In Vitro. Adv Exp Med Biol 2020; 1312: 1-17.
[http://dx.doi.org/10.1007/5584_2020_613] [PMID: 33385178]

[24]   Huang SXL, Green MD, de Carvalho AT, *et al*. The *in vitro* generation of lung and airway progenitor cells from human pluripotent stem cells. Nat Protoc 2015; 10(3): 413-25.
[http://dx.doi.org/10.1038/nprot.2015.023] [PMID: 25654758]

[25]   Miller AJ, Dye BR, Ferrer-Torres D, *et al*. Generation of lung organoids from human pluripotent stem cells in vitro. Nat Protoc 2019; 14(2): 518-40.
[http://dx.doi.org/10.1038/s41596-018-0104-8] [PMID: 30664680]

[26]   Rodrigues Toste de Carvalho AL, Liu HY, Chen YW, Porotto M, Moscona A, Snoeck HW. The *in vitro* multilineage differentiation and maturation of lung and airway cells from human pluripotent stem cell–derived lung progenitors in 3D. Nat Protoc 2021; 16(4): 1802-29.
[http://dx.doi.org/10.1038/s41596-020-00476-z] [PMID: 33649566]

[27]   McCracken KW, Aihara E, Martin B, *et al*. Erratum: Wnt/β-catenin promotes gastric fundus specification in mice and humans. Nature 2017; 543(7643): 136.
[http://dx.doi.org/10.1038/nature21381] [PMID: 28117442]

[28]   McCracken KW, Catá EM, Crawford CM, *et al*. Modelling human development and disease in pluripotent stem-cell-derived gastric organoids. Nature 2014; 516(7531): 400-4.
[http://dx.doi.org/10.1038/nature13863] [PMID: 25363776]

[29]   Broda TR, McCracken KW, Wells JM. Generation of human antral and fundic gastric organoids from pluripotent stem cells. Nat Protoc 2019; 14(1): 28-50.
[http://dx.doi.org/10.1038/s41596-018-0080-z] [PMID: 30470820]

[30]   Bayha E, Jørgensen MC, Serup P, Grapin-Botton A, Grapin-Botton A. Retinoic acid signaling organizes endodermal organ specification along the entire antero-posterior axis. PLoS One 2009; 4(6)e5845
[http://dx.doi.org/10.1371/journal.pone.0005845] [PMID: 19516907]

[31]   Han L, Chaturvedi P, Kishimoto K, *et al*. Single cell transcriptomics identifies a signaling network coordinating endoderm and mesoderm diversification during foregut organogenesis. Nat Commun 2020; 11(1): 4158.
[http://dx.doi.org/10.1038/s41467-020-17968-x] [PMID: 32855417]

[32]   Huch M, Bonfanti P, Boj SF, *et al*. Unlimited *in vitro* expansion of adult bi-potent pancreas progenitors through the Lgr5/R-spondin axis. EMBO J 2013; 32(20): 2708-21.
[http://dx.doi.org/10.1038/emboj.2013.204] [PMID: 24045232]

[33]   Huch M, Dollé L. The plastic cellular states of liver cells: Are EpCAM and Lgr5 fit for purpose? Hepatology 2016; 64(2): 652-62.
[http://dx.doi.org/10.1002/hep.28469] [PMID: 26799921]

[34]   Vyas D, Baptista PM, Brovold M, *et al*. Self-assembled liver organoids recapitulate hepatobiliary organogenesis in vitro. Hepatology 2018; 67(2): 750-61.
[http://dx.doi.org/10.1002/hep.29483] [PMID: 28834615]

[35]   Saheli M, Sepantafar M, Pournasr B, *et al*. Three-dimensional liver-derived extracellular matrix hydrogel promotes liver organoids function. J Cell Biochem 2018; 119(6): 4320-33.
[http://dx.doi.org/10.1002/jcb.26622] [PMID: 29247536]

[36]   Hu H, Gehart H, Artegiani B, *et al*. Long-Term Expansion of Functional Mouse and Human Hepatocytes as 3D Organoids. Cell 2018; 175(6): 1591-1606.e19.
[http://dx.doi.org/10.1016/j.cell.2018.11.013] [PMID: 30500538]

[37]   Mun SJ, Ryu JS, Lee MO, *et al*. Generation of expandable human pluripotent stem cell-derived hepatocyte-like liver organoids. J Hepatol 2019; 71(5): 970-85.
[http://dx.doi.org/10.1016/j.jhep.2019.06.030] [PMID: 31299272]

[38]   Olgasi C, Cucci A, Follenzi A. iPSC-Derived Liver Organoids: A Journey from Drug Screening, to Disease Modeling, Arriving to Regenerative Medicine. Int J Mol Sci 2020; 21(17): 6215.
[http://dx.doi.org/10.3390/ijms21176215] [PMID: 32867371]

[39]   Boj SF, Hwang CI, Baker LA, *et al.* Organoid models of human and mouse ductal pancreatic cancer. Cell 2015; 160(1-2): 324-38.
[http://dx.doi.org/10.1016/j.cell.2014.12.021] [PMID: 25557080]

[40]   Huang L, Holtzinger A, Jagan I, *et al.* Ductal pancreatic cancer modeling and drug screening using human pluripotent stem cell– and patient-derived tumor organoids. Nat Med 2015; 21(11): 1364-71.
[http://dx.doi.org/10.1038/nm.3973] [PMID: 26501191]

[41]   Wang W, Jin S, Ye K. Development of Islet Organoids from H9 Human Embryonic Stem Cells in Biomimetic 3D Scaffolds. Stem Cells Dev 2017; 26(6): 394-404.
[http://dx.doi.org/10.1089/scd.2016.0115] [PMID: 27960594]

[42]   Marti-Figueroa CR, Ashton RS. The case for applying tissue engineering methodologies to instruct human organoid morphogenesis. Acta Biomater 2017; 54: 35-44.
[http://dx.doi.org/10.1016/j.actbio.2017.03.023] [PMID: 28315813]

[43]   Tao T, Wang Y, Chen W, *et al.* Engineering human islet organoids from iPSCs using an organ-on-chip platform. Lab Chip 2019; 19(6): 948-58.
[http://dx.doi.org/10.1039/C8LC01298A] [PMID: 30719525]

[44]   Koike H, Iwasawa K, Ouchi R, *et al.* Modelling human hepato-biliary-pancreatic organogenesis from the foregut–midgut boundary. Nature 2019; 574(7776): 112-6.
[http://dx.doi.org/10.1038/s41586-019-1598-0] [PMID: 31554966]

[45]   Patel SN, Ishahak M, Chaimov D, *et al.* Organoid microphysiological system preserves pancreatic islet function within 3D matrix. Sci Adv 2021; 7(7)eaba5515
[http://dx.doi.org/10.1126/sciadv.aba5515] [PMID: 33579705]

[46]   Soltanian A, Ghezelayagh Z, Mazidi Z, *et al.* Generation of functional human pancreatic organoids by transplants of embryonic stem cell derivatives in a 3D-printed tissue trapper. J Cell Physiol 2019; 234(6): 9564-76.
[http://dx.doi.org/10.1002/jcp.27644] [PMID: 30362564]

[47]   Angus HCK, Butt AG, Schultz M, Kemp RA. Intestinal Organoids as a Tool for Inflammatory Bowel Disease Research. Front Med (Lausanne) 2020; 6: 334.
[http://dx.doi.org/10.3389/fmed.2019.00334] [PMID: 32010704]

[48]   Onozato D, Ogawa I, Kida Y, *et al.* Generation of Budding-Like Intestinal Organoids from Human Induced Pluripotent Stem Cells. J Pharm Sci 2021; 110(7): 2637-50.
[http://dx.doi.org/10.1016/j.xphs.2021.03.014] [PMID: 33794275]

[49]   Uchida H, Machida M, Miura T, *et al.* A xenogeneic-free system generating functional human gut organoids from pluripotent stem cells. JCI Insight 2017; 2(1)e86492
[http://dx.doi.org/10.1172/jci.insight.86492] [PMID: 28097227]

[50]   Wu H, Uchimura K, Donnelly EL, Kirita Y, Morris SA, Humphreys BD. Comparative Analysis and Refinement of Human PSC-Derived Kidney Organoid Differentiation with Single-Cell Transcriptomics. Cell Stem Cell 2018; 23(6): 869-881.e8.
[http://dx.doi.org/10.1016/j.stem.2018.10.010] [PMID: 30449713]

[51]   Romero-Guevara R, Ioannides A, Xinaris C. Kidney Organoids as Disease Models: Strengths, Weaknesses and Perspectives. Front Physiol 2020; 11563981
[http://dx.doi.org/10.3389/fphys.2020.563981] [PMID: 33250772]

[52]   Little MH, Combes AN. Kidney organoids: accurate models or fortunate accidents. Genes Dev 2019; 33(19-20): 1319-45.
[http://dx.doi.org/10.1101/gad.329573.119] [PMID: 31575677]

[53] Jin DC, Yun SR, Lee SW, *et al.* Current characteristics of dialysis therapy in Korea: 2016 registry data focusing on diabetic patients. Kidney Res Clin Pract 2018; 37(1): 20-9.
[http://dx.doi.org/10.23876/j.krcp.2018.37.1.20] [PMID: 29629274]

[54] Hale LJ, Howden SE, Phipson B, *et al.* 3D organoid-derived human glomeruli for personalised podocyte disease modelling and drug screening. Nat Commun 2018; 9(1): 5167.
[http://dx.doi.org/10.1038/s41467-018-07594-z] [PMID: 30514835]

[55] Calandrini C, Schutgens F, Oka R, *et al.* An organoid biobank for childhood kidney cancers that captures disease and tissue heterogeneity. Nat Commun 2020; 11(1): 1310.
[http://dx.doi.org/10.1038/s41467-020-15155-6] [PMID: 32161258]

[56] Taguchi A, Nishinakamura R. Higher-Order Kidney Organogenesis from Pluripotent Stem Cells. Cell Stem Cell 2017; 21(6): 730-746.e6.
[http://dx.doi.org/10.1016/j.stem.2017.10.011] [PMID: 29129523]

[57] Takasato M, Er PX, Chiu HS, *et al.* Kidney organoids from human iPS cells contain multiple lineages and model human nephrogenesis. Nature 2015; 526(7574): 564-8.
[http://dx.doi.org/10.1038/nature15695] [PMID: 26444236]

[58] Zhao X, Xu Z, Xiao L, *et al.* Review on the Vascularization of Organoids and Organoids-on-a-Chip. Front Bioeng Biotechnol 2021; 9637048
[http://dx.doi.org/10.3389/fbioe.2021.637048] [PMID: 33912545]

[59] Freedman BS, Brooks CR, Lam AQ, *et al.* Modelling kidney disease with CRISPR-mutant kidney organoids derived from human pluripotent epiblast spheroids. Nat Commun 2015; 6(1): 8715.
[http://dx.doi.org/10.1038/ncomms9715] [PMID: 26493500]

[60] Peters DJM, Breuning MH. Autosomal dominant polycystic kidney disease: modification of disease progression. Lancet 2001; 358(9291): 1439-44.
[http://dx.doi.org/10.1016/S0140-6736(01)06531-X] [PMID: 11705510]

[61] Wragg NM, Burke L, Wilson SL. A critical review of current progress in 3D kidney biomanufacturing: advances, challenges, and recommendations. Renal Replacement Therapy 2019; 5(1)

[62] Homan KA, Kolesky DB, Skylar-Scott MA, *et al.* Bioprinting of 3D Convoluted Renal Proximal Tubules on Perfusable Chips. Sci Rep 2016; 6(1): 34845.
[http://dx.doi.org/10.1038/srep34845] [PMID: 27725720]

[63] Bisht B, Hope A, Mukherjee A, Paul MK. Advances in the Fabrication of Scaffold and 3D Printing of Biomimetic Bone Graft. Ann Biomed Eng 2021; 49(4): 1128-50.
[http://dx.doi.org/10.1007/s10439-021-02752-9] [PMID: 33674908]

[64] Scheinpflug J, Pfeiffenberger M, Damerau A, *et al.* Journey into Bone Models: A Review. Genes (Basel) 2018; 9(5): 247.
[http://dx.doi.org/10.3390/genes9050247] [PMID: 29748516]

[65] Park Y, Cheong E, Kwak JG, Carpenter R, Shim JH, Lee J. Trabecular bone organoid model for studying the regulation of localized bone remodeling. Sci Adv 2021; 7(4)eabd6495
[http://dx.doi.org/10.1126/sciadv.abd6495] [PMID: 33523925]

[66] Cancedda R, Giannoni P, Mastrogiacomo M. A tissue engineering approach to bone repair in large animal models and in clinical practice. Biomaterials 2007; 28(29): 4240-50.
[http://dx.doi.org/10.1016/j.biomaterials.2007.06.023] [PMID: 17644173]

[67] Abreu CM, Gama L, Krasemann S, *et al.* Microglia Increase Inflammatory Responses in iPSC-Derived Human BrainSpheres. Front Microbiol 2018; 9: 2766.
[http://dx.doi.org/10.3389/fmicb.2018.02766] [PMID: 30619100]

[68] Comprehensive molecular characterization of human colon and rectal cancer. Nature 2012; 487(7407): 330-7.

[http://dx.doi.org/10.1038/nature11252] [PMID: 22810696]

[69] Akiva A, Melke J, Ansari S, *et al.* An Organoid for Woven Bone. Adv Funct Mater 2021; 31(17)2010524
[http://dx.doi.org/10.1002/adfm.202010524]

[70] Liu C, Carrera R, Flamini V, *et al.* Effects of mechanical loading on cortical defect repair using a novel mechanobiological model of bone healing. Bone 2018; 108: 145-55.
[http://dx.doi.org/10.1016/j.bone.2017.12.027] [PMID: 29305998]

[71] Fu S, Ni P, Wang B, *et al.* Injectable and thermo-sensitive PEG-PCL-PEG copolymer/collagen/n-HA hydrogel composite for guided bone regeneration. Biomaterials 2012; 33(19): 4801-9.
[http://dx.doi.org/10.1016/j.biomaterials.2012.03.040] [PMID: 22463934]

[72] Brady RT, O'Brien FJ, Hoey DA. Mechanically stimulated bone cells secrete paracrine factors that regulate osteoprogenitor recruitment, proliferation, and differentiation. Biochem Biophys Res Commun 2015; 459(1): 118-23.
[http://dx.doi.org/10.1016/j.bbrc.2015.02.080] [PMID: 25721667]

[73] Nilsson Hall G, Mendes LF, Gklava C, Geris L, Luyten FP, Papantoniou I. Developmentally Engineered Callus Organoid Bioassemblies Exhibit Predictive *in Vivo* Long Bone Healing. Adv Sci (Weinh) 2020; 7(2)1902295
[http://dx.doi.org/10.1002/advs.201902295] [PMID: 31993293]

[74] Lee GY, Kenny PA, Lee EH, Bissell MJ. Three-dimensional culture models of normal and malignant breast epithelial cells. Nat Methods 2007; 4(4): 359-65.
[http://dx.doi.org/10.1038/nmeth1015] [PMID: 17396127]

[75] Pauli C, Hopkins BD, Prandi D, *et al.* Personalized *in Vitro* and *in Vivo* Cancer Models to Guide Precision Medicine. Cancer Discov 2017; 7(5): 462-77.
[http://dx.doi.org/10.1158/2159-8290.CD-16-1154] [PMID: 28331002]

[76] Simian M, Hirai Y, Navre M, Werb Z, Lochter A, Bissell MJ. The interplay of matrix metalloproteinases, morphogens and growth factors is necessary for branching of mammary epithelial cells. Development 2001; 128(16): 3117-31.
[http://dx.doi.org/10.1242/dev.128.16.3117] [PMID: 11688561]

[77] Koledova Z, Zhang X, Streuli C, *et al.* SPRY1 regulates mammary epithelial morphogenesis by modulating EGFR-dependent stromal paracrine signaling and ECM remodeling. Proc Natl Acad Sci USA 2016; 113(39): E5731-40.
[http://dx.doi.org/10.1073/pnas.1611532113] [PMID: 27621461]

[78] Zeng YA, Nusse R. Wnt proteins are self-renewal factors for mammary stem cells and promote their long-term expansion in culture. Cell Stem Cell 2010; 6(6): 568-77.
[http://dx.doi.org/10.1016/j.stem.2010.03.020] [PMID: 20569694]

[79] Nusse R, Varmus HE. Many tumors induced by the mouse mammary tumor virus contain a provirus integrated in the same region of the host genome. Cell 1982; 31(1): 99-109.
[http://dx.doi.org/10.1016/0092-8674(82)90409-3] [PMID: 6297757]

[80] Yu Q, Verheyen E, Zeng Y. Mammary Development and Breast Cancer: A Wnt Perspective. Cancers (Basel) 2016; 8(7): 65.
[http://dx.doi.org/10.3390/cancers8070065] [PMID: 27420097]

[81] Rosenbluth JM, Schackmann RCJ, Gray GK, *et al.* Organoid cultures from normal and cancer-prone human breast tissues preserve complex epithelial lineages. Nat Commun 2020; 11(1): 1711.
[http://dx.doi.org/10.1038/s41467-020-15548-7] [PMID: 32249764]

[82] Berger AJ, Renner CM, Hale I, *et al.* Scaffold stiffness influences breast cancer cell invasion *via* EGFR-linked Mena upregulation and matrix remodeling. Matrix Biol 2020; 85-86: 80-93.
[http://dx.doi.org/10.1016/j.matbio.2019.07.006] [PMID: 31323325]

[83] Hu WF, Chahrour MH, Walsh CA. The diverse genetic landscape of neurodevelopmental disorders.

Annu Rev Genomics Hum Genet 2014; 15(1): 195-213.
[http://dx.doi.org/10.1146/annurev-genom-090413-025600] [PMID: 25184530]

[84] Silbereis JC, Pochareddy S, Zhu Y, Li M, Sestan N. The Cellular and Molecular Landscapes of the Developing Human Central Nervous System. Neuron 2016; 89(2): 248-68.
[http://dx.doi.org/10.1016/j.neuron.2015.12.008] [PMID: 26796689]

[85] Sakaguchi H, Kadoshima T, Soen M, *et al.* Generation of functional hippocampal neurons from self-organizing human embryonic stem cell-derived dorsomedial telencephalic tissue. Nat Commun 2015; 6(1): 8896.
[http://dx.doi.org/10.1038/ncomms9896] [PMID: 26573335]

[86] Jo J, Xiao Y, Sun AX, *et al.* Midbrain-like Organoids from Human Pluripotent Stem Cells Contain Functional Dopaminergic and Neuromelanin-Producing Neurons. Cell Stem Cell 2016; 19(2): 248-57.
[http://dx.doi.org/10.1016/j.stem.2016.07.005] [PMID: 27476966]

[87] Monzel AS, Smits LM, Hemmer K, *et al.* Derivation of Human Midbrain-Specific Organoids from Neuroepithelial Stem Cells. Stem Cell Reports 2017; 8(5): 1144-54.
[http://dx.doi.org/10.1016/j.stemcr.2017.03.010] [PMID: 28416282]

[88] Ozone C, Suga H, Eiraku M, *et al.* Functional anterior pituitary generated in self-organizing culture of human embryonic stem cells. Nat Commun 2016; 7(1): 10351.
[http://dx.doi.org/10.1038/ncomms10351] [PMID: 26762480]

[89] Qian X, Nguyen HN, Song MM, *et al.* Brain-Region-Specific Organoids Using Mini-bioreactors for Modeling ZIKV Exposure. Cell 2016; 165(5): 1238-54.
[http://dx.doi.org/10.1016/j.cell.2016.04.032] [PMID: 27118425]

[90] Muguruma K, Nishiyama A, Kawakami H, Hashimoto K, Sasai Y. Self-organization of polarized cerebellar tissue in 3D culture of human pluripotent stem cells. Cell Rep 2015; 10(4): 537-50.
[http://dx.doi.org/10.1016/j.celrep.2014.12.051] [PMID: 25640179]

[91] Giandomenico SL, Sutcliffe M, Lancaster MA. Generation and long-term culture of advanced cerebral organoids for studying later stages of neural development. Nat Protoc 2020.
[PMID: 33328611]

[92] Giandomenico SL, Mierau SB, Gibbons GM, *et al.* Cerebral organoids at the air–liquid interface generate diverse nerve tracts with functional output. Nat Neurosci 2019; 22(4): 669-79.
[http://dx.doi.org/10.1038/s41593-019-0350-2] [PMID: 30886407]

[93] Tung TC, Kü SH. Experimental studies on the development of the pronephric duct in anuran embryos. J Anat 1944; 78(Pt 1-2): 52-7.
[PMID: 17104942]

[94] Wilson HVON. Some phenomena of coalescence and regeneration in sponges. J Elisha Mitchell Sci Soc 1907; 23(4): 161-74.

[95] Zwilling E. Some aspects of differentiation: disaggregation and reaggregation of early chick embryos. Natl Cancer Inst Monogr 1960; 2: 19-39.
[PMID: 13847984]

[96] Zwilling E. Some aspects of differentiation: disaggregation and reaggregation of early chick embryos. Natl Cancer Inst Monogr 1960; 2: 19-39.
[PMID: 13847984]

[97] Martin GR. Isolation of a pluripotent cell line from early mouse embryos cultured in medium conditioned by teratocarcinoma stem cells. Proc Natl Acad Sci USA 1981; 78(12): 7634-8.
[http://dx.doi.org/10.1073/pnas.78.12.7634] [PMID: 6950406]

[98] Takahashi K, Yamanaka S. Induction of pluripotent stem cells from mouse embryonic and adult fibroblast cultures by defined factors. Cell 2006; 126(4): 663-76.
[http://dx.doi.org/10.1016/j.cell.2006.07.024] [PMID: 16904174]

[99]   Thomson JA, Itskovitz-Eldor J, Shapiro SS, *et al.* Embryonic stem cell lines derived from human blastocysts. Science 1998; 282(5391): 1145-7.
[http://dx.doi.org/10.1126/science.282.5391.1145] [PMID: 9804556]

[100] Lancaster MA, Renner M, Martin CA, *et al.* Cerebral organoids model human brain development and microcephaly. Nature 2013; 501(7467): 373-9.
[http://dx.doi.org/10.1038/nature12517] [PMID: 23995685]

[101] Sato T, Vries RG, Snippert HJ, *et al.* Single Lgr5 stem cells build crypt-villus structures *in vitro* without a mesenchymal niche. Nature 2009; 459(7244): 262-5.
[http://dx.doi.org/10.1038/nature07935] [PMID: 19329995]

[102] Xia Y, Sancho-Martinez I, Nivet E, Rodriguez Esteban C, Campistol JM, Izpisua Belmonte JC. The generation of kidney organoids by differentiation of human pluripotent cells to ureteric bud progenitor–like cells. Nat Protoc 2014; 9(11): 2693-704.
[http://dx.doi.org/10.1038/nprot.2014.182] [PMID: 25340442]

[103] Zhong X, Gutierrez C, Xue T, *et al.* Generation of three-dimensional retinal tissue with functional photoreceptors from human iPSCs. Nat Commun 2014; 5(1): 4047.
[http://dx.doi.org/10.1038/ncomms5047] [PMID: 24915161]

[104] Marton RM, Paşca SP. Organoid and Assembloid Technologies for Investigating Cellular Crosstalk in Human Brain Development and Disease. Trends Cell Biol 2020; 30(2): 133-43.
[http://dx.doi.org/10.1016/j.tcb.2019.11.004] [PMID: 31879153]

[105] Esk C, Lindenhofer D, Haendeler S, *et al.* A human tissue screen identifies a regulator of ER secretion as a brain-size determinant. Science 2020; 370(6519): 935-41.
[http://dx.doi.org/10.1126/science.abb5390] [PMID: 33122427]

[106] Trujillo CA, Muotri AR. Brain Organoids and the Study of Neurodevelopment. Trends Mol Med 2018; 24(12): 982-90.
[http://dx.doi.org/10.1016/j.molmed.2018.09.005] [PMID: 30377071]

[107] Birey F, Andersen J, Makinson CD, *et al.* Assembly of functionally integrated human forebrain spheroids. Nature 2017; 545(7652): 54-9.
[http://dx.doi.org/10.1038/nature22330] [PMID: 28445465]

[108] Grebenyuk S, Ranga A. Engineering Organoid Vascularization. Front Bioeng Biotechnol 2019; 7: 39.
[http://dx.doi.org/10.3389/fbioe.2019.00039] [PMID: 30941347]

[109] Peterson AB, Xu L, Daugherty J, Breiding MJ. Centers for Disease Control and Prevention (2019). Surveillance Report of Traumatic Brain Injury-related Emergency Department Visits, Hospitalizations, and Deaths—United States, 2014. Centers for Disease Control and Prevention, US Department of Health and Human Services Report 2019.

[110] Dewan MC, Rattani A, Gupta S, *et al.* Estimating the global incidence of traumatic brain injury. J Neurosurg 2019; 130(4): 1080-97.
[http://dx.doi.org/10.3171/2017.10.JNS17352] [PMID: 29701556]

[111] Chhibber T, Bagchi S, Lahooti B, *et al.* CNS organoids: an innovative tool for neurological disease modeling and drug neurotoxicity screening. Drug Discov Today 2020; 25(2): 456-65.
[http://dx.doi.org/10.1016/j.drudis.2019.11.010] [PMID: 31783130]

[112] Lai JD, Berlind JE, Fricklas G, Maria NS, Jacobs R, Yu V, *et al.* A model of traumatic brain injury using human iPSC-derived cortical brain organoids. BioRxiv 2020.
[http://dx.doi.org/10.1101/2020.07.05.180299]

[113] Koo B, Choi B, Park H, Yoon KJ. Past, Present, and Future of Brain Organoid Technology. Mol Cells 2019; 42(9): 617-27.
[PMID: 31564073]

[114] Kim MS, Kim DH, Kang HK, Kook MG, Choi SW, Kang KS. Modeling of Hypoxic Brain Injury

through 3D Human Neural Organoids. Cells 2021; 10(2): 234.
[http://dx.doi.org/10.3390/cells10020234] [PMID: 33504071]

[115]   Paşca AM, Park JY, Shin HW, *et al.* Human 3D cellular model of hypoxic brain injury of prematurity. Nat Med 2019; 25(5): 784-91.
[http://dx.doi.org/10.1038/s41591-019-0436-0] [PMID: 31061540]

[116]   Daviaud N, Chevalier C, Friedel RH, Zou H. Distinct Vulnerability and Resilience of Human Neuroprogenitor Subtypes in Cerebral Organoid Model of Prenatal Hypoxic Injury. Front Cell Neurosci 2019; 13: 336.
[http://dx.doi.org/10.3389/fncel.2019.00336] [PMID: 31417360]

[117]   Hor JH, Soh ESY, Tan LY, *et al.* Cell cycle inhibitors protect motor neurons in an organoid model of Spinal Muscular Atrophy. Cell Death Dis 2018; 9(11): 1100.
[http://dx.doi.org/10.1038/s41419-018-1081-0] [PMID: 30368521]

[118]   Wang Z, Wang SN, Xu TY, *et al.* Cerebral organoids transplantation improves neurological motor function in rat brain injury. CNS Neurosci Ther 2020; 26(7): 682-97.
[http://dx.doi.org/10.1111/cns.13286] [PMID: 32087606]

[119]   Chen X, Sun G, Tian E, *et al.* Modeling Sporadic Alzheimer's Disease in Human Brain Organoids under Serum Exposure. Adv Sci (Weinh) 2021; 8(18)2101462
[http://dx.doi.org/10.1002/advs.202101462] [PMID: 34337898]

[120]   Smits LM, Reinhardt L, Reinhardt P, Glatza M, Monzel AS, Stanslowsky N, *et al.* Modeling Parkinson's disease in midbrain-like organoids. npj. Parkinsons Dis 2019; 5(1)

[121]   Dharmalingam P, Venkatakrishnan K, Tan B. Predicting Metastasis from Cues of Metastatic Cancer Stem-like Cells-3D-Ultrasensitive Metasensor at a Single-Cell Level. ACS Nano 2021; 15(6): 9967-86.
[http://dx.doi.org/10.1021/acsnano.1c01436] [PMID: 34081852]

[122]   Sim J, Lee HJ, Jeong B, Park MH. Poly(Ethylene Glycol)-Poly(l-Alanine)/Hyaluronic Acid Complex as a 3D Platform for Understanding Cancer Cell Migration in the Tumor Microenvironment. Polymers (Basel) 2021; 13(7): 1042.
[http://dx.doi.org/10.3390/polym13071042] [PMID: 33810521]

[123]   Bose B, Kapoor S, Nihad M. Induced Pluripotent Stem Cell Derived Human Lung Organoids to Map and Treat the SARS-CoV2 Infections In Vitro. Cell Biology and Translational Medicine, Volume 11. Adv Exp Med Biol 2020; 1-17.

[124]   Kuo CJ, Curtis C. Organoids reveal cancer dynamics. Nature 2018; 556(7702): 441-2.
[http://dx.doi.org/10.1038/d41586-018-03841-x] [PMID: 29686366]

[125]   Muthuswamy SK. Organoid Models of Cancer Explode with Possibilities. Cell Stem Cell 2018; 22(3): 290-1.
[http://dx.doi.org/10.1016/j.stem.2018.02.010] [PMID: 29499146]

[126]   van de Wetering M, Francies HE, Francis JM, *et al.* Prospective derivation of a living organoid biobank of colorectal cancer patients. Cell 2015; 161(4): 933-45.
[http://dx.doi.org/10.1016/j.cell.2015.03.053] [PMID: 25957691]

[127]   Fujii M, Shimokawa M, Date S, *et al.* A Colorectal Tumor Organoid Library Demonstrates Progressive Loss of Niche Factor Requirements during Tumorigenesis. Cell Stem Cell 2016; 18(6): 827-38.
[http://dx.doi.org/10.1016/j.stem.2016.04.003] [PMID: 27212702]

[128]   Sato T, Stange DE, Ferrante M, *et al.* Long-term expansion of epithelial organoids from human colon, adenoma, adenocarcinoma, and Barrett's epithelium. Gastroenterology 2011; 141(5): 1762-72.
[http://dx.doi.org/10.1053/j.gastro.2011.07.050] [PMID: 21889923]

[129]   Bartfeld S, Bayram T, van de Wetering M, *et al. in vitro* expansion of human gastric epithelial stem cells and their responses to bacterial infection. Gastroenterology 2015; 148(1): 126-136.e6.

[http://dx.doi.org/10.1053/j.gastro.2014.09.042] [PMID: 25307862]

[130] Broutier L, Mastrogiovanni G, Verstegen MMA, *et al.* Human primary liver cancer–derived organoid cultures for disease modeling and drug screening. Nat Med 2017; 23(12): 1424-35.
[http://dx.doi.org/10.1038/nm.4438] [PMID: 29131160]

[131] Turco MY, Gardner L, Hughes J, *et al.* Long-term, hormone-responsive organoid cultures of human endometrium in a chemically defined medium. Nat Cell Biol 2017; 19(5): 568-77.
[http://dx.doi.org/10.1038/ncb3516] [PMID: 28394884]

[132] Sachs N, de Ligt J, Kopper O, *et al.* A Living Biobank of Breast Cancer Organoids Captures Disease Heterogeneity. Cell 2018; 172(1-2): 373-386.e10.
[http://dx.doi.org/10.1016/j.cell.2017.11.010] [PMID: 29224780]

[133] Weeber F, van de Wetering M, Hoogstraat M, *et al.* Preserved genetic diversity in organoids cultured from biopsies of human colorectal cancer metastases. Proc Natl Acad Sci USA 2015; 112(43): 13308-11.
[http://dx.doi.org/10.1073/pnas.1516689112] [PMID: 26460009]

[134] Gao D, Vela I, Sboner A, *et al.* Organoid cultures derived from patients with advanced prostate cancer. Cell 2014; 159(1): 176-87.
[http://dx.doi.org/10.1016/j.cell.2014.08.016] [PMID: 25201530]

[135] Drost J, Karthaus WR, Gao D, *et al.* Organoid culture systems for prostate epithelial and cancer tissue. Nat Protoc 2016; 11(2): 347-58.
[http://dx.doi.org/10.1038/nprot.2016.006] [PMID: 26797458]

[136] Engel RM, Chan WH, Nickless D, *et al.* Patient-Derived Colorectal Cancer Organoids Upregulate Revival Stem Cell Marker Genes following Chemotherapeutic Treatment. J Clin Med 2020; 9(1): 128.
[http://dx.doi.org/10.3390/jcm9010128] [PMID: 31906589]

[137] Vlachogiannis G, Hedayat S, Vatsiou A, *et al.* Patient-derived organoids model treatment response of metastatic gastrointestinal cancers. Science 2018; 359(6378): 920-6.
[http://dx.doi.org/10.1126/science.aao2774] [PMID: 29472484]

[138] Drost J, van Jaarsveld RH, Ponsioen B, *et al.* Sequential cancer mutations in cultured human intestinal stem cells. Nature 2015; 521(7550): 43-7.
[http://dx.doi.org/10.1038/nature14415] [PMID: 25924068]

[139] Verissimo CS, Overmeer RM, Ponsioen B, *et al.* Targeting mutant RAS in patient-derived colorectal cancer organoids by combinatorial drug screening. eLife 2016; 5e18489
[http://dx.doi.org/10.7554/eLife.18489] [PMID: 27845624]

[140] Sakamoto N, Feng Y, Stolfi C, *et al.* BRAF^{V600E} cooperates with CDX2 inactivation to promote serrated colorectal tumorigenesis. eLife 2017; 6e20331
[http://dx.doi.org/10.7554/eLife.20331] [PMID: 28072391]

[141] Ganesh K, Wu C, O'Rourke KP, *et al.* A rectal cancer organoid platform to study individual responses to chemoradiation. Nat Med 2019; 25(10): 1607-14.
[http://dx.doi.org/10.1038/s41591-019-0584-2] [PMID: 31591597]

[142] Yao Y, Xu X, Yang L, *et al.* Patient-Derived Organoids Predict Chemoradiation Responses of Locally Advanced Rectal Cancer. Cell Stem Cell 2020; 26(1): 17-26.e6.
[http://dx.doi.org/10.1016/j.stem.2019.10.010] [PMID: 31761724]

[143] Ooft SN, Weeber F, Dijkstra KK, *et al.* Patient-derived organoids can predict response to chemotherapy in metastatic colorectal cancer patients. Sci Transl Med 2019; 11(513)eaay2574
[http://dx.doi.org/10.1126/scitranslmed.aay2574] [PMID: 31597751]

[144] Kondo J, Endo H, Okuyama H, *et al.* Retaining cell–cell contact enables preparation and culture of spheroids composed of pure primary cancer cells from colorectal cancer. Proc Natl Acad Sci USA 2011; 108(15): 6235-40.
[http://dx.doi.org/10.1073/pnas.1015938108] [PMID: 21444794]

[145] Cristobal A, van den Toorn HWP, van de Wetering M, Clevers H, Heck AJR, Mohammed S. Personalized Proteome Profiles of Healthy and Tumor Human Colon Organoids Reveal Both Individual Diversity and Basic Features of Colorectal Cancer. Cell Rep 2017; 18(1): 263-74.
[http://dx.doi.org/10.1016/j.celrep.2016.12.016] [PMID: 28052255]

[146] Katsuda T, Kawamata M, Hagiwara K, *et al.* Conversion of Terminally Committed Hepatocytes to Culturable Bipotent Progenitor Cells with Regenerative Capacity. Cell Stem Cell 2017; 20(1): 41-55.
[http://dx.doi.org/10.1016/j.stem.2016.10.007] [PMID: 27840021]

[147] Huch M, Gehart H, van Boxtel R, *et al.* Long-term culture of genome-stable bipotent stem cells from adult human liver. Cell 2015; 160(1-2): 299-312.
[http://dx.doi.org/10.1016/j.cell.2014.11.050] [PMID: 25533785]

[148] Finnberg NK, Gokare P, Lev A, *et al.* Application of 3D tumoroid systems to define immune and cytotoxic therapeutic responses based on tumoroid and tissue slice culture molecular signatures. Oncotarget 2017; 8(40): 66747-57.
[http://dx.doi.org/10.18632/oncotarget.19965] [PMID: 28977993]

[149] Zumwalde NA, Haag JD, Sharma D, *et al.* Analysis of Immune Cells from Human Mammary Ductal Epithelial Organoids Reveals Vδ2+ T Cells That Efficiently Target Breast Carcinoma Cells in the Presence of Bisphosphonate. Cancer Prev Res (Phila) 2016; 9(4): 305-16.
[http://dx.doi.org/10.1158/1940-6207.CAPR-15-0370-T] [PMID: 26811335]

[150] Behjati S, Huch M, van Boxtel R, *et al.* Genome sequencing of normal cells reveals developmental lineages and mutational processes. Nature 2014; 513(7518): 422-5.
[http://dx.doi.org/10.1038/nature13448] [PMID: 25043003]

[151] Matano M, Date S, Shimokawa M, *et al.* Modeling colorectal cancer using CRISPR-Cas9–mediated engineering of human intestinal organoids. Nat Med 2015; 21(3): 256-62.
[http://dx.doi.org/10.1038/nm.3802] [PMID: 25706875]

[152] Roy K, Agarwal S, Banerjee R, Paul MK, Purbey PK. COVID-19 and gut immunomodulation. World J Gastroenterol 2021; 27(46): 7925-42.
[http://dx.doi.org/10.3748/wjg.v27.i46.7925] [PMID: 35046621]

[153] Helms J, Kremer S, Merdji H, *et al.* Neurologic Features in Severe SARS-CoV-2 Infection. N Engl J Med 2020; 382(23): 2268-70.
[http://dx.doi.org/10.1056/NEJMc2008597] [PMID: 32294339]

[154] Ramani A, Müller L, Ostermann PN, *et al.* SARS-CoV-2 targets neurons of 3D human brain organoids. EMBO J 2020; 39(20)e106230
[http://dx.doi.org/10.15252/embj.2020106230] [PMID: 32876341]

[155] Chen R, Wang K, Yu J, *et al.* The Spatial and Cell-Type Distribution of SARS-CoV-2 Receptor ACE2 in the Human and Mouse Brains. Front Neurol 2021; 11573095
[http://dx.doi.org/10.3389/fneur.2020.573095] [PMID: 33551947]

[156] Zhang BZ, Chu H, Han S, *et al.* SARS-CoV-2 infects human neural progenitor cells and brain organoids. Cell Res 2020; 30(10): 928-31.
[http://dx.doi.org/10.1038/s41422-020-0390-x] [PMID: 32753756]

[157] Han Y, Duan X, Yang L, *et al.* Identification of SARS-CoV-2 inhibitors using lung and colonic organoids. Nature 2021; 589(7841): 270-5.
[http://dx.doi.org/10.1038/s41586-020-2901-9] [PMID: 33116299]

[158] Sanyal R. Organoid Technology and the COVID Pandemic. SARS-CoV-2 Origin and COVID-19 Pandemic Across the Globe2021

[159] Zang R, Castro MFG, McCune BT, *et al.* TMPRSS2 and TMPRSS4 promote SARS-CoV-2 infection of human small intestinal enterocytes. Sci Immunol 2020; 5(47)eabc3582
[http://dx.doi.org/10.1126/sciimmunol.abc3582] [PMID: 32404436]

[160] Lamers MM, Beumer J, van der Vaart J, *et al.* SARS-CoV-2 productively infects human gut enterocytes. Science 2020; 369(6499): 50-4.
[http://dx.doi.org/10.1126/science.abc1669] [PMID: 32358202]

[161] Zhou J, Li C, Liu X, *et al.* Infection of bat and human intestinal organoids by SARS-CoV-2. Nat Med 2020; 26(7): 1077-83.
[http://dx.doi.org/10.1038/s41591-020-0912-6] [PMID: 32405028]

[162] Lancaster MA, Huch M. Disease modelling in human organoids. Dis Model Mech 2019; 12(7)dmm039347
[http://dx.doi.org/10.1242/dmm.039347] [PMID: 31383635]

[163] Perez-Lanzon M, Kroemer G, Maiuri MC. Organoids for Modeling Genetic Diseases. Int Rev Cell Mol Biol 2018; 337: 49-81.
[http://dx.doi.org/10.1016/bs.ircmb.2017.12.006] [PMID: 29551162]

[164] Baldassari S, Musante I, Iacomino M, Zara F, Salpietro V, Scudieri P. Brain Organoids as Model Systems for Genetic Neurodevelopmental Disorders. Front Cell Dev Biol 2020; 8590119
[http://dx.doi.org/10.3389/fcell.2020.590119] [PMID: 33154971]

[165] Hoffmann A, Ziller M, Spengler D. Progress in iPSC-Based Modeling of Psychiatric Disorders. Int J Mol Sci 2019; 20(19): 4896.
[http://dx.doi.org/10.3390/ijms20194896] [PMID: 31581684]

[166] Inak G, Rybak-Wolf A, Lisowski P, *et al.* Defective metabolic programming impairs early neuronal morphogenesis in neural cultures and an organoid model of Leigh syndrome. Nat Commun 2021; 12(1): 1929.
[http://dx.doi.org/10.1038/s41467-021-22117-z] [PMID: 33771987]

[167] Liu C, Qin T, Huang Y, Li Y, Chen G, Sun C. Drug screening model meets cancer organoid technology. Transl Oncol 2020; 13(11)100840
[http://dx.doi.org/10.1016/j.tranon.2020.100840] [PMID: 32822897]

[168] Breyer MD, Look AT, Cifra A. From bench to patient: model systems in drug discovery. Dis Model Mech 2015; 8(10): 1171-4.
[http://dx.doi.org/10.1242/dmm.023036] [PMID: 26438689]

[169] Pound P, Ritskes-Hoitinga M. Is it possible to overcome issues of external validity in preclinical animal research? Why most animal models are bound to fail. J Transl Med 2018; 16(1): 304.
[http://dx.doi.org/10.1186/s12967-018-1678-1] [PMID: 30404629]

[170] Jensen C, Teng Y. Is It Time to Start Transitioning From 2D to 3D Cell Culture? Front Mol Biosci 2020; 7: 33.
[http://dx.doi.org/10.3389/fmolb.2020.00033] [PMID: 32211418]

[171] Kondo J, Inoue M. Application of Cancer Organoid Model for Drug Screening and Personalized Therapy. Cells 2019; 8(5): 470.
[http://dx.doi.org/10.3390/cells8050470] [PMID: 31108870]

[172] Li L, Knutsdottir H, Hui K, *et al.* Human primary liver cancer organoids reveal intratumor and interpatient drug response heterogeneity. JCI Insight 2019; 4(2)e121490
[http://dx.doi.org/10.1172/jci.insight.121490] [PMID: 30674722]

[173] Kondo J, Ekawa T, Endo H, *et al.* High-throughput screening in colorectal cancer tissue-originated spheroids. Cancer Sci 2019; 110(1): 345-55.
[http://dx.doi.org/10.1111/cas.13843] [PMID: 30343529]

[174] Nuciforo S, Fofana I, Matter MS, *et al.* Organoid Models of Human Liver Cancers Derived from Tumor Needle Biopsies. Cell Rep 2018; 24(5): 1363-76.
[http://dx.doi.org/10.1016/j.celrep.2018.07.001] [PMID: 30067989]

[175] Saito Y, Muramatsu T, Kanai Y, *et al.* Establishment of Patient-Derived Organoids and Drug

Screening for Biliary Tract Carcinoma. Cell Rep 2019; 27(4): 1265-1276.e4.
[http://dx.doi.org/10.1016/j.celrep.2019.03.088] [PMID: 31018139]

[176]   Wilkinson DC, Alva-Ornelas JA, Sucre JMS, *et al.* Development of a Three-Dimensional Bioengineering Technology to Generate Lung Tissue for Personalized Disease Modeling. Stem Cells Transl Med 2017; 6(2): 622-33.
[http://dx.doi.org/10.5966/sctm.2016-0192] [PMID: 28191779]

[177]   Sucre JMS, Vijayaraj P, Aros CJ, *et al.* Posttranslational modification of β-catenin is associated with pathogenic fibroblastic changes in bronchopulmonary dysplasia. Am J Physiol Lung Cell Mol Physiol 2017; 312(2): L186-95.
[http://dx.doi.org/10.1152/ajplung.00477.2016] [PMID: 27941077]

[178]   van Ineveld RL, Ariese HCR, Wehrens EJ, Dekkers JF, Rios AC. Single-Cell Resolution Three-Dimensional Imaging of Intact Organoids. Journal of Visualized Experiments 2020.

[179]   Rakotoson I, Delhomme B, Djian P, *et al.* Fast 3-D Imaging of Brain Organoids With a New Single-Objective Planar-Illumination Two-Photon Microscope. Front Neuroanat 2019; 13: 77.
[http://dx.doi.org/10.3389/fnana.2019.00077] [PMID: 31481880]

[180]   Bolhaqueiro ACF, van Jaarsveld RH, Ponsioen B, Overmeer RM, Snippert HJ, Kops GJPL. Live imaging of cell division in 3D stem-cell organoid cultures. Methods Cell Biol 2018; 145: 91-106.
[http://dx.doi.org/10.1016/bs.mcb.2018.03.016] [PMID: 29957217]

[181]   Okkelman IA, Neto N, Papkovsky DB, Monaghan MG, Dmitriev RI. A deeper understanding of intestinal organoid metabolism revealed by combining fluorescence lifetime imaging microscopy (FLIM) and extracellular flux analyses. Redox Biol 2020; 30101420
[http://dx.doi.org/10.1016/j.redox.2019.101420] [PMID: 31935648]

[182]   Mazzucchelli S, Piccotti F, Allevi R, *et al.* Establishment and Morphological Characterization of Patient-Derived Organoids from Breast Cancer. Biol Proced Online 2019; 21(1): 12.
[http://dx.doi.org/10.1186/s12575-019-0099-8] [PMID: 31223292]

[183]   Qu Y, Yucer N, Garcia VJ, Giuliano AE, Cui X. hiPSC-Based Tissue Organoid Regeneration. Tissue Regeneration 2018.

[184]   Ariyachet C, Tovaglieri A, Xiang G, *et al.* Reprogrammed Stomach Tissue as a Renewable Source of Functional β Cells for Blood Glucose Regulation. Cell Stem Cell 2016; 18(3): 410-21.
[http://dx.doi.org/10.1016/j.stem.2016.01.003] [PMID: 26908146]

[185]   Okamoto R, Shimizu H, Suzuki K, *et al.* Organoid-based regenerative medicine for inflammatory bowel disease. Regen Ther 2020; 13: 1-6.
[http://dx.doi.org/10.1016/j.reth.2019.11.004] [PMID: 31970266]

[186]   Dong X, Xu S-B, Chen X, Tao M, Tang X-Y, Fang K-H, *et al.* Human cerebral organoids establish subcortical projections in the mouse brain after transplantation. Mol Psychiatry 2020.
[PMID: 33051604]

[187]   Eiraku M, Takata N, Ishibashi H, *et al.* Self-organizing optic-cup morphogenesis in three-dimensional culture. Nature 2011; 472(7341): 51-6.
[http://dx.doi.org/10.1038/nature09941] [PMID: 21475194]

[188]   Hohwieler M, Illing A, Hermann PC, *et al.* Human pluripotent stem cell-derived acinar/ductal organoids generate human pancreas upon orthotopic transplantation and allow disease modelling. Gut 2017; 66(3): 473-86.
[http://dx.doi.org/10.1136/gutjnl-2016-312423] [PMID: 27633923]

[189]   Bisht B, Hope A, Paul MK. From papyrus leaves to bioprinting and virtual reality: history and innovation in anatomy. Anat Cell Biol 2019; 52(3): 226-35.
[http://dx.doi.org/10.5115/acb.18.213] [PMID: 31598350]

[190]   Ong CS, Yesantharao P, Huang CY, *et al.* 3D bioprinting using stem cells. Pediatr Res 2018; 83(1-2): 223-31.

[http://dx.doi.org/10.1038/pr.2017.252] [PMID: 28985202]

[191] Moroni L, Burdick JA, Highley C, *et al.* Biofabrication strategies for 3D *in vitro* models and regenerative medicine. Nat Rev Mater 2018; 3(5): 21-37.
[http://dx.doi.org/10.1038/s41578-018-0006-y] [PMID: 31223488]

[192] Rothenbücher TSP, Gürbüz H, Pereira MP, Heiskanen A, Emneus J, Martinez-Serrano A. Next generation human brain models: engineered flat brain organoids featuring gyrification. Biofabrication 2021; 13(1)011001
[http://dx.doi.org/10.1088/1758-5090/abc95e] [PMID: 33724233]

[193] Reid JA, Palmer XL, Mollica PA, Northam N, Sachs PC, Bruno RD. A 3D bioprinter platform for mechanistic analysis of tumoroids and chimeric mammary organoids. Sci Rep 2019; 9(1): 7466.
[http://dx.doi.org/10.1038/s41598-019-43922-z] [PMID: 31097753]

[194] Skylar-Scott MA, Uzel SGM, Nam LL, *et al.* Biomanufacturing of organ-specific tissues with high cellular density and embedded vascular channels. Sci Adv 2019; 5(9)eaaw2459
[http://dx.doi.org/10.1126/sciadv.aaw2459] [PMID: 31523707]

[195] Brassard JA, Nikolaev M, Hübscher T, Hofer M, Lutolf MP. Recapitulating macro-scale tissue self-organization through organoid bioprinting. Nat Mater 2021; 20(1): 22-9.
[http://dx.doi.org/10.1038/s41563-020-00803-5] [PMID: 32958879]

[196] Cetnar AD, Tomov ML, Ning L, Jing B, Theus AS, Kumar A, *et al.* Patient-Specific 3D Bioprinted Models of Developing Human Heart. Adv Healthc Mater 2020.
[PMID: 33274834]

[197] Rachamalla H, Mukherjee A. K. Paul M. Nanotechnology Application and Intellectual Property Right Prospects of Mammalian Cell Culture. Cell Culture. Biochemistry 2021. [Working Title].

[198] Choudhury D, Ashok A, Naing MW. Commercialization of Organoids. Trends Mol Med 2020; 26(3): 245-9.
[http://dx.doi.org/10.1016/j.molmed.2019.12.002] [PMID: 31982341]

[199] Steeg R, Neubauer JC, Müller SC, Ebneth A, Zimmermann H. The EBiSC iPSC bank for disease studies. Stem Cell Res (Amst) 2020; 49102034
[http://dx.doi.org/10.1016/j.scr.2020.102034] [PMID: 33099110]

[200] Kimura S, Tsuji T. Mechanical and Immunological Regulation in Wound Healing and Skin Reconstruction. Int J Mol Sci 2021; 22(11): 5474.
[http://dx.doi.org/10.3390/ijms22115474] [PMID: 34067386]

[201] Schreurs RRCE, Baumdick ME, Drewniak A, Bunders MJ. In vitro co-culture of human intestinal organoids and lamina propria-derived CD4+ T cells. STAR Protocols 2021; 2(2)

[202] Xia T, Du WL, Chen XY, Zhang YN. Organoid models of the tumor microenvironment and their applications. J Cell Mol Med 2021; 25(13): 5829-41.
[http://dx.doi.org/10.1111/jcmm.16578] [PMID: 34033245]

[203] Ormel PR, Vieira de Sá R, van Bodegraven EJ, *et al.* Microglia innately develop within cerebral organoids. Nat Commun 2018; 9(1): 4167.
[http://dx.doi.org/10.1038/s41467-018-06684-2] [PMID: 30301888]

[204] Grebenyuk S, Ranga A. Engineering Organoid Vascularization. Front Bioeng Biotechnol 2019; 7: 39.
[http://dx.doi.org/10.3389/fbioe.2019.00039] [PMID: 30941347]

[205] Homan KA, Gupta N, Kroll KT, *et al.* Flow-enhanced vascularization and maturation of kidney organoids in vitro. Nat Methods 2019; 16(3): 255-62.
[http://dx.doi.org/10.1038/s41592-019-0325-y] [PMID: 30742039]

[206] Lim GS, Hor JH, Ho NRY, *et al.* Microhexagon gradient array directs spatial diversification of spinal motor neurons. Theranostics 2019; 9(2): 311-23.
[http://dx.doi.org/10.7150/thno.29755] [PMID: 30809276]

[207]   Zhang YS, Aleman J, Shin SR, *et al.* Multisensor-integrated organs-on-chips platform for automated and continual in situ monitoring of organoid behaviors. Proc Natl Acad Sci USA 2017; 114(12): E2293-302.
[http://dx.doi.org/10.1073/pnas.1612906114] [PMID: 28265064]

[208]   Iordachescu A, Hughes EAB, Joseph S, Hill EJ, Grover LM, Metcalfe AD. Trabecular bone organoids: a micron-scale 'humanised' prototype designed to study the effects of microgravity and degeneration. npj Microgravity 2021; 7(1)

[209]   Kumar V, Varghese S. *Ex vivo* Tumor-on-a-Chip Platforms to Study Intercellular Interactions within the Tumor Microenvironment. Adv Healthc Mater 2018.
[PMID: 30516355]

CHAPTER 2

Organoid Technology: Disease Modeling, Drug Discovery, and Personalized Medicine

Jyotirmoi Aich[1], Sangeeta Ballav[2], Isha Zafar[2], Aqdas Khan[1], Shubhi Singh[1], Manash K. Paul[3,4], Shine Devrajan[1] and Soumya Basu[2,*]

[1] *School of Biotechnology and Bioinformatics, Dr. D. Y. Patil Deemed to be University, CBD Belapur, Navi Mumbai, Maharashtra, 400 614, India*

[2] *Cancer and Translational Research Centre, Dr. D. Y. Patil Biotechnology and Bioinformatics Institute, Dr. D. Y. Patil Vidyapeeth, Pune, Maharashtra, 411 033, India*

[3] *Department of Radiation Biology and Toxicology, Manipal School of Life Sciences, Manipal Academy of Higher Education, Manipal, 576104, India*

[4] *Department of Pulmonary and Critical Care Medicine, David Geffen School of Medicine, University of California Los Angeles, Los Angeles, California, 90095, USA*

Abstract: The last three decades have witnessed revolutionary growth in the fields of biomedical science and pioneering the same in regenerative medicine and disease modeling. Historically, biological research has been performed using 2-dimensional animal cell culture, but now we are switching to more intricate 3-dimensional models for better replicability of experimental results. Organoids are stem cell-derived 3D cell cultures that are the cornerstone of this new development. They retain the significant features of biological organs and have opened up new, previously not-thought-of avenues to steer research in personalized healthcare and disease modeling. The current chapter encapsulates how organoids came into the picture, addresses the current research occurring worldwide, and discusses futuristic aspects and applications. The significance of organoids in disease modeling is discussed in detail, and the following aspects, such as disease modeling in congenital conditions, cancer, infectious diseases, gene editing, and futuristic microfluidics, were elucidated. This chapter also covers the role of organoids in drug discovery. Drug discovery is a very time and money-intensive process, and many attempts have been made over the years to bring about change in the same. It has been noted that the development of many new drugs is being hindered due to the complexity of the human genome. This point has been elaborately discussed in this present chapter, along with the potential of organoids as a solution in high-throughput drug screening and personalized treatment. The chapter concludes with a look at how the COVID-19 pandemic has underpinned the use of organoids in drug research and disease modeling, and finally, it provides a summary of future research directions.

---

* **Corresponding author Soumya Basu**: Cancer and Translational Research Centre, Dr. D.Y. Patil Biotechnology and Bioinformatics Institute, Dr. D. Y. Patil Vidyapeeth, Pune, Maharashtra, 411 033, India; E-Mail: soumya.bs@gmail.com

Keywords: Drug discovery, Disease modeling, Organoids, 2-dimensional culture, 3-dimensional culture.

INTRODUCTION AND THE BASICS OF ORGANOIDS

The past decades have witnessed significant research and development in biomedical science concerning regenerative medicine, disease modeling, drug discovery, and drug repurposing. The advent of cutting-edge technologies in 3-dimensional (3D) biology, like spheroids, organoids, organs-on-chip, 3D hydrogels, 3D co-cultures, and 3D bioprinting, has delivered a powerful boost towards precision and personalized medicine. Earlier, 2-dimensional (2D) monolayer cultures were prominently used, but the existing lacunae and the concomitant development of 3D culture systems have limited their use. Adherent 2D cultures grow as stretched and flat adherent monolayers of cells in contrast to their original *in-vivo* morphology. Proliferation, apoptosis, differentiation, gene, and protein expression are some of the essential cellular processes that may be influenced by this altered morphology [1, 2]. 2D monocultures of cell lines often fail to simulate the *In vivo* cellular functions, tissue-specific cell-cell, and cell-matrix interactions as observed in 3D cultures [3].

Though 2D-based cell culture systems were the main-stay for drug discovery, the inherent drawbacks of the said system have led to the delayed progress of drug molecules through the clinical pipeline, and the majority of the pharmaceutical compounds failed in human clinical trials. Despite promising results in 2D cultures, the same could not be reproduced in clinical settings. Notable differences between 2D cultured *in-vitro* systems and actual *in-vivo* conditions may lead to differential responses to treatment [4]. These limitations have been overcome mainly by transitioning to 3D cell culture models. These 3D models have arisen methodologically and conceptually because of the classical re-aggregation experiments that demonstrated that cells segregated from embryonic organs have the capability to re-aggregate and re-constitute the original organ structure. This chapter intends to focus on organoids as 3D model systems to study disease modeling, drug discovery, and personalized medicine.

An inclusive definition will refer to 'organoids' as Minuscule replicas of corresponding biological organs or mini-organs, that are cultured as 3D structures derived either from pluripotent or organ-specific stem cells, supported on 3D gels constituting the extracellular matrix (ECM), fortified with specific nutrients, signaling molecules, and growth factors so that they harbor the ability to replicate their biological counterparts in terms of architecture complexity and physiological functions, as well as having the multi-lineage commitment and differentiation potential [5]. Till now, multiple human organoids like brain, kidney, lung, stomach, liver, pancreas, endometrium, prostate, pancreas, thyroid, and retinal organoids have been produced [6] in different studies.

Before proceeding toward the various applications of organoids, it is imperative to understand how the microenvironment influences organoids. Engineering tissue-specific balanced and requisite microenvironment conditions is paramount to ensure the successful growth and development of an accurate organoid. The imbalance of any essential component may skew the organoid maturation and result in aberrant histomorphology, altered size, and cellular composition [7]. Organoids can be classified according to the type of stem cells [adult stem cells (ASCs) or pluripotent stem cells (PSCs)] used in their formation (Fig. **1**). PSCs (which give rise to three primary layers in the early embryo stage) can be re-activated by tuning and reprogramming them to form differentiated somatic cells. Induced Pluripotent Stem Cells (iPSCs) have great differentiation potential and have been progressively used for organoid cultures. Organoids may be described as "blank states" devoid of signals and need molecular cues to facilitate their growth and differentiation. Thus, delineating the role of extrinsic biochemical and physical cues is crucial for the development process. This enables a quantitative platform for the study of biological processes as significant experimental control can be established through their use [8].

The role of physical cues is to support cell attachment and maintain the structure for survivability [9]. Two types of techniques can be used for organoid development: i) Scaffold matrices and ii) Scaffold-free techniques. The matrigel derived from Engelbreth–Holm–Swarm mouse sarcomas is the most often utilized matrix. Four major basement membrane ECM proteins that make up the matrigel are: laminin (60 percent), collagen IV (30 percent), entactin (8 percent), and the heparin sulfate proteoglycan perlecan (2–3%) [10]. Scaffold-free techniques make optimal use of the self-aggregation properties of cells when kept in hanging drop microplates or low-adhesion plates to form spheroids and cell sheets [11]. Biochemical cues are the molecules that are responsible for the regulation (up or down-regulation) of the specific signaling pathways, which differ with the type of organoid cultured, the type of stem cell used (whether PSC or ASC), and the level of differentiation required [9]. Some of these biochemical cues are growth factors like insulin-like growth factor (IGF), vascular endothelial growth factor (VEGF), fibroblast growth factors (FGF), epidermal growth factor (EGF), hepatocyte growth factor (HGF), transforming growth factor (TGF), mitogen-activated protein kinase (MAPK), bone morphogenetic protein (BMP), RHO-associated protein kinase (ROCK), WNT-related integration site [7], and R-spondin (WNT regulator) [9]. Notch, Hedgehog, Wnt, and mTOR signaling pathways are frequently associated with and targeted in organoid technology. The physical and biochemical cues, including angiogenic growth factors, are transferred through

ECM to stem cells [12]. Manipulation of Wnt and hedgehog signaling pathways in retinal organoids using various molecules has recently been reviewed and reported [13]. In addition, *in-vivo* patterning of tissues also depends on mechanical forces (like shear stress, tension, compression, and hydrostatic pressure), giving rise to extending, bending, and twisting motions needed for accurate morphogenesis [14]. Organoids have already been explored for their widespread therapeutic and prophylactic applications. Stem cell-based human organoids have enabled and empowered disease modeling to be carried out with feasibility [6] as there are many prevalent diseases (pathogenic as well as genetic) that affect humans but do not perturb naturally in other species thereby needing artificial induction for disease simulation and modeling [15]. A recent example has been the proposed use of organoids to understand the pathophysiology and infectivity of severe acute respiratory syndrome coronavirus 2 (SARS-CoV2) using bronchial organoids [16]. The applications of organoids with respect to disease modeling and drug discovery are further elaborated in subsequent sections of this chapter. Further, the current chapter throws insights into high throughput drug screening for drug discovery and drug screening for personalized medicine.

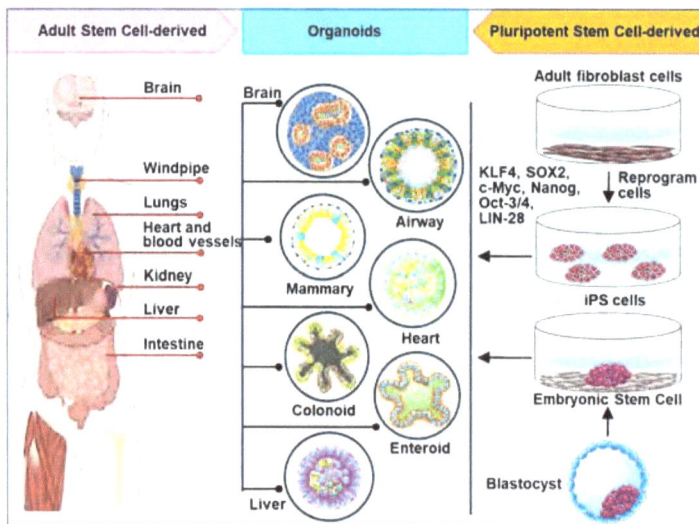

Fig. (1). Organoid Models of Diseases.

DISEASE MODELING AND ORGANOIDS

In recent years, organoid technology has been used to learn more about stem cell biology, organogenesis, and different human pathologies. Replication of various biological phenomena that occur in humans is not entirely feasible using animal models only. For example - human anatomy and physiology are comparatively very dissimilar from the *In vivo* mouse model [17]. So, organoids are made up of cell types that are similar to those found within the human body [18]. The latest developments in human patient-derived organoids have allowed us to proceed with disease modeling with great accuracy. The future potential can be seen in biomedical applications, translational medicine, and personalized therapy [19].

Organoids have become potent instruments for simulating many illnesses, providing notable benefits compared to conventional cell culture techniques and animal models. Organoids are used for the purpose of modeling a number of diseases, as elucidated subsequently.

Congenital Conditions

Using Cystic Fibrosis (CF) organoids, researchers can investigate various aspects of CF pathophysiology, including mucus production, ion transport defects, inflammation, and susceptibility to infection. In 2013, the first CF-patient-derived intestinal organoids were generated [20]. The researchers discovered that certain organoids act in response to the activation of cAMP by observing the swelling that takes place when a fluid is imported into the lumen. This method was adopted in the Netherlands, becoming a one-of-a-kind personalized medical treatment for CF patients [21]. One published report showed that CF could be modeled *In vitro* by using iPSCs from CF patients as the iPSCs got differentiated into liver cholangiocytes [22]. Another study demonstrated the use of cerebral organoids to model human microcephaly, a genetic condition caused by a mutation in the CDK5 regulatory subunit-associated protein 2 gene (CDK5RAP2) [23]. Leber's congenital amaurosis is a genetic disorder that affects the retina and causes hereditary blindness. A known genetic cause of this disorder was a mutation in the CE290 gene. To study the same, Parfitt *et al.* used iPSCs with the mutation and generated retinal organoids. At the initial stage, the organoids normally develop into optic cups. However, decreased ciliation and cilia lengths were observed in the resulting tissues [24].

Cancer Organoids

Since researchers were enlightened by the fact that HeLa cells could be grown *in vitro*, several cell lines have been established for most tumor types. However, there are certain downsides to the same, like – i) Cell lines do not have the tissue architecture of the organ in question. ii) A strong cellular selection is required. iii) They show widespread non-uniformity when compared between different laboratories, with significant dissimilarity in gene expression and proliferation along with a significant difference in drug response [25]. For these reasons, scientists are always on the lookout for alternatives. A major breakthrough was made with the discovery that healthy tissue from the colon and stomach could be expanded *In vitro* [26]. Consequently, several researchers began investigating organoid technology to mimic and model different aspects of cancer. Currently, organoids derived from tissue resections and biopsies have become more prevalent. The genetic and phenotypic features of the tumor of origin are also retained by cancer-derived organoids. There is a major possibility of opening of living biobanks of cancer-derived organoid culture from different tumor types like colorectal, gastric, breast, and bladder cancer, and this, in turn, can provide various possibilities for not just drug screening but also for the process of drug development [27 - 29]. Although, as of now, only two studies – colon and bladder cancer have shown the [30] promising potential of cancer organoids for predicting patient response [31, 32].

Infectious Diseases

A plethora of published reports have demonstrated the use of organoids as a model to study infectious diseases and their mechanisms. Norovirus [HuNoV] is a contagious virus that causes diarrhea and vomiting. Using small intestinal organoids, researchers have been able to model norovirus infection and transmission [33]. Intestinal organoids have also been used to decipher the small intestine as an alternative pathway for MERS-CoV, which causes severe human respiratory tract infections. *Helicobacter pylori* infections were also studied using gastric organoid models. On the other hand, lung organoids have been used to model influenza virus infection *in vitro*. Likewise, some researchers are currently using brain organoids that mimic inherited microcephaly to focus on the mechanisms of Zika-induced microcephaly. The study sheds light on how brain organoids are developed and how easy it is for independent researchers to opt for this advancement in technology. It is very interesting that preliminary findings of Zika's effects on brain organoids were published just 3 months after the World Health Organization (WHO) declared the Zika virus a major worldwide crisis in 2016. These observations describe the smaller overall size of infected patient-derived organoids and specific effects on neural progenitors, leading to cell death, decreased proliferation, and early differentiation [34]. At least five independent research groups have recently used these Zika virus-infected brain organoids to test therapeutic programs and ideas to fight the effect of the virus on neural progenitor cells [35].

Methods in Organoid Technology

The field of organoid research is seeing ongoing advancements in advanced techniques, which are progressively improving their effectiveness in disease modeling, drug exploration, and regenerative medicine. The area of organoid-based disease modeling has been significantly improved by several techniques, such as CRISPR-Cas9 technology, microfluidics and organ-on-a-chip systems, coculture and organoid fusion, bioprinting and biomaterial engineering, and multi-omics technologies.

Gene Editing in Organoids

Gene editing in organoids was initiated by the Clevers laboratory. With the help of CRISPR/Cas9 and organoid technology, the researchers set the right mutations and corrected the chloride channel function of intestinal organoids sourced from a patient suffering from cystic fibrosis [36]. This major breakthrough in organoid gene editing opened many different approaches and new avenues. In colon organoids, gene editing has allowed us to work with the step-wise repetition of tumorigenesis *In vitro* and identify cancer signatures in microsatellite-unstable tumors. Another noteworthy development is the creation of the first human brain organoid cancer model for neuroectodermal tumors [37].

Microfluidics and Organoids: A Quantum Leap

Microfluidic organoids or organ-on-a-chip platforms represent an upcoming group of models that generalize three-dimensional tissue structure and physiology. Microfluidics technology is used in many different ways because it offers us more control and ways to change fluid flow with a high degree of precision. The amalgamation of microfluidics with cell biology led to organoids or organs-on-a-chip [38]. Organs-on-chips are small integrated chips that have transparent 3D polymeric microchannels. These channels perform the following important functions of organ structures - i) 3D microarchitecture, which is made up of various tissue types; ii) Stimulation induced by mechanical and biomechanical forces; and iii) Lastly, the multi-tissue integrations [39, 40].

A few important fundamentals, like the supply of nutrients, shear stress, and flow parameters, can easily be monitored in a microfluidic organ-on-a-chip [41]. Cells can now also be stimulated with the help of mechanical or electrical stimuli and sometimes even through both. With the recent advancements in microfluidic technology, organ-on-a-chip will be upgraded to barriers-on-chip and body-on-a-chip models very soon [42]. Another important aspect of organs-on-chips is that they can be utilized to replicate the current physiological intricacies of the human body. This is also known as "body-on-a-chip". Due to these applications, the ADME of drugs through different organs can be studied [43]. The combination of microfluidic and organoid technology is an upcoming tool for studying disease development and progression. An example would be the use of this technology in personalized therapeutic strategies for cancer treatment. With the help of this, well-grounded, fast, robust, cost-effective, and reproducible results can be formulated [41].

The gut-on-a-chip is an interesting example of organ-on-a-chip models, where gut organoids are derived from intestinal stem cells or patient biopsies and grown inside a microfluidic device. Nutrition, shear stress, and oxygen tension may be managed by perfusing cell culture media into microfluidic channels [41]. Gut-o--a-chip models are currently being used to examine intestinal barrier function by measuring epithelial permeability and are effective in studying intestinal permeability diseases, including IBD and leaky gut syndrome. Gut-on-a-chip models may be colonized with commensal or pathogenic microbes to study host-microbe interplay, followed by drug screening absorption, and toxicity. By inserting inflammatory stimuli or patient-derived immune cells into microfluidic channels, gut-on-a-chip models are used to study intestinal inflammation and IBD. These models can study immune cell recruitment, cytokine production, and tissue damage caused by inflammation [43]. Thus, gut-on-a-chip models combine organoid technology and microfluidic culture system to study intestinal biology, disease processes, and pharmaceutical responses in a physiologically realistic environment.

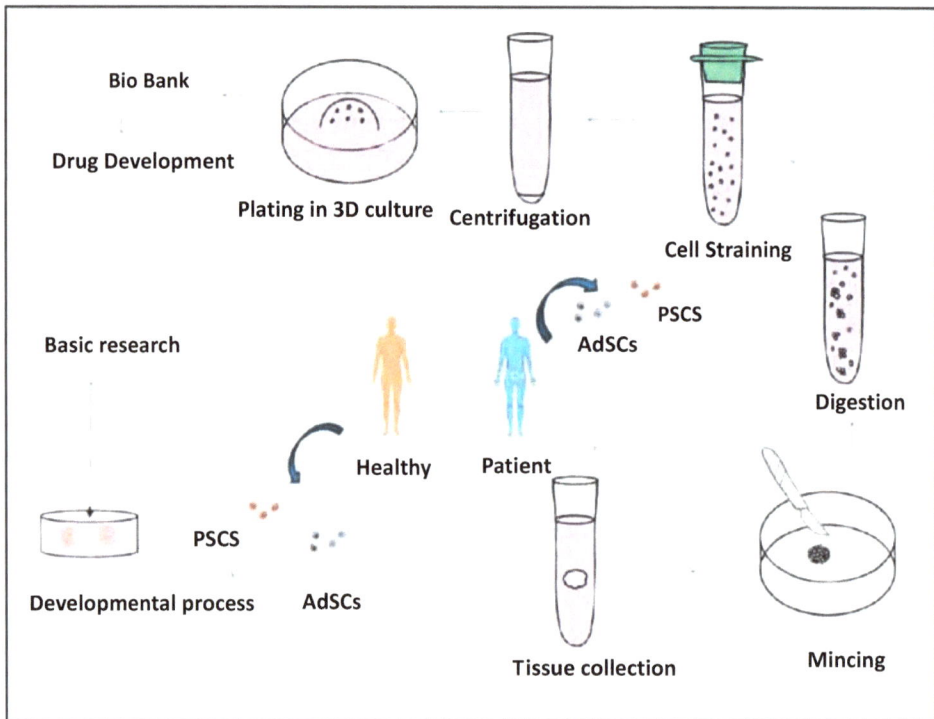

Fig. (2). Schematic diagram of human organoids for drug development. The most commonly used organoids are, i) Pluripotent stem cells (PSCs) for basic research, and ii) Human adult stem cells (ASCs) to understand the mechanism of human disease and drug screening.

ORGANOID MODEL FOR DRUG DISCOVERY

Since the last decade, organoid technology has gained momentum in the literature to reduce the gap between *In vitro* and *In vivo* studies to test drug responses. Various patient-derived organoids have undergone screening for different chemotherapeutic drugs [44 - 46], and the information derived from the same has provided a logical basis for personalized medicine for different patients, as depicted in Fig. (**2**) [47, 48].

One of the most prominent liver organoid culture protocols by Broutier *et al.* [49] demonstrated exogenous addition of Rho-kinase inhibitors, dexamethasone inhibitors, and removal of R-spondin-1, Noggin, and Wnt3a suggested that primary liver tumors harbored different gene mutation patterns depicting differential sensitivities to therapeutic drugs. The subsequent accumulation has been reported in the literature, suggesting that human EGF receptor-targeted breast cancer organoids appeared as pillars for drug screening [50]. Furthermore, the role of patient-derived organoids (PDO) is becoming more apparent in the literature, supporting the notion that PDO is used in high-throughput drug screening. For example, using 56 PDOs not only marks a hierarchical subset of biomarker identification and drug screening but also targets the pathway for further modulation by secreted factors, which led to the identification of inhibitors as a potential therapeutic agent [51, 52]. Such studies also highlight the use of PDO in a wide range of biomedical utilities in developing personalized medicine [53]. It is becoming increasingly evident that genome sequencing of patients, followed by high-throughput screening of drugs using the organoid model, facilitates the transition of 3D cultures over 2D cultures in identifying personalized therapeutic strategies [54, 55]. A growing body of evidence has shown the combination of organoid technologies with small-micelle-mediated human organ efficient clearing and labeling (SHANEL) to obtain transparent organoids, providing an approach toward quickly identifying drug targets [56].

Overall, the applications of such culture systems are elaborated; delineating micro-changes at the cellular level to be utilized for large-scale drug screens and drug development [57]. However, the structural and biological complexity of 3D organoid models in precision medicine has many challenges, and their relevance to current clinical therapeutics needs to be explored [58]. Table **1** presents some of the significant organoid studies with associated protocols.

Table 1. Details of significant organoid studies with associated protocols.

Source of Organoid	Disease	Drugs Used	Significance of Study	Reference
Human mammary epithelial-derived organoids	Breast cancer	Afatinib AZD8055 everolimus GDC-0068 gefitinib pictilisib	Described organoids bank available for *In vitro* personalized drug screening.	[52]
Stem cell-based islet organoids	Diabetes	Galunisertib	Described disease models and regenerative therapies for diabetes.	[53]
Synovial and chondral organoids	Rheumatoid arthritis		Demonstrated that chondral organoids stimulate a higher rate of physiological cartilage architecture.	[54]
Hepatocyte derived organoids	Liver disease	GSK3β- and ROCK1	Described cell-based therapies in regenerative medicine.	[49]
Human iPSC-derived cerebral organoids	Alzheimer's disease (AD)	FDA-approved AD repositioning drugs	Demonstrated AD drug development by integrating mathematical modeling and the pathological features of human ICOs.	[57]

HIGH THROUGHPUT DRUG SCREENING FOR DRUG DISCOVERY

In the past few years, many clinically established drugs have been stalled due to the complexity and heterogeneity of the human genome. Despite the elusiveness of these drugs, efforts continue to develop novel methods for accurate and individualized drug screening. Fueled by the recognition of organoid cultures, scientists became more attentive to screening drugs on tissue models due to their effectiveness and time-saving features. The governing factors of an organoid culture environment include complex cellular interactions, tissue-specific architecture, self-renewal ability, and preservation of original tissue characteristics [59]. Pertaining to these factors, to date, many organoid-based models have been created from many types of cells, such as adult, embryonic, primary, and stem cells. Thereafter, profuse research studies owe the significance of these organoid models in varying applications that have inspired the ability of organoid models for drug development and personalized treatment strategies. In an instance, the amalgamation of vorinostat and buparlisib was found to be potent for uterine carcinosarcoma patients possessing mutations in phosphatase and tensin homolog (PTEN) and phosphatidylinositol-4,5-bisphosphate-3-kinase catalytic subunit alpha (PIK3CA). In addition, administering another drug combination, buparlisib, and olaparib exhibited better anti-cancer effects in the endometrial adenocarcinoma organoid with mutations in PIK3CA and PTEN [60]. In another instance, mammary tumor organoid culture has been exploited to examine its potentiality with doxorubicin and latrunculin A. The authors used non-adhesive alginate to develop the organoid culture by encapsulating the mouse tumor pieces produced with a droplet microfluidic device. The results revealed that organoids manifested different responses to these two drugs, mainly due to differential uptake patterns influenced by the luminal size and pressures of organoids [61].

Organoid technology has eased the outlook in therapeutics and drug development research to understand the characteristic features of tissue-specificity, heterogeneity, organoid stability, and drug sensitivity. One such example demonstrates a novel, robust, and miniaturized assay that identifies the sensitivity of drugs on organoids obtained from patients through surgery. The method involved the application of simplified geometry that consists of mini-ring (seeding cells around the rim of the wells) [62]. This high-throughput drug screening approach has proven feasible in profiling the potential and actionable drugs on the constructed organoids. One of the important advantages of using this mini-ring approach is that it requires a very small number of cells for seeding, which narrows down the expansion to *In vitro* or *in vivo*. Recently, Pleguezuelos-Manzano and colleagues established human intestinal organoids from tissue biopsies of epithelial cells of adult stem cells. These intestinal organoids have proved futile in retaining the characteristic features of the original tissue along with genome stability [63]. Au *et al.* reported the development of microfluidic organoids prepared from liver tissue for high-throughput screening of drugs. This digital system consists of arrays of hepatic organoids that manifested better contractile behavior, which is fibroblast-dependent, and recapitulated the physiological behavior, preferably more nuanced than that of the *In vivo* liver tissue. These features of microfluidic organoids provide an advantage in detecting early-stage hepatotoxicity [64].

DRUG SCREENING FOR PERSONALIZED MEDICINE

Organoids hold great potential for successful personalized treatment due to their novelty and unique ability to mimic the functional and structural motifs of real organs. This highly dynamic, cutting-edge technology has filled critical gaps in biomedical research wherein researchers are able to study the *In vitro* lineage specification, organogenesis, and tissue homeostasis. Over the last few years, PDOs have paved the way for comprehending various diseases as they offer a better screening of drugs with respect to penetration ability, binding activity, and radiation sensitivity in individual probes [65]. Considering these, PDOs play a decisive role in therapeutics and biomedical research. A recently published report by Tiriac *et al.* exemplarily established 66 pancreatic cancer PDO libraries from fine-needle biopsies, surgical resection, and rapid autopsy pancreatic ductal adenocarcinoma (PDAC) specimens that could mimic the mutational spectrum and transcriptional subtypes of primary pancreatic cancer. To confirm the reliability of the developed PDO library, they conducted a case study, which showed that longitudinal PDO generation paralleled patient outcomes [66]. Along the line, the authors developed a drug-testing pipeline composed of PDAC-specific PDO and termed pharmacotyping. This novel approach illustrates the sensitivity of drugs for each PDO within a given timeframe and serves as guidance for precision medicine in pancreatic cancer [67].

More and more advances are achieved in diabetes, which has widened the spectrum of personalized medicine targets. In this aspect, islet organoid and patient-derived β cells have gained considerable momentum to improve the heterogeneity and functionality of islets. Mainly, iPSCs reprogrammed either from fibroblasts/peripheral blood mononuclear cells (PBMCs) or ASCs are applied to derive personalized islet organoids and patient-derived β cells from pancreatic fragments [68]. With the advent of iPSC technology and personalized medicine, these organoids provide informative disease models for recapitulating structural elements of specific patients, which are further used to examine drug reactions upon screening. In an instance, Millman *et al.* reported an efficient drug screening organoid model wherein the authors generated functional stem cell-derived β-cells (SC-β cells) from type 1 diabetic (T1D) patients. They treated the SC-β cells with three anti-diabetic compounds, namely, tolbutamide, liraglutide, and LY2608204, eliciting a surge in insulin release [69].

Besides these, one of the most challenging and risky infectious diseases is SARS-CoV-2. The global pandemic is prompting the advent of many novel therapeutic strategies for coronavirus 2019 (COVID-19) and associated models to mimic the viral infection on a dish. For COVID-19 therapeutics development, there are two *In vitro* approaches to screen drugs: 2-dimensional and 3-dimensional cell culture-based platforms. Hence, researchers are investigating comprehensive knowledge behind the pathogenesis and trying to develop COVID-19 prophylactic approaches that could screen the drugs for primary preventive measures. A study conducted by Ramezankhani *et al.* provides detailed insights into recapitulating virus-body interaction. According to their report, organoids and microfluidic devices are currently the most promising and reliable platforms [70]. Most models proposed so far comprise monkey kidney-derived cell lines that have susceptibility to SARS-CoV infection and mimic the exact replication process of the virus due to the high expression of ACE-2 in kidney tissue [71]. Table **2** summarizes different advantages and disadvantages of various applications of organoids.

Table 2. Pros and cons of various applications of organoids.

Application	Pros	Cons
Fundamental research	- Capable of self-renewal - Requires a small amount of tissue - Permits co-cultures - Resembles tissue architecture, complexity, and gene expression of native tissue.	- Relative accessibility to hospitals or tissue networks is required. - Optimization protocols are not globally standardized. - Static condition
Disease models	- Provides better physiological characteristics. - Permits comparison between multiple cases for heterogeneity. - Amenable to high-throughput screening (HTS).	- Relative accessibility to hospitals or tissue networks is required - Not feasible, as it requires proper skills - HTS is costly.
Precision and personalized medicine	- Patient-specific - Potential for genome editing.	- Requires selective procedures for corrected clones.

CONCLUSION

Despite current limitations, organoid technology, as both models and methods, holds a benchmark for evaluating preclinical drug candidates, identifying biomarkers, and proposing individualized therapy. It aids as an extraordinary platform for drug screening and personalized medicine. Screening of drugs using organoid models has efficiently improved the therapy response prediction and potentially served in determining better drug candidates for individual patients. Future efforts must work on the reliability, feasibility, and optimization of these developed organoid models that could refine the clinical application.

ABBREVIATIONS

2D	2-Dimensional
3D	3-Dimensional
ASCs	Adult stem cells
AD	Alzheimer's disease
BMP	Bone morphogenetic protein
CF	Cystic Fibrosis
CDK5RAP2	CDK5 regulatory subunit-associated protein 2
COVID-19	Coronavirus 2019
ECM	Extracellular matrix
EGF	Epidermal growth factor
FGF	Fibroblast growth factor

HGF	Hepatocyte growth factor
HuNoV	Norovirus
HTS	High-throughput screening
iPSCs	Induced pluripotent stem cells
IGF	Insulin-like growth factor
MAPK	Mitogen-activated protein kinase
PSCs	Pluripotent stem cells
PDO	Patient-derived organoids
PTEN	Phosphatase and tensin homolog
PIK3CA	Phosphatidylinositol-4-5-bisphosphate-3-kinase catalytic subunit alpha
PDAC	Pancreatic ductal adenocarcinoma
PBMCs	Peripheral blood mononuclear cells
ROCK	RHO-associated protein kinase
SHANEL	Small-micelle-mediated human organ efficient clearing and labeling
SC-β	cells Stem cell-derived β-cells
T1D	Type 1 diabetes
TGF	Transforming growth factor
VEGF	Vascular endothelial growth factor
WHO	World Health Organization

ACKNOWLEDGEMENTS

Intramural Grants, Dr. D. Y. Patil Vidyapeeth (DPU), Pimpri, Pune, India to S. Basu [DPU/644-43/2021].

REFERENCES

[1]   Tibbitt MW, Anseth KS. Hydrogels as extracellular matrix mimics for 3D cell culture. Biotechnol Bioeng 2009; 103(4): 655-63.
[http://dx.doi.org/10.1002/bit.22361] [PMID: 19472329]

[2]   Edmondson R, Broglie JJ, Adcock AF, Yang L. Three-dimensional cell culture systems and their applications in drug discovery and cell-based biosensors. Assay Drug Dev Technol 2014; 12(4): 207-18.
[http://dx.doi.org/10.1089/adt.2014.573] [PMID: 24831787]

[3]   Yin X, Mead BE, Safaee H, Langer R, Karp JM, Levy O. Engineering stem cell organoids. Cell Stem Cell 2016; 18(1): 25-38.
[http://dx.doi.org/10.1016/j.stem.2015.12.005] [PMID: 26748754]

[4]   Breslin S, O'Driscoll L. Three-dimensional cell culture: the missing link in drug discovery. Drug Discov Today 2013; 18(5-6): 240-9.
[http://dx.doi.org/10.1016/j.drudis.2012.10.003] [PMID: 23073387]

[5]   Perkhofer L, Frappart PO, Müller M, Kleger A. Importance of organoids for personalized medicine.

Per Med 2018; 15(6): 461-5.
[http://dx.doi.org/10.2217/pme-2018-0071] [PMID: 30418092]

[6]   Kim J, Koo BK, Knoblich JA. Human organoids: model systems for human biology and medicine. Nat Rev Mol Cell Biol 2020; 21(10): 571-84.
[http://dx.doi.org/10.1038/s41580-020-0259-3] [PMID: 32636524]

[7]   Tortorella I, Argentati C, Emiliani C, Martino S, Morena F. The role of physical cues in the development of stem cell-derived organoids. Eur Biophys J 2021; 1-3.
[PMID: 34120215]

[8]   Di Lullo E, Kriegstein AR. The use of brain organoids to investigate neural development and disease. Nat Rev Neurosci 2017; 18(10): 573-84.
[http://dx.doi.org/10.1038/nrn.2017.107] [PMID: 28878372]

[9]   2018. Providing the Right Cues for Organoid Development | Tempo Bioscience, https://www.tempo-bioscience.com/blog/providing-the-right-cues-for-organoid-development/

[10]   Aisenbrey EA, Murphy WL. Synthetic alternatives to Matrigel. Nat Rev Mater 2020; 5(7): 539-51.
[http://dx.doi.org/10.1038/s41578-020-0199-8] [PMID: 32953138]

[11]   Langhans SA. Three-dimensional *In vitro* cell culture models in drug discovery and drug repositioning. Front Pharmacol 2018; 9: 6.
[http://dx.doi.org/10.3389/fphar.2018.00006] [PMID: 29410625]

[12]   Lee HN, Choi YY, Kim JW, *et al.* Effect of biochemical and biomechanical factors on vascularization of kidney organoid-on-a-chip. Nano Converg 2021; 8(1): 35.
[http://dx.doi.org/10.1186/s40580-021-00285-4] [PMID: 34748091]

[13]   Wagstaff E, Heredero Berzal A, Boon C, Quinn P, ten Asbroek A, Bergen A. The role of small molecules and their effect on the molecular mechanisms of early retinal organoid development. Int J Mol Sci 2021; 22(13): 7081.
[http://dx.doi.org/10.3390/ijms22137081] [PMID: 34209272]

[14]   Kim S, Uroz M, Bays JL, Chen CS. Harnessing Mechanobiology for Tissue Engineering. Dev Cell 2021; 56(2): 180-91.
[http://dx.doi.org/10.1016/j.devcel.2020.12.017] [PMID: 33453155]

[15]   Weinhart M, Hocke A, Hippenstiel S, Kurreck J, Hedtrich S. 3D organ models—Revolution in pharmacological research? Pharmacol Res 2019; 139: 446-51.
[http://dx.doi.org/10.1016/j.phrs.2018.11.002] [PMID: 30395949]

[16]   Zhou J, Li C, Liu X, *et al.* Infection of bat and human intestinal organoids by SARS-CoV-2. Nat Med 2020; 26(7): 1077-83.
[http://dx.doi.org/10.1038/s41591-020-0912-6] [PMID: 32405028]

[17]   Dutta D, Heo I, Clevers H. Disease Modeling in Stem Cell-Derived 3D Organoid Systems. Trends Mol Med 2017; 23(5): 393-410.
[http://dx.doi.org/10.1016/j.molmed.2017.02.007] [PMID: 28341301]

[18]   Kim J, Koo BK, Knoblich JA. Human organoids: model systems for human biology and medicine. Nat Rev Mol Cell Biol 2020; 21(10): 571-84.
[http://dx.doi.org/10.1038/s41580-020-0259-3] [PMID: 32636524]

[19]   Sachs N, Papaspyropoulos A, Zomer-van Ommen DD, *et al.* Long-term expanding human airway organoids for disease modeling. EMBO J 2019; 38(4)e100300
[http://dx.doi.org/10.15252/embj.2018100300] [PMID: 30643021]

[20]   Dekkers JF, Wiegerinck CL, de Jonge HR, *et al.* A functional CFTR assay using primary cystic fibrosis intestinal organoids. Nat Med 2013; 19(7): 939-45.
[http://dx.doi.org/10.1038/nm.3201] [PMID: 23727931]

[21]   Lancaster MA, Renner M, Martin CA, *et al.* Cerebral organoids model human brain development and

microcephaly. Nature 2013; 501(7467): 373-9.
[http://dx.doi.org/10.1038/nature12517] [PMID: 23995685]

[22] Sampaziotis F, Cardoso de Brito M, Madrigal P, *et al.* Cholangiocytes derived from human induced pluripotent stem cells for disease modeling and drug validation. Nat Biotechnol 2015; 33(8): 845-52.
[http://dx.doi.org/10.1038/nbt.3275] [PMID: 26167629]

[23] Berkers G, van Mourik P, Vonk AM, *et al.* Rectal Organoids Enable Personalized Treatment of Cystic Fibrosis. Cell Rep 2019; 26(7): 1701-1708.e3.
[http://dx.doi.org/10.1016/j.celrep.2019.01.068] [PMID: 30759382]

[24] Parfitt DA, Lane A, Ramsden CM, *et al.* Identification and Correction of Mechanisms Underlying Inherited Blindness in Human iPSC-Derived Optic Cups. Cell Stem Cell 2016; 18(6): 769-81.
[http://dx.doi.org/10.1016/j.stem.2016.03.021] [PMID: 27151457]

[25] Alley MC, Scudiero DA, Monks A, *et al.* Feasibility of drug screening with panels of human tumor cell lines using a microculture tetrazolium assay. Cancer Res 1988; 48(3): 589-601.
[PMID: 3335022]

[26] Ben-David U, Siranosian B, Ha G, *et al.* Genetic and transcriptional evolution alters cancer cell line drug response. Nature 2018; 560(7718): 325-30.
[http://dx.doi.org/10.1038/s41586-018-0409-3] [PMID: 30089904]

[27] Sachs N, de Ligt J, Kopper O, *et al.* A Living Biobank of Breast Cancer Organoids Captures Disease Heterogeneity. Cell 2018; 172(1-2): 373-386.e10.
[http://dx.doi.org/10.1016/j.cell.2017.11.010] [PMID: 29224780]

[28] Yan HHN, Siu HC, Law S, *et al.* A Comprehensive Human Gastric Cancer Organoid Biobank Captures Tumor Subtype Heterogeneity and Enables Therapeutic Screening. Cell Stem Cell 2018; 23(6): 882-897.e11.
[http://dx.doi.org/10.1016/j.stem.2018.09.016] [PMID: 30344100]

[29] van de Wetering M, Francies HE, Francis JM, *et al.* Prospective derivation of a living organoid biobank of colorectal cancer patients. Cell 2015; 161(4): 933-45.
[http://dx.doi.org/10.1016/j.cell.2015.03.053] [PMID: 25957691]

[30] Zhou J, Li C, Zhao G, *et al.* Human intestinal tract serves as an alternative infection route for Middle East respiratory syndrome coronavirus. Sci Adv 2017; 3(11)eaao4966
[http://dx.doi.org/10.1126/sciadv.aao4966] [PMID: 29152574]

[31] Vlachogiannis G, Hedayat S, Vatsiou A, *et al.* Patient-derived organoids model treatment response of metastatic gastrointestinal cancers. Science 2018; 359(6378): 920-6.
[http://dx.doi.org/10.1126/science.aao2774] [PMID: 29472484]

[32] Lee SH, Hu W, Matulay JT, *et al.* Tumor Evolution and Drug Response in Patient-Derived Organoid Models of Bladder Cancer. Cell 2018; 173(2): 515-528.e17.
[http://dx.doi.org/10.1016/j.cell.2018.03.017] [PMID: 29625057]

[33] Ettayebi K, Crawford SE, Murakami K, *et al.* Replication of human noroviruses in stem cell–derived human enteroids. Science 2016; 353(6306): 1387-93.
[http://dx.doi.org/10.1126/science.aaf5211] [PMID: 27562956]

[34] Garcez PP, Loiola EC, Madeiro da Costa R, *et al.* Zika virus impairs growth in human neurospheres and brain organoids. Science 2016; 352(6287): 816-8.
[http://dx.doi.org/10.1126/science.aaf6116] [PMID: 27064148]

[35] Watanabe M, Buth JE, Vishlaghi N, *et al.* Self-Organized Cerebral Organoids with Human-Specific Features Predict Effective Drugs to Combat Zika Virus Infection. Cell Rep 2017; 21(2): 517-32.
[http://dx.doi.org/10.1016/j.celrep.2017.09.047] [PMID: 29020636]

[36] Schwank G, Koo BK, Sasselli V, *et al.* Functional repair of CFTR by CRISPR/Cas9 in intestinal stem cell organoids of cystic fibrosis patients. Cell Stem Cell 2013; 13(6): 653-8.
[http://dx.doi.org/10.1016/j.stem.2013.11.002] [PMID: 24315439]

[37]   Drost J, van Boxtel R, Blokzijl F, *et al.* Use of CRISPR-modified human stem cell organoids to study the origin of mutational signatures in cancer. Science 2017; 358(6360): 234-8.
[http://dx.doi.org/10.1126/science.aao3130] [PMID: 28912133]

[38]   Esch EW, Bahinski A, Huh D. Organs-on-chips at the frontiers of drug discovery. Nat Rev Drug Discov 2015; 14(4): 248-60.
[http://dx.doi.org/10.1038/nrd4539] [PMID: 25792263]

[39]   Huh D, Kim HJ, Fraser JP, *et al.* Microfabrication of human organs-on-chips. Nat Protoc 2013; 8(11): 2135-57.
[http://dx.doi.org/10.1038/nprot.2013.137] [PMID: 24113786]

[40]   Human Organs-on-Chips . [Cited 2022 Jan 7]. Available from: https://wyss.harvard.edu/technology/human-organs-on-chips/

[41]   Duzagac F, Saorin G, Memeo L, Canzonieri V, Rizzolio F. Microfluidic Organoids-on-a-Chip: Quantum Leap in Cancer Research. Cancers (Basel) 2021; 13(4): 737.
[http://dx.doi.org/10.3390/cancers13040737] [PMID: 33578886]

[42]   Sung JH, Wang YI, Narasimhan Sriram N, *et al.* Recent Advances in Body-on-a-Chip Systems. Anal Chem 2019; 91(1): 330-51.
[http://dx.doi.org/10.1021/acs.analchem.8b05293] [PMID: 30472828]

[43]   Maschmeyer I, Lorenz AK, Schimek K, *et al.* A four-organ-chip for interconnected long-term co-culture of human intestine, liver, skin and kidney equivalents. Lab Chip 2015; 15(12): 2688-99.
[http://dx.doi.org/10.1039/C5LC00392J] [PMID: 25996126]

[44]   Kondo J, Inoue M. Application of Cancer Organoid Model for Drug Screening and Personalized Therapy. Cells 2019; 17; 8(5): 470.
[http://dx.doi.org/10.3390/cells8050470]

[45]   Kim J, Koo BK, Knoblich JA. Human organoids: model systems for human biology and medicine. Nat Rev Mol Cell Biol 2020; 21(10): 571-84.
[http://dx.doi.org/10.1038/s41580-020-0259-3] [PMID: 32636524]

[46]   Takahashi T. Organoids for Drug Discovery and Personalized Medicine. Annu Rev Pharmacol Toxicol 2019; 59(1): 447-62.
[http://dx.doi.org/10.1146/annurev-pharmtox-010818-021108] [PMID: 30113875]

[47]   Xu H, Jiao Y, Qin S, Zhao W, Chu Q, Wu K. Organoid technology in disease modelling, drug development, personalized treatment and regeneration medicine. Exp Hematol Oncol 2018; 7(1): 30.
[http://dx.doi.org/10.1186/s40164-018-0122-9] [PMID: 30534474]

[48]   Nie X, Liang Z, Li K, *et al.* Novel organoid model in drug screening: Past, present, and future. Liver Res 2021; 5(2): 72-8.
[http://dx.doi.org/10.1016/j.livres.2021.05.003]

[49]   Liu L, Yu L, Li Z, Li W, Huang W. Patient-derived organoid (PDO) platforms to facilitate clinical decision making. J Transl Med 2021; 19(1): 40.
[http://dx.doi.org/10.1186/s12967-020-02677-2] [PMID: 33478472]

[50]   Sachs N, de Ligt J, Kopper O, *et al.* A Living Biobank of Breast Cancer Organoids Captures Disease Heterogeneity. Cell 2018; 11; 172(1-2): 373-86.
[http://dx.doi.org/10.1016/j.cell.2017.11.010]

[51]   Vlachogiannis G, Hedayat S, Vatsiou A, *et al.* Patient-derived organoids model treatment response of metastatic gastrointestinal cancers. Science 2018; 359(6378): 920-6.
[http://dx.doi.org/10.1126/science.aao2774] [PMID: 29472484]

[52]   Sachs N, de Ligt J, Kopper O, *et al.* A Living Biobank of Breast Cancer Organoids Captures Disease Heterogeneity. Cell 2018; 172(1-2): 373-386.e10.
[http://dx.doi.org/10.1016/j.cell.2017.11.010] [PMID: 29224780]

[53]   Zhang X, Ma Z, Song E, Xu T. Islet organoid as a promising model for diabetes. Protein Cell 2021;
        10: 1-19.
        [PMID: 33751396]

[54]   Rothbauer M, Byrne RA, Schobesberger S, *et al.* Establishment of a human three-dimensional chip-
        based chondro-synovial coculture joint model for reciprocal cross talk studies in arthritis research. Lab
        Chip 2021; 21(21): 4128-43.
        [http://dx.doi.org/10.1039/D1LC00130B] [PMID: 34505620]

[55]   Nuciforo S, Heim MH. Organoids to model liver disease. JHEP Reports 2021; 3(1)100198
        [http://dx.doi.org/10.1016/j.jhepr.2020.100198] [PMID: 33241206]

[56]   Lehmann R, Lee CM, Shugart EC, *et al.* Human organoids: a new dimension in cell biology. Mol Biol
        Cell 2019; 30(10): 1129-37.
        [http://dx.doi.org/10.1091/mbc.E19-03-0135] [PMID: 31034354]

[57]   Park JC, Jang SY, Lee D, *et al.* A logical network-based drug-screening platform for Alzheimer's
        disease representing pathological features of human brain organoids. Nat Commun 2021; 12(1): 280.
        [http://dx.doi.org/10.1038/s41467-020-20440-5] [PMID: 33436582]

[58]   Sachs N, de Ligt J, Kopper O, *et al.* A Living Biobank of Breast Cancer Organoids Captures Disease
        Heterogeneity. Cell 2018; 172(1-2): 373-386.e10.
        [http://dx.doi.org/10.1016/j.cell.2017.11.010] [PMID: 29224780]

[59]   Xu H, Jiao Y, Qin S, Zhao W, Chu Q, Wu K. Organoid technology in disease modelling, drug
        development, personalized treatment and regeneration medicine. Exp Hematol Oncol 2018; 7(1): 30.
        [http://dx.doi.org/10.1186/s40164-018-0122-9] [PMID: 30534474]

[60]   Pauli C, Hopkins BD, Prandi D, *et al.* Personalized *In Vitro* and *In Vivo* Cancer Models to Guide
        Precision Medicine. Cancer Discov 2017; 7(5): 462-77.
        [http://dx.doi.org/10.1158/2159-8290.CD-16-1154] [PMID: 28331002]

[61]   Fang G, Lu H, Rodriguez de la Fuente L, *et al.* Mammary Tumor Organoid Culture in Non-Adhesive
        Alginate for Luminal Mechanics and High-Throughput Drug Screening. Adv Sci (Weinh) 2021;
        8(21)2102418
        [http://dx.doi.org/10.1002/advs.202102418] [PMID: 34494727]

[62]   Phan N, Hong JJ, Tofig B, *et al.* A simple high-throughput approach identifies actionable drug
        sensitivities in patient-derived tumor organoids. Commun Biol 2019; 2(1): 78.
        [http://dx.doi.org/10.1038/s42003-019-0305-x] [PMID: 30820473]

[63]   Pleguezuelos-Manzano C, Puschhof J, van den Brink S, Geurts V, Beumer J, Clevers H. Establishment
        and Culture of Human Intestinal Organoids Derived from Adult Stem Cells. Curr Protoc Immunol
        2020; 130(1)e106
        [http://dx.doi.org/10.1002/cpim.106] [PMID: 32940424]

[64]   Au SH, Chamberlain MD, Mahesh S, Sefton MV, Wheeler AR. Hepatic organoids for microfluidic
        drug screening. Lab Chip 2014; 14(17): 3290-9.
        [http://dx.doi.org/10.1039/C4LC00531G] [PMID: 24984750]

[65]   Graff CP, Wittrup KD. Theoretical analysis of antibody targeting of tumor spheroids: importance of
        dosage for penetration, and affinity for retention. Cancer Res 2003; 63(6): 1288-96.
        [PMID: 12649189]

[66]   Tiriac H, Belleau P, Engle DD, *et al.* Organoid Profiling Identifies Common Responders to
        Chemotherapy in Pancreatic Cancer. Cancer Discov 2018; 8(9): 1112-29.
        [http://dx.doi.org/10.1158/2159-8290.CD-18-0349] [PMID: 29853643]

[67]   Seufferlein T, Kleger A. Organoidomics — falling star or new galaxy in pancreatic cancer? Nat Rev
        Gastroenterol Hepatol 2018; 15(10): 586-7.
        [http://dx.doi.org/10.1038/s41575-018-0052-3] [PMID: 30046146]

[68]   Millman JR, Xie C, Van Dervort A, Gürtler M, Pagliuca FW, Melton DA. Generation of stem cell-derived β-cells from patients with type 1 diabetes. Nat Commun 2016; 7(1): 11463.
[http://dx.doi.org/10.1038/ncomms11463] [PMID: 27163171]

[69]   Millman JR, Xie C, Van Dervort A, Gürtler M, Pagliuca FW, Melton DA. Generation of stem cell-derived β-cells from patients with type 1 diabetes. Nat Commun 2016; 7(1): 11463.
[http://dx.doi.org/10.1038/ncomms11463] [PMID: 27163171]

[70]   Ramezankhani R, Solhi R, Chai YC, *et al.* Organoid and microfluidics-based platforms for drug screening in COVID-19. Drug Discov Today 2021; S1359-6446(21): 00565-1.

[71]   Kaye M, Druce J, Tran T, *et al.* SARS-associated coronavirus replication in cell lines. Emerg Infect Dis 2006; 12(1): 128-33.
[http://dx.doi.org/10.3201/eid1201.050496] [PMID: 16494729]

CHAPTER 3

Intestinal Organoid Bioengineering, Disease Modeling, and Drug Discovery

Janvie Manhas[1,*]

[1] *Department of Biochemistry, All India Institute of Medical Sciences, New Delhi, India*

Abstract: The intestinal organoid system is a unique *ex-vivo* representation of the complex and dynamic mammalian intestinal epithelium. Intestinal organoids are three-dimensional, crypt-villus structures with a central lumen that can be sourced from adult intestinal stem cells, embryonic stem cells as well as induced pluripotent stem cells. They serve as a bona fide model for not only understanding intestinal biology and development but also for disease modeling, regenerative therapeutics, and drug discovery. Organoids help bridge the gap in existing model systems by incorporating complex, spatial, and biological parameters such as cell-cell interactions, cell-matrix interactions, gut-microbe interactions, and other components of intestinal *in-vivo* physiology and pathology. In this chapter, we discuss the basic strategies to generate intestinal organoids and how different bioengineering approaches can be used to effectively model both genetic and infectious intestinal diseases to enhance their utility in research and therapeutics.

Keywords: Colon cancer, Colonoids, Cystic fibrosis, Disease modeling, Enteroids, Inflammatory bowel disease (IBD) modeling, Intestinal organoids, Intestinal stem cells, Organoid culture, Organoid engineering.

INTRODUCTION

Multicellular organisms orchestrate cellular and tissue renewal by harnessing the regenerative and multi-differentiation potential of adult somatic stem cells for the maintenance of homeostasis. The human intestinal epithelium in particular, is well known to utilize the regenerative program of the intestinal stem cells (ISCs) in order to prevent the accumulation of genetic mutations and to maintain intestinal health and function. The intestinal epithelium is a crypt-villus structure that plays a crucial role in the absorption of water, nutrients and electrolytes as well as drug transport, metabolism, and hormone excretion. Due to this complex array of functionalities, it is constantly exposed to substances of varied compositions and

---

[*] **Corresponding author Janvie Manhas**: Department of Biochemistry, All India Institute of Medical Sciences, New Delhi, India; E-mail: manhasjanvi11@gmail.com

Manash K. Paul (Ed.)

toxicities. However, it has a protective mechanism in the form of its disposable epithelium, which sheds off intestinal epithelial cells cyclically. The ISCs then replenish and replace the intestinal epithelium every 3-4 days. These ISCs are housed at the bottom of the intestinal crypt and produce progenitor cells called transit-amplifying cells (TA cells) which proliferate and migrate up along the crypt-villus axis to differentiate into mature intestinal epithelial cells, which constitute the intestinal surface epithelium. These ISCs are armed with long-term self-renewal capacity and multipotent differentiation, which continue to regenerate the crypt-villus structure of the entire intestinal epithelium during the individual's lifetime [1].

Even though it has been difficult to recapitulate the multifactorial and complex intestinal crypt-villus mechanism in the laboratory, several preclinical models have been developed during this endeavor. These include the conventional models such as cell lines [2 - 6], Using chamber [7 - 9], and everted sac [10 - 12], as well as the more sophisticated models such as InTESTine™ [13, 14], microfluidic gut-on-chip [15 - 17], and intestinal organoids (IOs). These intestinal model systems have their advantages, applications, and limitations [18]. They have contributed to enhancing our understanding of intestinal epithelium and its myriad functions and mechanisms in health and disease. In this chapter, we will take a deep dive to understand the stem cell-derived *In vitro* model of the intestinal epithelium, IOs, which closely mimics the physiological 3D intestinal crypt-villus morphology. Previously published reports have used analogous terms such as 'enteroids', 'tumoroids', 'mini-gut organoids', colonoids' for *In vitro* cultured-3D-intestinal structures which we refer to as IOs in this chapter. Kindly note that 'intestinal' in the term intestinal organoids, encompasses both small and large intestine (colon) for the purpose of this chapter.

ESTABLISHMENT OF INTESTINAL ORGANOID CULTURE SYSTEM

Knowledge of the developmental biology of intestinal crypt-villus structure, the micro environmental factors, and signaling pathways involved in the same have equipped us with tools necessary to instruct somatic ISCs to self-renew, differentiate *ex-vivo*, and self-organize into three-dimensional (3D) structures that preserve the structural, molecular and functional identity of the original intestinal tissue. The cornerstone of intestinal organoid culture is the ability to isolate stem cell containing units, intestinal crypts, or a single population of stem cells and grow them with appropriate 'niche' culture conditions providing cell-cell and cell-extracellular matrix (ECM) cues to self-assemble into the parent tissue of origin in the culture dish.

Research and Development

Although some culture systems have been described, it has been challenging to establish *in vitro* propagation of adult somatic stem cell-derived organoids without inducing genetic transformations. One of the first of these organoid-type cultures was the rat intestinal crypt culture system, established and propagated for 1-2 weeks by seeding crypt epithelial cells with subepithelial fibroblasts on a Type I collagen-coated surface [19]. Notably, this population of rat intestinal crypts were not able to survive without the companion fibroblasts, which proved the importance of 'niche' signaling for the establishment of organoid cultures. Another study reported long-lasting maintenance of IOs by culturing neonatal intestinal epithelium within a niche comprising mesenchymal fibroblasts providing Wnt signaling cues using a novel, air-liquid interface culture system [20]. Although the existence of multipotent cells in intestinal epithelium has been known for a long time, confirmatory evidence of leucine-rich repeat-containing G-protein coupled receptor 5 (Lgr5) positive crypt base columnar cells possessing the requisite properties of stemness, long-term self-renewal, and multipotent differentiation came from elegant fate-mapping experiments in mice [21]. In these experiments, lineage tracing of Lgr5+ expressing cells was visualized with a Cre-activated reporter (Lgr5-EGFP-IRES-CreER) [22]. Independent research groups also reported Bmi1, Hopx, and mTert expressing label-retaining cells as ISCs [23, 24]. Another decade of research using gene-targeting experiments in mice revealed that intestinal stem cell plasticity was determined more by niche signaling pathways than cell-intrinsic properties like cell position or marker expression [25 - 27]. This stem cell niche signaling in the intestinal epithelium was found to be dominated by Wnt signaling, the master regulator of proliferation and undifferentiated status of ISCs [28]. The binding of Wnt ligand to Frizzled and LRP5/6 receptor complexes suppress the APC/GSK-3/Axin leading to the nuclear translocation of β-catenin, where it forms the Tcf4/β-catenin complex. The Tcf4/β-catenin induces downstream expression of Wnt target genes such as Lgr5. R-spondin is the ligand of the Lgr5 receptor, which enables sustained activation of Wnt signaling. R-spondin-treated mice have been observed to develop crypt hyperplasia and high ISC numbers due to enhanced Wnt activation [29, 30]. In contrast, inhibition of bone morphogenetic protein signaling (BMP) is necessary for ISC renewal [31, 32]. It has been observed that BMP inhibitors such as Noggin or Gremlin induce crypt formation and increase ISCs. Notch signaling is another player in ISCs maintenance and regulates secretory lineage differentiation [33 - 35]. EGF signaling is also essential for intestinal stem cell renewal, and PI3K/Akt signaling induces hyperplastic changes in the intestinal epithelium [36 - 38].

To achieve 3D structural organization and mimic extracellular matrix engagement through integrin signaling, the culture and maintenance of ISCs or isolated intestinal crypts is done on Matrigel [39]. Matrigel [40] is derived from a mouse origin basement membrane Engelbreth-Holm-Swarm (EHS) tumor and supplies ECM components such as laminin, proteoglycans, collagen type IV, entactin, and growth factors FGF, TGF-β required for IO culture. An optimal human ECM substitute for Matrigel to support human organoid culture is yet to be identified or synthesized and is an area of active research.

Adult Intestinal Stem Cell-derived Organoids

The self-renewing and self-organizing nature of ISCs was the key to establishing the first IO system in mice, where a single Lgr5+ adult intestinal stem cell formed a 3D organoid system recapitulating the *in vivo* intestinal epithelium-like villus-crypt morphology [41]. The system employed a unique mesenchymal-free, feeder layer-free, niche-mimicking habitat comprising Matrigel, R-spondin, Noggin, EGF, *etc.* This cocktail of niche factors can be selected based on previous research (*see section 2.1*) and experimentation. R-spondin is a Wnt agonist that has been shown to induce crypt hyperplasia [42]. Noggin is a BMP antagonist that induces crypt proliferation [32]. Rho-associated protein kinase (ROCK) inhibitor Y-27632 inhibits anoikis in stem cells and can be useful for organoid passaging [43]. Mouse colonic organoids also require Wnt-3a supplementation, whereas human small intestinal and colonic organoids need TGF-β and p38 inhibition [44]. Some studies suggest the use of prostaglandin E2 and nicotinamide in addition to Wnt-3a for human colonic organoids [45]. To simplify media formulation for organoid culture as well as reduce costs, a novel cell line L-WRN was engineered whose supernatant can be used as conditioned media for organoid culture [46]. L-WRN cell line is derived from mouse L cells (fibroblasts) and secretes Wnt-3a, R-spondin, and Noggin to provide high-titer niche factors to stimulate Wnt signaling in organoid culture. Besides recapitulating the morphology and *in vivo* localization of individual cell types in the intestinal crypts, organoids also mimic a variety of absorptive and digestive functions of the intestinal crypts [47].

Fetal Intestinal Epithelium-derived Organoids

The developmental stage and the region of intestinal epithelium govern the culture media composition and conditions required for the growth of fetal-derived intestinal organoids. During early developmental stages, intestinal epithelium grows into a cystic-spheroid type organoid structure without requiring EGF or Wnt signaling [48]. Interestingly, from postnatal day 15 onwards, fetal IOs require EGF and Wnt ligand supplementation, similar to adult IOs [48, 49]. In fact, these fetal IO transform into adult-type IOs in presence of γ-secretase inhibitor or Wnt-3A, emphasizing the role of specific niche-derived signals in intestinal epithelium maturation. The cellular plasticity has been reported to be higher for fetal small IO as compared to adult small IO. For example, murine fetal small IOs, when transplanted into damaged murine adult colonic mucosa, differentiated into colonic epithelium while adult small intestinal organoids were not able to do the same [49, 50].

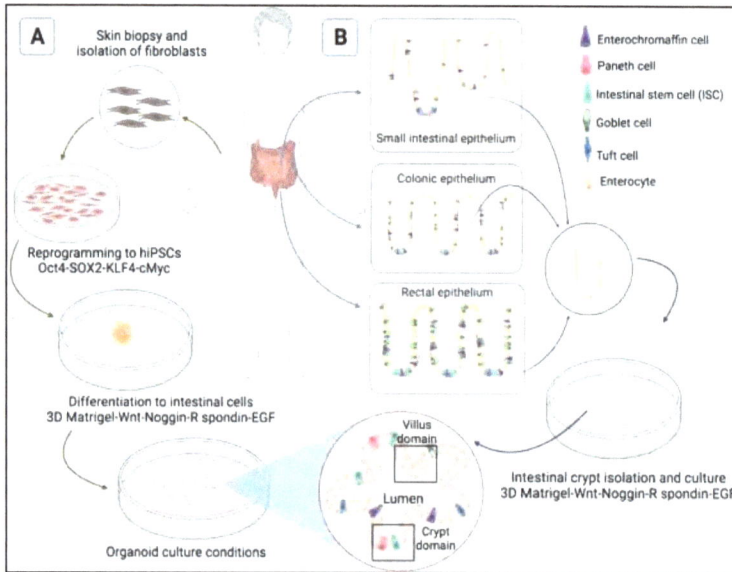

Fig. (1). Derivation of intestinal organoids from A) human-induced pluripotent stem cells and B) adult intestinal tissue by crypt isolation and organoid culture.

Human-Induced Pluripotent Stem Cell-derived Organoids

Besides somatic tissues, IOs have also been derived successfully from human iPSCs [51] (Fig. **1**). In this system, iPSCs are first treated with Activin to differentiate them into endodermal cells, followed by supplementation with Wnt3a and FGF for hindgut endodermal maturation. As both epithelium and mesenchymal cells are formed, this model preserves the niche-specific epithelial-mesenchymal interaction. Hindgut endoderm forms spheroids that attach as budding structures to the mesenchymal cells. Though this system favors IOs growth induced by paracrine R-spondin and Wnt secreted from the neighboring mesenchyme, this resembles fetal intestinal epithelium rather than the adult intestinal epithelium. It has been reported that providing an *in vivo* microenvironment such as transplantation of these human iPSC-derived IOs into kidney capsules can help in aging and maturation into adult-type IOs [52].

ENGINEERING INTESTINAL ORGANOIDS FOR DISEASE MODELING

Developmental Biology and Homeostasis in Intestine

The structure of the intestinal epithelium undergoes dynamic changes during development, tissue regeneration, and intestinal disease pathology. The most important unit of intestinal epithelial morphogenesis is the crypt-villus structure, which is responsible for the maintenance of intestinal homeostasis and function. Understanding the genesis and development of this crypt-villus unit helps in elucidating the physiological and pathological roles of the intestinal epithelium. Animal models have previously been used for the study of the crypt-villus axis. However, critical knowledge gaps still remain in the field due to inter-species differences and difficulties in real-time monitoring of intestinal morphogenesis. Intestinal cell lines based on 2D culture grow in flat monolayers attached to the culture dish, which is unsuitable for dynamic morphogenesis and crypt-villus axis development. With the recent development of 3D organoid culture methods and new culture media compositions, researchers are now able to closely study intestinal dynamic cellular morphogenesis, including crypt-villus formation *in vitro*.

Intestinal organoid technology enables *in vitro* creation of budding structures surrounding a central lumen. These budding structures are functionally and structurally similar to intestinal crypt domain enriched with Lgr5+ ISCs. Lgr5+ ISCs continuously divide to generate self-renewing stem cells, as well as terminally differentiated intestinal cells such as enterocytes, Paneth cells, enteroendocrine, and goblet cells. In addition, cytokines such as EGF, R-spondin 1, and Noggin are included in the intestinal organoid culture medium. The intestinal organoid model system has been used to elucidate the molecular mechanisms associated with intestinal self-organization *via* symmetry breaking [53]. The organoid model system has also been useful to study specific molecules or genes responsible for the maintenance of the crypt-villus compartment. For example, genetic ablation of Rac1 in intestinal organoids was shown to cause disruption of the crypt-villus region [54].

Modified Organoid Culture Systems

Organoid-based modeling of human diseases has provided us with a unique opportunity to study both epithelial autonomous dysregulation and microenvironmental factors that misguide normal cellular behavior independently or collectively (Fig. **2**). Epithelial dysregulation can be modeled in organoids by investigating genetic and epigenetic modifications, clonogenic growth, morphological abnormalities, and response to drugs or chemicals using a direct culture of disease-derived organoids or by genetic engineering organoids.

Intercellular crosstalk between epithelial and non-epithelial cells and the effect of non-cellular extrinsic factors can be reproduced in culture *ex vivo* by co-culturing different cell types or modulating culture conditions by exogenous interventions. Imitation of pathogenic tissue environmental conditions *in vitro* by using the below described modifications is a simple and powerful strategy to expand the versatility of organoid-based disease modeling systems.

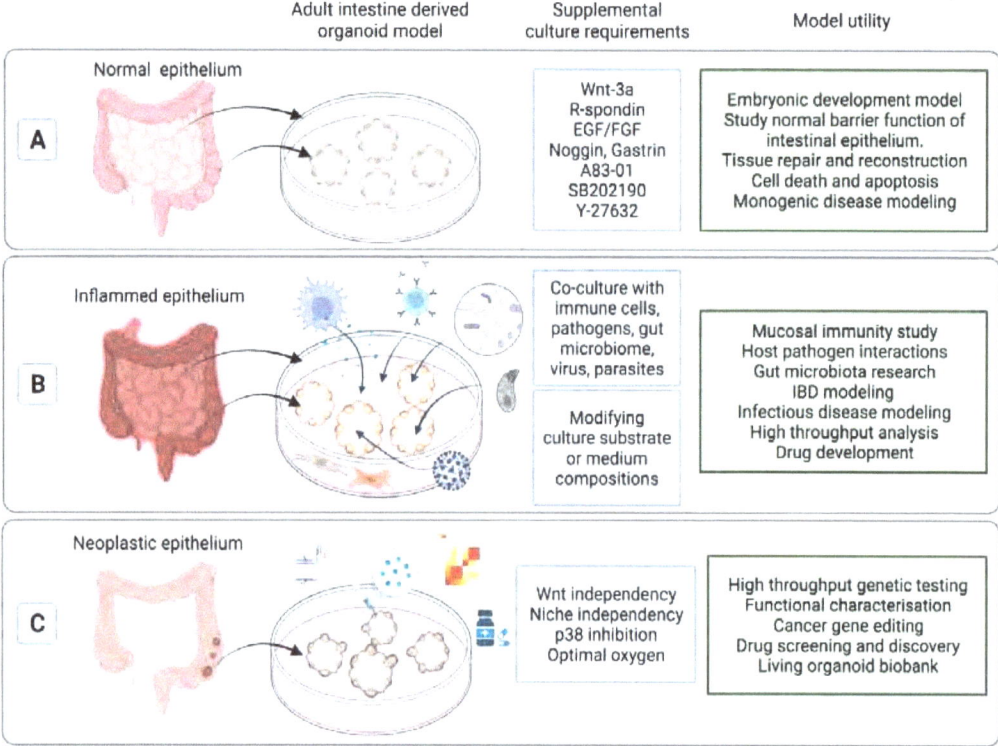

Fig. (2). Disease modeling using adult intestine-derived organoids. Culture requirements for growth and maintenance of A, Normal intestinal organoids, B, Intestinal organoids from inflammatory disease epithelium, C, Tumor-derived organoids, and examples of downstream applications.

Organoid Co-culture System to Model Diseases

In many human diseases, there is an interplay between epithelial and non-epithelial cells which can be very difficult to reproduce in a conventional organoid system with only epithelial elements. Organoid co-culture with non-epithelial cells has been developed to comprehend the cellular diversity involved in the intestinal tissue, which can be altered in disease. A coordinated relationship between enterocytes and immune cells is needed to maintain proper barrier function and mucosal immunity in human intestine. Hybrid culture of organoids with immune cells has helped to investigate complex immune interactions at the level of the intestinal epithelium, also allowing for long-term culture and studies targeting specific pathways in this interactome. To interrogate human innate immunological responses and cellular communications, a macrophage-enteroid model was developed, and it successfully recapitulated intestinal cell identity, signaling, and response to bacterial infection [55]. When co-cultured with human-derived intestinal enteroids, activated T-cells developed dendritic extensions as a possible sensing mechanism for environmental cues [56]. Intraepithelial lymphocytes (IELs) were co-cultured both within and outside intestinal organoids in the presence of cytokines such as IL-2, IL-7, and IL-15, and their dynamic interaction with enterocytes was observed [57]. Type 1 T-regulatory cells were found to suppress immune activation by secretion of IL-22 and promote the production of mucin-producing goblet cells in intestinal organoid culture [58]. The intestinal organoid co-culture platform is also being successfully used to evaluate the clinical activity of cancer immunotherapies and cell-based therapies [59, 60].

Host-pathogen Interaction Culture as an Infectious Disease Model

The gastrointestinal epithelial-mucosal barrier is an interface that facilitates two-way communication between epithelial cells and microorganisms. A healthy intestine maintains a balance between host-microbiome interactions by tuning this signaling across the epithelial-mucosal barrier. IO models with enteric pathogens such as bacteria, viruses, and parasites have been successfully exploited to study physiological and pathological mechanisms involved in the barrier function of the intestinal epithelium (Table **1**). Intestinal cell lines and animal models may not accurately mimic the complex interplay of host microbes at work in gastrointestinal health and associated diseases. Many of the commensal microbes or pathogens that reside in or infect the human intestine fail to grow in 2D cultures [61]. Most 3D intestinal organoid systems have a cellular alignment with a central lumen which can be used for physiological reconstruction of the host-pathogen interaction by microinjection of microbes/pathogens in this space [61 - 63].

Table 1. Pathogens that have been co-cultured with intestinal organoids for studying host-pathogen interactions and infectious disease modeling.

Pathogen	Intestinal Organoid (Host)	Reference
MICROBE		
Escherichia coli (E. coli) strains: *E. coli enterohemorrhagic* *E. coli enterotoxigenic* *E. coli enteropathogenic* *E. coli enteroaggregative*	Human small intestine Human large intestine	[73, 74]
		[55, 74]
		[55]
		[74, 75]

(Table 1) cont.....

Pathogen	Intestinal Organoid (Host)	Reference
Klebsiella pneumoniae	Human small intestine	[76]
Listeria monocytogenes	Human small intestine	[67]
Clostridium difficile	Mouse small intestine	[77]
Vibrio cholerae	Human small intestine	[68 - 70]
Shigella flexneri	Human large intestine	[71, 72]
Salmonella typhimurium	Mouse small intestine	[78]
	Human small intestine	[67]
VIRUS		
Human norovirus	Human small intestine	[62]
Human rotavirus	Human small intestine	[61, 79, 80]
Enterovirus	Human fetal small intestine	[81]
Human adenovirus	Human small intestine	[82]
Human astrovirus	Human small and large intestine	[83]
Middle east respiratory syndrome coronavirus (MERS-Cov)	Human large intestine	[84]
Severe acute respiratory syndrome coronavirus 2 (SARS-Cov-2)	Human small and large intestine	[85 - 87]
	Bat small intestine	[87]
PARASITE		
Cryptosporidium parvum	Mouse small intestine	[88]
	Human small intestine	[89]
Toxoplasma gondii	Cat and mouse small intestine	[90, 91]

Though extremely valuable for studying host-microbe interactions, the practice of 3D organoid co-culture with microinjection of microbes in the lumen is labor-intensive and requires expertise. To improve the accuracy and throughput of organoid microinjection, efforts are being made to automate this process [64]. The human intestine also has a rich repertoire of gut microbiota, most of which are obligate anaerobes. Using anaerobic culture systems and human IOs, this physiological aspect of intestinal host-microbiome interaction has also been recapitulated *in vitro* [65, 66]. Another interesting model system to study intestinal epithelium and pathogenic bacteria interaction is by inverting the cellular polarity of IOs by growing them in suspension cultures. This allows us with increased accessibility to the luminal side of the organoids and direct organoid-microbe co-culture and nutrient transport analysis independent of the challenges of luminal microinjection of bacteria [67]. These co-culture strategies have helped to model interactions between the intestinal epithelium and pathogenic bacteria such as *Escherichia coli, Klebsiella pneumoniae, Listeria monocytogenes, Clostridium difficile and Salmonella typhimurium* (see Table 1).

The mechanisms of secretory diarrhea in *Vibrio cholera* infection have also been studied using IOs [68 - 70]. It was observed that cholera toxin exposure to IOs inhibited NHE3 (Na(+)/H(+) exchanger) ion transporter activity, which causes fluid secretion responsible for the clinical phenotype in cholera. The pathogenesis of dysentery caused by *Shigella* has also been studied using human IOs [71, 72]. The advantage of organoids in infectious disease modeling is very useful when specific cell types are preferentially affected. For instance, *Enterohemorrhagic E. coli* colonization in human colonic organoids was enhanced when the organoids were differentiated towards enterocyte and goblet cell phenotype [73].

Organoids have also contributed to the study of virus-host cell interactions. ISC-derived 3D organoids have been used to grow human norovirus (HuNov) *in vitro*, providing a reproducible cultivation system for the study of HuNov-related enteric infections in humans. Using the organoid system, it was further established that genogroup II, genotype 4 (GII.4) HuNov could not replicate in IOs derived from patients with functional fucosyltransferase 2 (FUT2) deficiency [62]. This pathogen genotype-specific phenotypic host relationship was confirmed with genetic manipulation of FUT2 in human IOs and can be used to develop antiviral therapies [92]. It has also been reported that some strains of HuNov need bile supplementation for their replication in human IOs [62]. Rotavirus is a common cause of acute, fatal gastroenteritis infection in infants and young children. Human IOs have been reported to support the replication of rotaviruses directly from patient stool samples [93]. Using human IOs, it was observed that rotavirus infected both enterocytes and enteroendocrine cells in human intestinal organoids [79, 80]. Similarly, a strain of human adenovirus preferentially infected goblet cells as compared to other cell types in human IOs [82]. Human IOs have been a robust model system to study SARS-CoV-2 infections. Human IOs can be infected with different variants of SARS-Cov-2 viruses easily independent of the key host factors such as ACE-2 expression and have permitted scientists to directly study the SARS-CoV-2 virus-epithelial cell interactions [87, 94 - 96].

Intestinal parasites often have complex lifecycles with phases involving specific host tissues and species tropism. The organoid co-culture system can recapitulate this complex mechanism and provide a suitable model system for studying parasitic diseases. Toxoplasmosis is one of the most common parasitic human infections. *Toxoplasma gondii* has two main phases in its lifecycle: the sexual phase restricted to cat intestinal epithelium and an asexual phase that can be completed in any warm-blooded animal, including humans. Bovine and porcine small intestine-derived organoids have been infected with *Toxoplasma gondii* tachyzoites to model toxoplasmosis infection in intermediate hosts [91]. A model of feline small intestine-derived organoids has helped to unravel the tropism for the sexual stage of *Toxoplasma gondii* in cats. It has been identified using this IO model that an abundance of linoleic acid or deficiency of delta-6-desaturase, an enzyme vital for linoleic acid metabolism, is a critical factor for the sexual reproduction of *Toxoplasma gondii* [90]. Recently it has also been demonstrated that *Cryptosporidium parvum* can propagate within mouse and human IOs, preferentially in differentiated organoids rather than expanding organoids, and now researchers can model its life cycle with both the sexual and asexual stages *In vitro* [88, 89].

IOs can be a potential model for vaccines and antiviral therapy testing. As compared to the 2D Caco-2 cell line, it was observed that human IOs were more susceptible to rotavirus infection, and treatment of IOs with interferon-alpha or ribavirin inhibited viral replication in IOs by diverse antiviral mechanisms [61]. Subsequently, it was also reported that PI3K-Akt-mTOR signaling was responsible for the maintenance of rotavirus infection within IOs, and mTOR inhibitor rapamycin could be a promising antiviral therapy in human rotavirus infection [97].

Culture Milieu Modification to Mimic Diseases

Once an intestinal organoid culture is established, altering the culture medium composition according to the context of the disease under consideration is the simplest method to model diseases. For instance, to mimic inflammatory disease conditions, various cytokines or microbial products can be added to an organoid culture medium. To study the protective role of autophagy-related 16 like 1 (ATG16L1) protein, encoded by *Atg16l1* gene in humans, mouse intestinal organoids genetically lacking ATG16L1 were treated with TNF-α or interferon-γ and were found to upregulate apoptosis-related pathways [98, 99]. IL-22 promotes regeneration of intestinal stem cells and protects against genotoxic stress in mouse intestinal organoids [100, 101]. The advantage provided by the organoid model is the ability to isolate and study individual cytokine effects in an epithelial and genotype-specific context. Intestinal organoids grown in type I collagen gel have recapitulated the mechanotransduction-induced remodeling of the ECM during tissue repair [102]. Further, it has also been observed using organoids that ECM signaling and inflammatory insults gear the intestinal epithelium towards a regenerative fetal phenotype for repair and healing [103].

Interestingly, cancers often hijack stem cell niche pathways or embryonic signaling pathways to enable autonomous growth independent of growth factor requirements [104]. Human colorectal cancer cells accumulate genetic mutations in APC, which enables the expansion of colonic organoids in the absence of exogenous Wnt and R-spondin [105]. Another study shows that Braf^{V600E} mutation can constitutively switch on Wnt signaling in colonic organoids following which these organoid cultures grow independent of Wnt and R-spondin supplementation [106]. Exposing organoids to a cancer-specific culture milieu facilitates the selective expansion of patient-derived tumor organoids and helps to create efficient *In vitro* disease models.

Organoids Derived from Diseased Tissue

The relative ease in the availability of patient-derived intestinal epithelium supported by advances in endoscopic accessibility have inspired the use of organoids for disease models (Fig. **2**). IOs have been insightful in studying the pathophysiology of various human diseases, a few of which will be enumerated in this section.

Colorectal Cancer (CRC)

Colorectal cancer occurs as a result of multiple-hit mutations that cause genomic and epigenetic alterations in the intestinal epithelium [107]. The adenoma-carcinoma sequence is implicated in about 60% of all CRC pathogenesis [108, 109]. Murine IOs-derived from APC deficient mice models have been used to study tumorigenesis in CRC. IOs-derived from CRC-susceptible $Apc^{Min/+}$ mice were found to have upregulated Wnt signaling, which interferes with the differentiation process. These IOs were used to screen for epigenetically active compounds which can enhance organoid differentiation [110]. Using IOs derived from CRC-susceptible $Apc^{Min/+}$ mice as a model system, the role of Cohesin Rad21 in APC gene deficiency and carcinogenesis in familial adenomatous polyposis was established [111].

Organoids from human CRC cells can be developed from both surgical specimens and endoscopic biopsy tissues. The efficiency of tumor-derived organoid formation from fresh tumor tissue is about 30% to 90% and largely depends on sample volume, purity, origin, and culture conditions [112]. Transcriptomic profiling of organoids cultured from human CRC cells has been known to mirror biological features, including the genotype and molecular fingerprint of the original tumor [108, 113]. Tumor organoids also recapitulate the accurate morphology, histological grade, or differentiation status of the tumor. Tumor organoid from a single tumor cell is able to reconstitute intratumoral heterogeneity and mutational landscape [114]. However, sometimes patient-derived tumor organoids can accumulate chromosomal mutations and replication errors, which have been observed in tumor organoids with chromosomal instability and mismatch repair inability [115, 116]. It is therefore recommended to cryo-preserve early passage organoids and avoid the long-term culture of tumor organoids to enable experimental reproducibility [116]. Nevertheless, human tumor-derived organoids have demonstrated molecular and phenotypic credibility as a physiologically relevant *In vitro* substitute for *In vivo* tumor patient. The first living organoid biobank was established with 22 primary CRC patients and 19 normal epithelial controls with a 90% success rate [117]. Analysis of these IOs revealed aberrant activation of the Wnt pathway in patient-derived tumor organoids, which explains the advantage of Wnt and R-spondin supplementation in culture medium leading to enhanced tumor organoid expansion and purity. Living organoid biobanks with rare histological subtypes of CRC, premalignant subtypes, and metastatic CRC have been successfully utilized in personalized medicine programs and drug screening [112, 113].

Inflammatory Bowel Disease (IBD)

Inflammatory bowel disease (IBD) is a long-lasting, relapsing-remitting group of inflammatory disorders of the intestinal epithelium, which includes ulcerative colitis, Crohn's disease, and a rare form of microscopic colitis. IBDs frequently have a chronic disease course with acute exacerbations, hospital admissions requiring surgical interventions, and cause a tremendous negative impact on the quality of life. Unfortunately, the exact pathogenesis of IBD still remains elusive due to scarcity of models that recapitulate the development of the disease. Using intestinal organoid technology combines human origin, non-interference by malignant genetics, and the ability to study all epithelial cell types present in normal epithelium, including rare cells such as M, Tuft, and endocrine cells [118 - 120]. Accessibility to *In vitro* organoid cultures with these rare cell types implicated in IBD pathogenesis is a game-changing strategy in IBD research, especially when these cells are difficult to study in other models [121]. Furthermore, this model is simple to generate from mucosal biopsies of IBD patients and healthy controls [122 - 124]. Co-culture organoid systems adding non-epithelial immune elements such as cytokines or fibroblasts, enteric nerves, macrophages, or T-cells help to modulate the stem cell niche and generate accurate models for IBD [55 - 57, 125].

The luminal contents of the bowel have been reported to be indispensable for IBD pathogenesis [73, 126, 127]. This includes interaction with microbes like bacteria and viruses that alter the inflammatory milieu of the enterocytes (*also see sections 3.2.2 and 3.2.3*). Rare forms of IBD with very early onset in childhood (VEO-IBD) have been reported to be caused by gene mutations linked to epithelial integrity, immune system, cellular migration, oxidative stress as well as carbohydrate metabolism [128 - 130]. Intestinal organoids retain the genomic and transcriptomic profile of the patient's original tissue and can be used to model intestinal host-microbe interactions, mucosal barrier alterations, and bioengineered to faithfully represent the intestinal niche in IBD.

Monogenic Disorders

Microvillus inclusion disease (MVID) is a congenital, neonatal, intestinal epithelial disorder characterized by defective apical vesicular transport [131]. Myosin 5B and Syntaxin-3 gene mutations have been reported in the majority of the cases of MVID. MVID is characterized by a loss of apical microvilli and diffuse villus atrophy, giving rise to clinical symptoms associated with malabsorption and malnutrition. It has been shown that MVID enteropathy and loss of microvilli phenotype can be efficiently recapitulated using intestinal organoids [132, 133].

Multiple intestinal atresia (MIA) is a rare disease characterized by disruption of the epithelial barrier along the entire gastrointestinal tract. Multiple mutations in the tetratricopeptide repeat domain 7A (TTC7A) were found to be responsible for this disease [134]. Using intestinal organoid cultures from patient biopsies, it was demonstrated that MIA organoids show an inversion of the apicobasal polarity of epithelial cells. TTC7A deficiency results in increased Rho kinase activity which disrupts intestinal cell polarity and results in disease phenotype. Pharmacological inhibition of Rho kinase was found to restore normal apicobasal polarity in MIA organoids suggesting a link between intestinal epithelial function and Rho signaling pathway.

Diacylglycerol-acyltransferase-1 (DGAT1) deficiency mutations in the DGAT1 gene have been associated with rare, inherited congenital diarrheal disorders characterized by life-threatening dehydration and nutrient malabsorption. The DGAT1 enzyme catalyzes the conversion of diacylglycerol (DG) and fatty acyl CoA to triacylglycerol (TG) in humans. In the human intestine, DGAT1 is the only highly expressed enzyme. The mechanisms for intestinal dysregulation and aberrations in absorption or transport of nutrients and electrolytes have remained unclear. However, it has been recently reported using patient-derived intestinal organoids that loss of DGAT1 results in increased sensitivity to lipid-induced toxicity in enterocytes and development of the disease phenotype [135]. Further experiments also proved that aberrant lipid metabolism in DGAT1 deficient enterocytes can be rescued by isoenzyme DGAT-2 expression and can be used as a therapeutic modality. The intestinal organoid model system provided a novel link between gut epithelial lipid metabolism dysfunction and congenital diarrheal disorders [135].

Cystic fibrosis (CF) is an autosomal recessive disease caused by mutations in the CF transmembrane conductance regulator (CFTR) gene. The CFTR gene encodes for a chloride channel and is the only channel that uses a cAMP-dependent mechanism in the intestine. Due to the aberrant function of the CFTR channel, CF patients develop a highly viscous buildup of mucus and impaired intraluminal fluid secretion in the lung and gastrointestinal tract, predisposing them to serious and recurrent infections. The use of IOs in CF was first reported in 2012 when researchers used murine intestinal crypt-derived organoids to evaluate the physiology and transport activity of CFTR [136]. Subsequently, human intestinal organoids derived from ISCs collected from rectal biopsy tissue have been developed for clinical and pre-clinical testing and disease modeling. Rectal biopsy is an easy procedure that can be performed in all age groups, including newborns, without the need for hospitalization or anesthesia [137]. IOs generated are a functional expression of the broad spectrum of CFTR mutations seen in CF [138]. These CF patient-derived IOs can be expanded in culture, stored in liquid nitrogen, and continuously cultured for >6 months to evaluate CFTR modulator drug responses. In this regard, a sophisticated microscopic assay called Forskolin-induced swelling (FIS) assay (Box 1) was developed to model CFTR function *In vitro* [139, 140]. Currently, FIS is considered a standard assay for quantifying residual CFTR function, disease severity, and genotype-phenotype correlations, which can then be used to identify patient-specific drugs [141, 142]. F508del is the most dominant CFTR mutation in CF. CF-derived IOs show attenuated FIS in organoids with CFTR F508del. Using FIS assay to model and evaluate potential therapies, it was shown that CFTR potentiator, Ivacaftor (VX-770), and CFTR corrector, Lumacaftor (VX-809), when used as a single treatment, could restore normal function in CFTR F508del derived IO [141, 143].

Forskolin is a root extract of the Indian plant *Coleus forskohlii*. It has been used for centuries in the traditional Indian system of Ayurvedic medicine to treat various diseases such as heart disease and respiratory disorders [144]. Its anti-inflammatory and blood-pressure lowering effects have been attributed to its ability to increase cyclic AMP levels in bronchial muscles and vascular smooth muscles [145]. In the intestinal cells, CFTR is the only channel that operates on a cAMP-dependent mechanism. Incubation with forskolin increases intracellular cyclic AMP concentration and induces extensive swelling in wild-type IOs within 60 min, which can be observed by microscopy. Forskolin-induced swelling (FIS) phenotype is absent in CF organoids with abnormally functioning CFTR or CFTR-deleted organoids. This swelling of IOs is quantified microscopically, and a relative increase in size is calculated over time using 10-minute intervals.

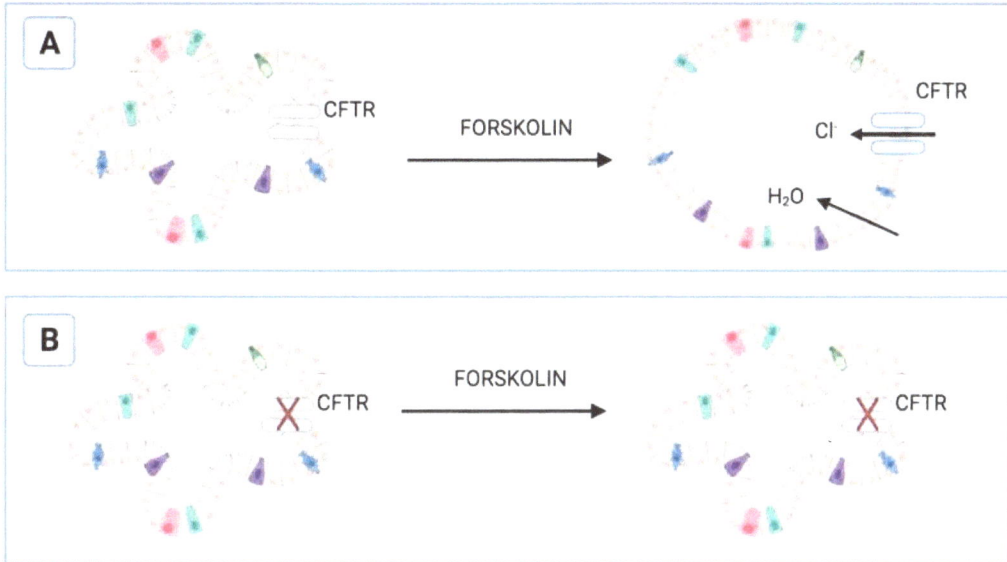

Box 1. Forskolin-Induced Swelling (FIS) assay in Intestinal organoids derived from A) normal intestinal epithelium and B) Cystic fibrosis patient's intestinal epithelium. When the CFTR channel is defective, treatment of forskolin does not induce swelling due to defective luminal secretion of chloride and water.

Engineered Biomaterials and ECM Scaffolds

The dependency of organoid culture system on natural ECM has been a major hurdle for producing clinical-grade organoids compatible with good manufacturing practice (GMP) regulations. The standard basement membrane matrix (BMM) used for organoid culture is derived from Englebreth-Hol--Swarm mouse sarcoma, a benign murine tumor. Although BMM is of non-human origin, its biochemical compatibility with the native basement membrane encouraged its use in 3D cell culture. Type I collagen extracted from bovine or porcine tissues can also support 3D cell and tissue culture. Despite the wide usage of these two ECMs in organoid culture, there is a growing demand for biosimilar, non-xenobiotic culture substrates for human organoids. Biochemical and physical cues from ECM influence cellular proliferation and differentiation [146]. This has motivated efforts in the field to produce synthetic culture scaffolds and bioengineer hydrogels that mimic natural ECM with the ability to support organoid development and maturation. Synthetic ECMs with tunable biophysical and molecular characteristics have been shown to help in the development of human tissue-derived organoids, starting with single stem cells [147]. These synthetic scaffolds consisting of polyethylene glycol hydrogels modified with ECM-binding peptides and different concentrations of integrin-binding motifs crosslinked by peptides amenable to degradation by matrix metalloproteases showed robust support for human duodenal and colon organoid culture [148].

Modeling metastatic cancers with specific tropism and distinct mechanical and biophysical properties is also possible now using designer hydrogels tailored to recreate the metastatic niche [149]. Together with advanced technologies like robotically controlled release of biomolecules, microfluidics, and 3D bioprinting of vascular structures, using engineered materials and biomimetic scaffolds have an exciting potential to generate *in vitro* preclinical platforms for drug discovery and precision oncology.

Genetic Editing of Intestinal Organoids

The establishment of patient-derived intestinal organoids may be difficult in extremely rare diseases or where the gene defect can cause prenatal mortality. To analyze genotype-phenotype correlations in intestinal diseases, various genetic engineering approaches to manipulate gene expression in intestinal organoids have been utilized. Besides conventional gene overexpression and knockdown assays, CRISPR (clusters of irregularly spaced short palindromic repeats)-Cas9 (CRISPR-associated protein) based genome editing has been pivotal in engineering intestinal organoids [150].

CFTR locus in Cystic fibrosis has been modified by homologous recombination *via* CRISPR-Cas9 gene editing, and a normal CFTR was inserted into diseased organoids from CF patients [151]. It was reported that the phenotype, morphology, and function (measured by FIS assay) of CFTR gene-edited CF organoids were significantly similar to wild-type healthy rectal organoids. These repaired organoids, in theory, may have the potential to be engrafted into the patient as a gene therapy model for CF. However, since CF is a multiorgan disease, genetic engineering of organoids could be more useful to screen and select drugs to restore wild type activity in mutant CFTR-derived IO.

CRISPR-Cas9 engineering approach has also been used to genetically induce human colon cancer in intestinal organoids using the classical genetic model of colon carcinogenesis, the adenoma-carcinoma sequence. Activation of oncogenic hotspot mutations (KRAS and PIK3CA) along with disruption of tumor suppressor genes (APC, TP53, and SMAD4) in human normal colonic organoids was reported to induce tumorigenesis in these organoids [152]. Likewise, deletion of cellular DNA repair genes, MLH1, NTHL1, and XPC in human colonic organoids incremented the mutational signature towards cancer development [116, 153]. Perturbation of p53 lead to the acquisition of carcinogenic features in colonic organoids [154]. Genetically engineered colonic organoids with BRAF^{V600E} mutation were treated with TGF-β to investigate developmental drivers and epithelial-to-mesenchymal (EMT) transition in colorectal cancer [155]. A rare subtype of colonic polyp, traditional serrated adenoma, was also genetically modeled and studied in human colonic organoids using complex genetic rearrangements introduced by a CRISPR-Cas9-based approach [156]. Cancer stem cells from patient-derived colon cancer organoids can be targeted and labeled using CRISPR-Cas9 based genetic editing for evaluating their extensive de-differentiation and self-renewal capacity [157]. This strategy to build model systems of human intestinal cancers using organoids along with gene editing allows researchers to study the direct causality of gene defects in cancer and is instrumental in the deconstruction of the complex, multi-hit, multi-genomic abnormalities in colon cancer into single genotype-phenotype elements.

CONCLUSION, LIMITATIONS AND FUTURE DIRECTIONS

Intestinal organoids are a high-fidelity tool for studying cell-intrinsic and environmental factors that control intestinal disease phenotype. Integrating genetic engineering and niche factors with organoid culture enable deeper probing of the complex interactions at play in health and disease. From a clinical perspective, gene-corrected organoids hold promise for regenerative gene therapies. The regenerative capacity of intestinal epithelium may be decreased or defective in disease states. Patients with absorptive dysfunction of the intestinal epithelium may benefit from organoid-based regenerative therapy. The massive expandability and phenotypic reliability of patient-derived intestinal organoids justify their use in regenerative medicine. MVID is characterized by protracted, watery diarrhea and malabsorption syndrome in infants, where the only treatment options available are total parenteral nutrition or small bowel transplantation [158]. Using gene-edited autologous organoids or replacement by human leukocyte antigen (HLA)-matched donor organoids could benefit these functional disorders of the intestinal epithelium. Organoid-based selection of responders to drugs has pushed the frontiers in translation of *in-vitro* drug testing to clinical decision making in cystic fibrosis [138, 159]. In animal studies, human intestinal organoids can repopulate the denuded rectal epithelium and maintain self-renewal and differentiation as xenografts over months without showing any signs of tumorigenesis [160].

Although organoid culture holds promise in future precision medicine, several uncertainties and limitations inherent to each of the cellular, extracellular, and medium components optimization must be addressed. Recent research has integrated state-of-the-art material science and protein design technologies to resolve these limitations. Mouse-derived ECM is now being replaced by synthetic hydrogels and biosimilar culture substrates, serum supplementation is being replaced by highly active, stable, and selective biomimetic compounds. Digital decoding of organoids using transcriptomic, proteomic, and epigenomic data can be used to address comparability and reproducibility issues.

With the global affordability of high throughput sequencing technology, systematic and high-definition comparison of organoids using single-cell sequencing may be the next standard for benchmarking the accuracy of organoids for clinical usage. Several scientific, technical, and ethical issues remain to be addressed about the future of organoid research. However, we are optimistic about the kaleidoscopic future of this technology and believe that organoid technology propelled by other advances in biomedical research will open new horizons in understanding intestinal epithelial disease pathogenesis and treatment.

REFERENCES

[1]   Bjerknes M, Cheng H. Intestinal epithelial stem cells and progenitors. Methods Enzymol 2006; 419: 337-83.
        [http://dx.doi.org/10.1016/S0076-6879(06)19014-X] [PMID: 17141062]

[2]   Walter E, Janich S, Roessler BJ, Hilfinger JM, Amidon GL. HT29-MTX/Caco-2 cocultures as an *in vitro* model for the intestinal epithelium: in vitro-in vivo correlation with permeability data from rats and humans. J Pharm Sci 1996; 85(10): 1070-6.
        [http://dx.doi.org/10.1021/js960110x] [PMID: 8897273]

[3]   Walter E, Janich S, Roessler BJ, Hilfinger JM, Amidon GL. HT29-MTX/Caco-2 cocultures as an *in vitro* model for the intestinal epithelium: in vitro-in vivo correlation with permeability data from rats and humans. J Pharm Sci 1996; 85(10): 1070-6.
        [http://dx.doi.org/10.1021/js960110x] [PMID: 8897273]

[4]   van Breemen RB, Li Y. Caco-2 cell permeability assays to measure drug absorption. Expert Opin Drug Metab Toxicol 2005; 1(2): 175-85.
        [http://dx.doi.org/10.1517/17425255.1.2.175] [PMID: 16922635]

[5]   Madara JL, Dharmsathaphorn K. Occluding junction structure-function relationships in a cultured epithelial monolayer. J Cell Biol 1985; 101(6): 2124-33.
        [http://dx.doi.org/10.1083/jcb.101.6.2124] [PMID: 3934178]

[6]   Lozoya-Agullo I, Araújo F, González-Álvarez I, *et al.* Usefulness of Caco-2/HT29-MTX and Caco-2/HT29-MTX/Raji B Coculture Models To Predict Intestinal and Colonic Permeability Compared to Caco-2 Monoculture. Mol Pharm 2017; 14(4): 1264-70.
        [http://dx.doi.org/10.1021/acs.molpharmaceut.6b01165] [PMID: 28263609]

[7]   Ussing HH, Zerahn K. Active transport of sodium as the source of electric current in the short-circuited isolated frog skin. Acta Physiol Scand 1951; 23(2-3): 110-27.
        [http://dx.doi.org/10.1111/j.1748-1716.1951.tb00800.x] [PMID: 14868510]

[8]   Lomasney KW, Hyland NP. The application of Ussing chambers for determining the impact of microbes and probiotics on intestinal ion transport. Can J Physiol Pharmacol 2013; 91(9): 663-70.
        [http://dx.doi.org/10.1139/cjpp-2013-0027] [PMID: 23984904]

[9]   Thomson A, Smart K, Somerville MS, *et al.* The Ussing chamber system for measuring intestinal permeability in health and disease. BMC Gastroenterol 2019; 19(1): 98.
        [http://dx.doi.org/10.1186/s12876-019-1002-4] [PMID: 31221083]

[10]   Alam MA, Al-Jenoobi FI, Al-mohizea AM. Everted gut sac model as a tool in pharmaceutical research: limitations and applications. J Pharm Pharmacol 2012; 64(3): 326-36.
        [http://dx.doi.org/10.1111/j.2042-7158.2011.01391.x] [PMID: 22309264]

[11]   Masiiwa WL, Gadaga LL. Intestinal Permeability of Artesunate-Loaded Solid Lipid Nanoparticles Using the Everted Gut Method. J Drug Deliv 2018; 2018: 1-9.
        [http://dx.doi.org/10.1155/2018/3021738] [PMID: 29854465]

[12]   Wilson TH, Wiseman G. The use of sacs of everted small intestine for the study of the transference of substances from the mucosal to the serosal surface. J Physiol 1954; 123(1): 116-25.
[http://dx.doi.org/10.1113/jphysiol.1954.sp005036] [PMID: 13131249]

[13]   Westerhout J, Steeg E, Grossouw D, *et al.* A new approach to predict human intestinal absorption using porcine intestinal tissue and biorelevant matrices. Eur J Pharm Sci 2014; 63: 167-77.
[http://dx.doi.org/10.1016/j.ejps.2014.07.003] [PMID: 25046168]

[14]   Stevens LJ, van Lipzig MMH, Erpelinck SLA, *et al.* A higher throughput and physiologically relevant two-compartment human *ex vivo* intestinal tissue system for studying gastrointestinal processes. Eur J Pharm Sci 2019; 137: 104989.
[http://dx.doi.org/10.1016/j.ejps.2019.104989] [PMID: 31301485]

[15]   Donkers JM, Eslami Amirabadi H, van de Steeg E. Intestine-on-a-chip: Next level in vitro research model of the human intestine. Curr Opin Toxicol 2021; 25: 6-14.
[http://dx.doi.org/10.1016/j.cotox.2020.11.002]

[16]   Marrero D, Pujol-Vila F, Vera D, *et al.* Gut-on-a-chip: Mimicking and monitoring the human intestine. Biosens Bioelectron 2021; 181: 113156.
[http://dx.doi.org/10.1016/j.bios.2021.113156] [PMID: 33761417]

[17]   Verhulsel M, Simon A, Bernheim-Dennery M, *et al.* Developing an advanced gut on chip model enabling the study of epithelial cell/fibroblast interactions. Lab Chip 2021; 21(2): 365-77.
[http://dx.doi.org/10.1039/D0LC00672F] [PMID: 33306083]

[18]   Rahman S, Ghiboub M, Donkers JM, *et al.* The Progress of Intestinal Epithelial Models from Cell Lines to Gut-On-Chip. Int J Mol Sci 2021; 22(24): 13472.
[http://dx.doi.org/10.3390/ijms222413472] [PMID: 34948271]

[19]   Evans GS, Flint N, Somers AS, Eyden B, Potten CS. The development of a method for the preparation of rat intestinal epithelial cell primary cultures. J Cell Sci 1992; 101(1): 219-31.
[http://dx.doi.org/10.1242/jcs.101.1.219] [PMID: 1569126]

[20]   Ootani A, Li X, Sangiorgi E, *et al.* Sustained *in vitro* intestinal epithelial culture within a Wnt-dependent stem cell niche. Nat Med 2009; 15(6): 701-6.
[http://dx.doi.org/10.1038/nm.1951] [PMID: 19398967]

[21]   Barker N, van Es JH, Kuipers J, *et al.* Identification of stem cells in small intestine and colon by marker gene Lgr5. Nature 2007; 449(7165): 1003-7.
[http://dx.doi.org/10.1038/nature06196] [PMID: 17934449]

[22]   Barker N, van Es JH, Kuipers J, *et al.* Identification of stem cells in small intestine and colon by marker gene Lgr5. Nature 2007; 449(7165): 1003-7.
[http://dx.doi.org/10.1038/nature06196] [PMID: 17934449]

[23]   Montgomery RK, Carlone DL, Richmond CA, *et al.* Mouse telomerase reverse transcriptase (mTert) expression marks slowly cycling intestinal stem cells. Proc Natl Acad Sci USA 2011; 108(1): 179-84.
[http://dx.doi.org/10.1073/pnas.1013004108] [PMID: 21173232]

[24]   Takeda N, Jain R, LeBoeuf MR, Wang Q, Lu MM, Epstein JA. Interconversion between intestinal stem cell populations in distinct niches. Science 2011; 334(6061): 1420-4.
[http://dx.doi.org/10.1126/science.1213214] [PMID: 22075725]

[25]   Itzkovitz S, Lyubimova A, Blat IC, *et al.* Single-molecule transcript counting of stem-cell markers in the mouse intestine. Nat Cell Biol 2012; 14(1): 106-14.
[http://dx.doi.org/10.1038/ncb2384] [PMID: 22119784]

[26]   Kim TH, Li F, Ferreiro-Neira I, *et al.* Broadly permissive intestinal chromatin underlies lateral inhibition and cell plasticity. Nature 2014; 506(7489): 511-5.
[http://dx.doi.org/10.1038/nature12903] [PMID: 24413398]

[27]   Muñoz J, Stange DE, Schepers AG, *et al.* The Lgr5 intestinal stem cell signature: robust expression of

proposed quiescent '+4' cell markers. EMBO J 2012; 31(14): 3079-91.
[http://dx.doi.org/10.1038/emboj.2012.166] [PMID: 22692129]

[28]   van de Wetering M, Sancho E, Verweij C, *et al.* The beta-catenin/TCF-4 complex imposes a crypt progenitor phenotype on colorectal cancer cells. Cell 2002; 111(2): 241-50.
[http://dx.doi.org/10.1016/S0092-8674(02)01014-0] [PMID: 12408868]

[29]   Kim KA, Kakitani M, Zhao J, *et al.* Mitogenic influence of human R-spondin1 on the intestinal epithelium. Science 2005; 309(5738): 1256-9.
[http://dx.doi.org/10.1126/science.1112521] [PMID: 16109882]

[30]   Koo BK, Spit M, Jordens I, *et al.* Tumour suppressor RNF43 is a stem-cell E3 ligase that induces endocytosis of Wnt receptors. Nature 2012; 488(7413): 665-9.
[http://dx.doi.org/10.1038/nature11308] [PMID: 22895187]

[31]   Davis H, Irshad S, Bansal M, *et al.* Aberrant epithelial GREM1 expression initiates colonic tumorigenesis from cells outside the stem cell niche. Nat Med 2015; 21(1): 62-70.
[http://dx.doi.org/10.1038/nm.3750] [PMID: 25419707]

[32]   Haramis APG, Begthel H, van den Born M, *et al.* De novo crypt formation and juvenile polyposis on BMP inhibition in mouse intestine. Science 2004; 303(5664): 1684-6.
[http://dx.doi.org/10.1126/science.1093587] [PMID: 15017003]

[33]   Milano J, McKay J, Dagenais C, *et al.* Modulation of notch processing by γ-secretase inhibitors causes intestinal goblet cell metaplasia and induction of genes known to specify gut secretory lineage differentiation. Toxicol Sci 2004; 82(1): 341-58.
[http://dx.doi.org/10.1093/toxsci/kfh254] [PMID: 15319485]

[34]   JH E, H C Notch and Wnt inhibitors as potential new drugs for intestinal neoplastic disease 2005.

[35]   JH E, N G, M B, H C, BA H. Intestinal stem cells lacking the Math1 tumour suppressor are refractory to Notch inhibitors. Nat Commun 2010; 1(18)

[36]   Feng Y, Bommer GT, Zhao J, *et al.* Mutant KRAS promotes hyperplasia and alters differentiation in the colon epithelium but does not expand the presumptive stem cell pool. Gastroenterology 2011; 141(3): 1003-1013.e10, 10.
[http://dx.doi.org/10.1053/j.gastro.2011.05.007] [PMID: 21699772]

[37]   Powell AE, Wang Y, Li Y, *et al.* The pan-ErbB negative regulator Lrig1 is an intestinal stem cell marker that functions as a tumor suppressor. Cell 2012; 149(1): 146-58.
[http://dx.doi.org/10.1016/j.cell.2012.02.042] [PMID: 22464327]

[38]   Wong VWY, Stange DE, Page ME, *et al.* Lrig1 controls intestinal stem-cell homeostasis by negative regulation of ErbB signalling. Nat Cell Biol 2012; 14(4): 401-8.
[http://dx.doi.org/10.1038/ncb2464] [PMID: 22388892]

[39]   Yui S, Nakamura T, Sato T, *et al.* Functional engraftment of colon epithelium expanded in vitro from a single adult Lgr5+ stem cell. Nat Med 2012; 18(4): 618-23.
[http://dx.doi.org/10.1038/nm.2695] [PMID: 22406745]

[40]   Kleinman HK, McGarvey ML, Hassell JR, *et al.* Basement membrane complexes with biological activity. Biochemistry 1986; 25(2): 312-8.
[http://dx.doi.org/10.1021/bi00350a005] [PMID: 2937447]

[41]   Sato T, Vries RG, Snippert HJ, *et al.* Single Lgr5 stem cells build crypt-villus structures *in vitro* without a mesenchymal niche. Nature 2009; 459(7244): 262-5.
[http://dx.doi.org/10.1038/nature07935] [PMID: 19329995]

[42]   Barker N, van Es JH, Kuipers J, *et al.* Identification of stem cells in small intestine and colon by marker gene Lgr5. Nature 2007; 449(7165): 1003-7.
[http://dx.doi.org/10.1038/nature06196] [PMID: 17934449]

[43]   Y-27632 preserves epidermal integrity in a human skin organ-culture (hSOC) system by regulating

AKT and ERK signaling pathways Journal of Dermatological Science Available from: https://wwwjdsjournalcom/article/S0923-1811(19)30338-X/fulltext

[44]   Sato T, Stange DE, Ferrante M, *et al.* Long-term expansion of epithelial organoids from human colon, adenoma, adenocarcinoma, and Barrett's epithelium. Gastroenterology 2011; 141(5): 1762-72.
[http://dx.doi.org/10.1053/j.gastro.2011.07.050] [PMID: 21889923]

[45]   Jung P, Sato T, Merlos-Suárez A, *et al.* Isolation and *in vitro* expansion of human colonic stem cells. Nat Med 2011; 17(10): 1225-7.
[http://dx.doi.org/10.1038/nm.2470] [PMID: 21892181]

[46]   Miyoshi H, Stappenbeck TS. *In vitro* expansion and genetic modification of gastrointestinal stem cells in spheroid culture. Nat Protoc 2013; 8(12): 2471-82.
[http://dx.doi.org/10.1038/nprot.2013.153] [PMID: 24232249]

[47]   Middendorp S, Schneeberger K, Wiegerinck CL, *et al.* Adult stem cells in the small intestine are intrinsically programmed with their location-specific function. Stem Cells 2014; 32(5): 1083-91.
[http://dx.doi.org/10.1002/stem.1655] [PMID: 24496776]

[48]   Mustata RC, Vasile G, Fernandez-Vallone V, *et al.* Identification of Lgr5-independent spheroid-generating progenitors of the mouse fetal intestinal epithelium. Cell Rep 2013; 5(2): 421-32.
[http://dx.doi.org/10.1016/j.celrep.2013.09.005] [PMID: 24139799]

[49]   Fordham RP, Yui S, Hannan NRF, *et al.* Transplantation of expanded fetal intestinal progenitors contributes to colon regeneration after injury. Cell Stem Cell 2013; 13(6): 734-44.
[http://dx.doi.org/10.1016/j.stem.2013.09.015] [PMID: 24139758]

[50]   Fukuda M, Mizutani T, Mochizuki W, *et al.* Small intestinal stem cell identity is maintained with functional Paneth cells in heterotopically grafted epithelium onto the colon. Genes Dev 2014; 28(16): 1752-7.
[http://dx.doi.org/10.1101/gad.245233.114] [PMID: 25128495]

[51]   Spence JR, Mayhew CN, Rankin SA, *et al.* Directed differentiation of human pluripotent stem cells into intestinal tissue in vitro. Nature 2011; 470(7332): 105-9.
[http://dx.doi.org/10.1038/nature09691] [PMID: 21151107]

[52]   Watson CL, Mahe MM, Múnera J, *et al.* An *in vivo* model of human small intestine using pluripotent stem cells. Nat Med 2014; 20(11): 1310-4.
[http://dx.doi.org/10.1038/nm.3737] [PMID: 25326803]

[53]   Serra D, Mayr U, Boni A, *et al.* Self-organization and symmetry breaking in intestinal organoid development. Nature 2019; 569(7754): 66-72.
[http://dx.doi.org/10.1038/s41586-019-1146-y] [PMID: 31019299]

[54]   Sumigray KD, Terwilliger M, Lechler T. Morphogenesis and Compartmentalization of the Intestinal Crypt. Dev Cell 2018; 45(2): 183-197.e5.
[http://dx.doi.org/10.1016/j.devcel.2018.03.024] [PMID: 29689194]

[55]   Noel G, Baetz NW, Staab JF, *et al.* A primary human macrophage-enteroid co-culture model to investigate mucosal gut physiology and host-pathogen interactions. Sci Rep 2017; 7(1): 45270.
[http://dx.doi.org/10.1038/srep45270] [PMID: 28345602]

[56]   Rogoz A, Reis BS, Karssemeijer RA, Mucida D. A 3-D enteroid-based model to study T-cell and epithelial cell interaction. J Immunol Methods 2015; 421: 89-95.
[http://dx.doi.org/10.1016/j.jim.2015.03.014] [PMID: 25841547]

[57]   Nozaki K, Mochizuki W, Matsumoto Y, *et al.* Co-culture with intestinal epithelial organoids allows efficient expansion and motility analysis of intraepithelial lymphocytes. J Gastroenterol 2016; 51(3): 206-13.
[http://dx.doi.org/10.1007/s00535-016-1170-8] [PMID: 26800996]

[58]   Cook L, Stahl M, Han X, *et al.* Suppressive and Gut-Reparative Functions of Human Type 1 T Regulatory Cells. Gastroenterology 2019; 157(6): 1584-98.

[http://dx.doi.org/10.1053/j.gastro.2019.09.002] [PMID: 31513797]

[59]	Schnalzger TE, de Groot MHP, Zhang C, *et al.* 3D model for CAR -mediated cytotoxicity using patient-derived colorectal cancer organoids. EMBO J 2019; 38(12): e100928.
[http://dx.doi.org/10.15252/embj.2018100928] [PMID: 31036555]

[60]	Dijkstra KK, Cattaneo CM, Weeber F, *et al.* Generation of Tumor-Reactive T Cells by Co-culture of Peripheral Blood Lymphocytes and Tumor Organoids. Cell 2018; 174(6): 1586-1598.e12.
[http://dx.doi.org/10.1016/j.cell.2018.07.009] [PMID: 30100188]

[61]	Yin Y, Bijvelds M, Dang W, *et al.* Modeling rotavirus infection and antiviral therapy using primary intestinal organoids. Antiviral Res 2015; 123: 120-31.
[http://dx.doi.org/10.1016/j.antiviral.2015.09.010] [PMID: 26408355]

[62]	Ettayebi K, Crawford SE, Murakami K, *et al.* Replication of human noroviruses in stem cell–derived human enteroids. Science 2016; 353(6306): 1387-93.
[http://dx.doi.org/10.1126/science.aaf5211] [PMID: 27562956]

[63]	Engevik MA, Engevik KA, Yacyshyn MB, *et al.* Human *Clostridium difficile* infection: inhibition of NHE3 and microbiota profile. Am J Physiol Gastrointest Liver Physiol 2015; 308(6): G497-509.
[http://dx.doi.org/10.1152/ajpgi.00090.2014] [PMID: 25552580]

[64]	Williamson IA, Arnold JW, Samsa LA, *et al.* A High-Throughput Organoid Microinjection Platform to Study Gastrointestinal Microbiota and Luminal Physiology. Cell Mol Gastroenterol Hepatol 2018; 6(3): 301-19.
[http://dx.doi.org/10.1016/j.jcmgh.2018.05.004] [PMID: 30123820]

[65]	Jalili-Firoozinezhad S, Gazzaniga FS, Calamari EL, *et al.* A complex human gut microbiome cultured in an anaerobic intestine-on-a-chip. Nat Biomed Eng 2019; 3(7): 520-31.
[http://dx.doi.org/10.1038/s41551-019-0397-0] [PMID: 31086325]

[66]	Sasaki N, Miyamoto K, Maslowski KM, Ohno H, Kanai T, Sato T. Development of a Scalable Coculture System for Gut Anaerobes and Human Colon Epithelium. Gastroenterology 2020; 159(1): 388-390.e5.
[http://dx.doi.org/10.1053/j.gastro.2020.03.021] [PMID: 32199883]

[67]	Co JY, Margalef-Català M, Li X, *et al.* Controlling Epithelial Polarity: A Human Enteroid Model for Host-Pathogen Interactions. Cell Rep 2019; 26(9): 2509-2520.e4.
[http://dx.doi.org/10.1016/j.celrep.2019.01.108] [PMID: 30811997]

[68]	Foulke-Abel J, In J, Kovbasnjuk O, *et al.* Human enteroids as an *ex-vivo* model of host–pathogen interactions in the gastrointestinal tract. Exp Biol Med (Maywood) 2014; 239(9): 1124-34.
[http://dx.doi.org/10.1177/1535370214529398] [PMID: 24719375]

[69]	Kane L, Hale C, Goulding D, *et al.* Using human iPSC derived small intestinal organoids as a model for enteric disease caused by Enterotoxigenic E. coli and Vibrio cholerae. Access Microbiol 2019; 1(1A): 754.
[http://dx.doi.org/10.1099/acmi.ac2019.po0482]

[70]	Ommen DDZ, Pukin AV, Fu O, van Ufford LHCQ, Janssens HM, Beekman JM, *et al.* Functional Characterization of Cholera Toxin Inhibitors Using Human Intestinal Organoids. 2016. ACS Publications. American Chemical Society; Available from: https://pubs.acs.org/doi/pdf/10.1021/acs.jmedchem.6b00770

[71]	Koestler BJ, Ward CM, Fisher CR, Rajan A, Maresso AW, Payne SM. Human Intestinal Enteroids as a Model System of *Shigella* Pathogenesis. Infect Immun 2019; 87(4): e00733-18.https://www.ncbi.nlm.nih.gov/labs/pmc/articles/PMC6434139/
[http://dx.doi.org/10.1128/IAI.00733-18] [PMID: 30642906]

[72]	Ranganathan S, Doucet M, Grassel CL, Delaine-Elias B, Zachos NC, Barry EM. Evaluating Shigella flexneri Pathogenesis in the Human Enteroid Model. Infection and Immunity Available from: https://journalsasmorg/doi/abs/101128/IAI00740-18 2019.

[http://dx.doi.org/10.1128/IAI.00740-18]

[73]   In J, Foulke-Abel J, Zachos NC, *et al.* Enterohemorrhagic *Escherichia coli* reduce mucus and intermicrovillar bridges in human stem cell-derived colonoids. Cell Mol Gastroenterol Hepatol 2016; 2(1): 48-62.e3.
[http://dx.doi.org/10.1016/j.jcmgh.2015.10.001] [PMID: 26855967]

[74]   VanDussen KL, Marinshaw JM, Shaikh N, *et al.* Development of an enhanced human gastrointestinal epithelial culture system to facilitate patient-based assays. Gut 2015; 64(6): 911-20.
[http://dx.doi.org/10.1136/gutjnl-2013-306651] [PMID: 25007816]

[75]   Rajan A, Vela L, Zeng XL, *et al.* Novel Segment- and Host-Specific Patterns of Enteroaggregative *Escherichia coli* Adherence to Human Intestinal Enteroids. MBio 2018; 9(1): e02419-17.
[http://dx.doi.org/10.1128/mBio.02419-17] [PMID: 29463660]

[76]   Nakamoto N, Sasaki N, Aoki R, *et al.* Gut pathobionts underlie intestinal barrier dysfunction and liver T helper 17 cell immune response in primary sclerosing cholangitis. Nat Microbiol 2019; 4(3): 492-503.
[http://dx.doi.org/10.1038/s41564-018-0333-1] [PMID: 30643240]

[77]   Saavedra PHV, Huang L, Ghazavi F, *et al.* Apoptosis of intestinal epithelial cells restricts Clostridium difficile infection in a model of pseudomembranous colitis. Nat Commun 2018; 9(1): 4846.
[http://dx.doi.org/10.1038/s41467-018-07386-5] [PMID: 30451870]

[78]   Zhang YG, Wu S, Xia Y, Sun J. *Salmonella* -infected crypt-derived intestinal organoid culture system for host-bacterial interactions. Physiol Rep 2014; 2(9): e12147.
[http://dx.doi.org/10.14814/phy2.12147] [PMID: 25214524]

[79]   Saxena K, Blutt SE, Ettayebi K, *et al.* Human Intestinal Enteroids: a New Model To Study Human Rotavirus Infection, Host Restriction, and Pathophysiology. J Virol 2016; 90(1): 43-56.
[http://dx.doi.org/10.1128/JVI.01930-15] [PMID: 26446608]

[80]   Chang-Graham AL, Danhof HA, Engevik MA, *et al.* Human Intestinal Enteroids With Inducible Neurogenin-3 Expression as a Novel Model of Gut Hormone Secretion. Cell Mol Gastroenterol Hepatol 2019; 8(2): 209-29.
[http://dx.doi.org/10.1016/j.jcmgh.2019.04.010] [PMID: 31029854]

[81]   Drummond CG, Bolock AM, Ma C, Luke CJ, Good M, Coyne CB. Enteroviruses infect human enteroids and induce antiviral signaling in a cell lineage-specific manner. Proc Natl Acad Sci USA 2017; 114(7): 1672-7.
[http://dx.doi.org/10.1073/pnas.1617363114] [PMID: 28137842]

[82]   Holly MK, Smith JG. Adenovirus Infection of Human Enteroids Reveals Interferon Sensitivity and Preferential Infection of Goblet Cells. J Virol 2018; 92(9): e00250-18.
[http://dx.doi.org/10.1128/JVI.00250-18] [PMID: 29467318]

[83]   Kolawole AO, Mirabelli C, Hill DR, *et al.* Astrovirus replication in human intestinal enteroids reveals multi-cellular tropism and an intricate host innate immune landscape. PLoS Pathog 2019; 15(10): e1008057.
[http://dx.doi.org/10.1371/journal.ppat.1008057] [PMID: 31671153]

[84]   Zhou J, Li C, Zhao G, *et al.* Human intestinal tract serves as an alternative infection route for Middle East respiratory syndrome coronavirus. Sci Adv 2017; 3(11): eaao4966.
[http://dx.doi.org/10.1126/sciadv.aao4966] [PMID: 29152574]

[85]   Lamers MM, Beumer J, van der Vaart J, *et al.* SARS-CoV-2 productively infects human gut enterocytes. Science 2020; 369(6499): 50-4.
[http://dx.doi.org/10.1126/science.abc1669] [PMID: 32358202]

[86]   Zang R, Castro MFG, McCune BT, *et al.* TMPRSS2 and TMPRSS4 promote SARS-CoV-2 infection of human small intestinal enterocytes. Sci Immunol 2020; 5(47): eabc3582.
[http://dx.doi.org/10.1126/sciimmunol.abc3582] [PMID: 32404436]

[87]   Zhou J, Li C, Liu X, *et al.* Infection of bat and human intestinal organoids by SARS-CoV-2. Nat Med 2020; 26(7): 1077-83.
        [http://dx.doi.org/10.1038/s41591-020-0912-6] [PMID: 32405028]

[88]   Wilke G, Funkhouser-Jones LJ, Wang Y, *et al.* A Stem-Cell-Derived Platform Enables Complete Cryptosporidium Development *In Vitro* and Genetic Tractability. Cell Host Microbe 2019; 26(1): 123-134.e8.
        [http://dx.doi.org/10.1016/j.chom.2019.05.007] [PMID: 31231046]

[89]   Heo I, Dutta D, Schaefer DA, *et al.* Modelling Cryptosporidium infection in human small intestinal and lung organoids. Nat Microbiol 2018; 3(7): 814-23.
        [http://dx.doi.org/10.1038/s41564-018-0177-8] [PMID: 29946163]

[90]   Martorelli Di Genova B, Wilson SK, Dubey JP, Knoll LJ. Intestinal delta-6-desaturase activity determines host range for Toxoplasma sexual reproduction. PLoS Biol 2019; 17(8): e3000364.
        [http://dx.doi.org/10.1371/journal.pbio.3000364] [PMID: 31430281]

[91]   Derricott H, Luu L, Fong WY, *et al.* Developing a 3D intestinal epithelium model for livestock species. Cell Tissue Res 2019; 375(2): 409-24.
        [http://dx.doi.org/10.1007/s00441-018-2924-9] [PMID: 30259138]

[92]   Haga K, Ettayebi K, Tenge VR, Karandikar UC, Lewis MA, Lin S-C, *et al.* Genetic Manipulation of Human Intestinal Enteroids Demonstrates the Necessity of a Functional Fucosyltransferase 2 Gene for Secretor-Dependent Human Norovirus Infection. mBio Available from: https://journalsasm org/doi/abs/101128/mBio00251-20 2020.

[93]   Finkbeiner SR, Zeng XL, Utama B, Atmar RL, Shroyer NF, Estes MK. Stem cell-derived human intestinal organoids as an infection model for rotaviruses. MBio 2012; 3(4): e00159-12.
        [http://dx.doi.org/10.1128/mBio.00159-12] [PMID: 22761392]

[94]   Geurts MH, van der Vaart J, Beumer J, Clevers H. The Organoid Platform: Promises and Challenges as Tools in the Fight against COVID-19. Stem Cell Reports 2021; 16(3): 412-8.
        [http://dx.doi.org/10.1016/j.stemcr.2020.11.009] [PMID: 33691146]

[95]   Hou YJ, Okuda K, Edwards CE, *et al.* SARS-CoV-2 Reverse Genetics Reveals a Variable Infection Gradient in the Respiratory Tract. Cell 2020; 182(2): 429-446.e14.
        [http://dx.doi.org/10.1016/j.cell.2020.05.042] [PMID: 32526206]

[96]   Wang M, Cao R, Zhang L, *et al.* Remdesivir and chloroquine effectively inhibit the recently emerged novel coronavirus (2019-nCoV) in vitro. Cell Res 2020; 30(3): 269-71.
        [http://dx.doi.org/10.1038/s41422-020-0282-0] [PMID: 32020029]

[97]   Yin Y, Dang W, Zhou X, *et al.* PI3K-Akt-mTOR axis sustains rotavirus infection *via* the 4E-BP1 mediated autophagy pathway and represents an antiviral target. Virulence 2018; 9(1): 83-98.
        [http://dx.doi.org/10.1080/21505594.2017.1326443] [PMID: 28475412]

[98]   Matsuzawa-Ishimoto Y, Shono Y, Gomez LE, *et al.* Autophagy protein ATG16L1 prevents necroptosis in the intestinal epithelium. J Exp Med 2017; 214(12): 3687-705.
        [http://dx.doi.org/10.1084/jem.20170558] [PMID: 29089374]

[99]   Pott J, Kabat AM, Maloy KJ. Intestinal Epithelial Cell Autophagy Is Required to Protect against TNF-Induced Apoptosis during Chronic Colitis in Mice. Cell Host Microbe 2018; 23(2): 191-202.e4.
        [http://dx.doi.org/10.1016/j.chom.2017.12.017] [PMID: 29358084]

[100]   Lindemans CA, Calafiore M, Mertelsmann AM, *et al.* Interleukin-22 promotes intestinal-stem-cell-mediated epithelial regeneration. Nature 2015; 528(7583): 560-4.
        [http://dx.doi.org/10.1038/nature16460] [PMID: 26649819]

[101]   Gronke K, Hernández PP, Zimmermann J, *et al.* Interleukin-22 protects intestinal stem cells against genotoxic stress. Nature 2019; 566(7743): 249-53.
        [http://dx.doi.org/10.1038/s41586-019-0899-7] [PMID: 30700914]

[102] Yui S, Azzolin L, Maimets M, *et al.* YAP/TAZ-Dependent Reprogramming of Colonic Epithelium Links ECM Remodeling to Tissue Regeneration. Cell Stem Cell 2018; 22(1): 35-49.e7.
[http://dx.doi.org/10.1016/j.stem.2017.11.001] [PMID: 29249464]

[103] Nusse YM, Savage AK, Marangoni P, *et al.* Parasitic helminths induce fetal-like reversion in the intestinal stem cell niche. Nature 2018; 559(7712): 109-13.
[http://dx.doi.org/10.1038/s41586-018-0257-1] [PMID: 29950724]

[104] Sato T, Clevers H. Growing self-organizing mini-guts from a single intestinal stem cell: Mechanism 2013.
[http://dx.doi.org/10.1126/science.1234852]

[105] Fujii M, Shimokawa M, Date S, *et al.* A Colorectal Tumor Organoid Library Demonstrates Progressive Loss of Niche Factor Requirements during Tumorigenesis. Cell Stem Cell 2016; 18(6): 827-38.
[http://dx.doi.org/10.1016/j.stem.2016.04.003] [PMID: 27212702]

[106] Tao Y, Kang B, Petkovich DA, *et al.* Aging-like Spontaneous Epigenetic Silencing Facilitates Wnt Activation, Stemness, and Braf^{V600E}-Induced Tumorigenesis. Cancer Cell 2019; 35(2): 315-328.e6.
[http://dx.doi.org/10.1016/j.ccell.2019.01.005] [PMID: 30753828]

[107] IJspeert JEG, Vermeulen L, Meijer GA, Dekker E. Serrated neoplasia—role in colorectal carcinogenesis and clinical implications. Nat Rev Gastroenterol Hepatol 2015; 12(7): 401-9.
[http://dx.doi.org/10.1038/nrgastro.2015.73] [PMID: 25963511]

[108] Lau HCH, Kranenburg O, Xiao H, Yu J. Organoid models of gastrointestinal cancers in basic and translational research. Nat Rev Gastroenterol Hepatol 2020; 17(4): 203-22.
[http://dx.doi.org/10.1038/s41575-019-0255-2] [PMID: 32099092]

[109] Shaashua L, Mayer S, Lior C, Lavon H, Novoselsky A, Scherz-Shouval R. Stromal Expression of the Core Clock Gene *Period 2* Is Essential for Tumor Initiation and Metastatic Colonization. Front Cell Dev Biol 2020; 8: 587697.
[http://dx.doi.org/10.3389/fcell.2020.587697] [PMID: 33123539]

[110] Cao L, Kuratnik A, Xu W, *et al.* Development of intestinal organoids as tissue surrogates: Cell composition and the Epigenetic control of differentiation. Mol Carcinog 2015; 54(3): 189-202.
[http://dx.doi.org/10.1002/mc.22089] [PMID: 24115167]

[111] Xu H, Yan Y, Deb S, *et al.* Cohesin Rad21 mediates loss of heterozygosity and is upregulated *via* Wnt promoting transcriptional dysregulation in gastrointestinal tumors. Cell Rep 2014; 9(5): 1781-97.
[http://dx.doi.org/10.1016/j.celrep.2014.10.059] [PMID: 25464844]

[112] Fujii M, Sato T. Somatic cell-derived organoids as prototypes of human epithelial tissues and diseases. Nat Mater 2021; 20(2): 156-69.
[http://dx.doi.org/10.1038/s41563-020-0754-0] [PMID: 32807924]

[113] Vlachogiannis G, Hedayat S, Vatsiou A, *et al.* Patient-derived organoids model treatment response of metastatic gastrointestinal cancers. Science 2018; 359(6378): 920-6.
[http://dx.doi.org/10.1126/science.aao2774] [PMID: 29472484]

[114] Roerink SF, Sasaki N, Lee-Six H, *et al.* Intra-tumour diversification in colorectal cancer at the single-cell level. Nature 2018; 556(7702): 457-62.
[http://dx.doi.org/10.1038/s41586-018-0024-3] [PMID: 29643510]

[115] Bolhaqueiro ACF, Ponsioen B, Bakker B, *et al.* Ongoing chromosomal instability and karyotype evolution in human colorectal cancer organoids. Nat Genet 2019; 51(5): 824-34.
[http://dx.doi.org/10.1038/s41588-019-0399-6] [PMID: 31036964]

[116] Drost J, van Boxtel R, Blokzijl F, *et al.* Use of CRISPR-modified human stem cell organoids to study the origin of mutational signatures in cancer. Science 2017; 358(6360): 234-8.
[http://dx.doi.org/10.1126/science.aao3130] [PMID: 28912133]

[117] van de Wetering M, Francies HE, Francis JM, *et al.* Prospective derivation of a living organoid biobank of colorectal cancer patients. Cell 2015; 161(4): 933-45.
[http://dx.doi.org/10.1016/j.cell.2015.03.053] [PMID: 25957691]

[118] Bellono NW, Bayrer JR, Leitch DB, *et al.* Enterochromaffin cells are gut chemosensors that couple to sensory neural pathways. Cell 2017; 170(1): 185-198.e16.
[http://dx.doi.org/10.1016/j.cell.2017.05.034] [PMID: 28648659]

[119] Howitt MR, Lavoie S, Michaud M, *et al.* Tuft cells, taste-chemosensory cells, orchestrate parasite type 2 immunity in the gut. Science 2016; 351(6279): 1329-33.
[http://dx.doi.org/10.1126/science.aaf1648] [PMID: 26847546]

[120] de Lau W, Kujala P, Schneeberger K, *et al.* Peyer's patch M cells derived from Lgr5(+) stem cells require SpiB and are induced by RankL in cultured "miniguts". Mol Cell Biol 2012; 32(18): 3639-47.
[http://dx.doi.org/10.1128/MCB.00434-12] [PMID: 22778137]

[121] Angus HCK, Butt AG, Schultz M, Kemp RA. Intestinal Organoids as a Tool for Inflammatory Bowel Disease Research. Front Med (Lausanne) 2020; 6: 334.
[http://dx.doi.org/10.3389/fmed.2019.00334] [PMID: 32010704]

[122] d'Aldebert E, Quaranta M, Sébert M, *et al.* Characterization of Human Colon Organoids From Inflammatory Bowel Disease Patients. Front Cell Dev Biol 2020; 8: 363.
https://www.frontiersin.org/article/10.3389/fcell.2020.00363
[http://dx.doi.org/10.3389/fcell.2020.00363] [PMID: 32582690]

[123] O'Connell L, Winter DC, Aherne CM. The Role of Organoids as a Novel Platform for Modeling of Inflammatory Bowel Disease. Front Pediatr 2021; 9: 624045. https://www.frontiersin.org/article/10.3389/fped.2021.624045
[http://dx.doi.org/10.3389/fped.2021.624045] [PMID: 33681101]

[124] Sarvestani SK, Signs S, Hu B, *et al.* Induced organoids derived from patients with ulcerative colitis recapitulate colitic reactivity. Nat Commun 2021; 12(1): 262.
[http://dx.doi.org/10.1038/s41467-020-20351-5] [PMID: 33431859]

[125] Pastuła A, Middelhoff M, Brandtner A, *et al.* Three-Dimensional Gastrointestinal Organoid Culture in Combination with Nerves or Fibroblasts: A Method to Characterize the Gastrointestinal Stem Cell Niche. Stem Cells Int 2016; 2016: 1-16.
[http://dx.doi.org/10.1155/2016/3710836] [PMID: 26697073]

[126] Ramani S, Crawford SE, Blutt SE, Estes MK. Human organoid cultures: transformative new tools for human virus studies. Curr Opin Virol 2018; 29: 79-86.
[http://dx.doi.org/10.1016/j.coviro.2018.04.001] [PMID: 29656244]

[127] Karve SS, Pradhan S, Ward DV, Weiss AA. Intestinal organoids model human responses to infection by commensal and Shiga toxin producing Escherichia coli. PLoS One 2017; 12(6): e0178966.
[http://dx.doi.org/10.1371/journal.pone.0178966] [PMID: 28614372]

[128] Moran CJ, Klein C, Muise AM, Snapper SB. Very early-onset inflammatory bowel disease: gaining insight through focused discovery. Inflamm Bowel Dis 2015; 21(5): 1166-75.
[http://dx.doi.org/10.1097/MIB.0000000000000329] [PMID: 25895007]

[129] Lehle AS, Farin HF, Marquardt B, *et al.* Intestinal Inflammation and Dysregulated Immunity in Patients With Inherited Caspase-8 Deficiency. Gastroenterology 2019; 156(1): 275-8.
[http://dx.doi.org/10.1053/j.gastro.2018.09.041] [PMID: 30267714]

[130] Schwerd T, Bryant RV, Pandey S, *et al.* NOX1 loss-of-function genetic variants in patients with inflammatory bowel disease. Mucosal Immunol 2018; 11(2): 562-74.
[http://dx.doi.org/10.1038/mi.2017.74] [PMID: 29091079]

[131] Vogel GF, Hess MW, Pfaller K, Huber LA, Janecke AR, Müller T. Towards understanding microvillus inclusion disease. Mol Cell Pediatr 2016; 3(1): 3.
[http://dx.doi.org/10.1186/s40348-016-0031-0] [PMID: 26830108]

[132] Mosa MH, Nicolle O, Maschalidi S, *et al.* Dynamic Formation of Microvillus Inclusions During Enterocyte Differentiation in *Munc18-2*–Deficient Intestinal Organoids. Cell Mol Gastroenterol Hepatol 2018; 6(4): 477-493.e1.
[http://dx.doi.org/10.1016/j.jcmgh.2018.08.001] [PMID: 30364784]

[133] Wiegerinck CL, Janecke AR, Schneeberger K, *et al.* Loss of syntaxin 3 causes variant microvillus inclusion disease. Gastroenterology 2014; 147(1): 65-68.e10.
[http://dx.doi.org/10.1053/j.gastro.2014.04.002] [PMID: 24726755]

[134] Bigorgne AE, Farin HF, Lemoine R, *et al.* TTC7A mutations disrupt intestinal epithelial apicobasal polarity. J Clin Invest 2014; 124(1): 328-37.
[http://dx.doi.org/10.1172/JCI71471] [PMID: 24292712]

[135] van Rijn JM, Ardy RC, Kuloğlu Z, *et al.* Intestinal Failure and Aberrant Lipid Metabolism in Patients With DGAT1 Deficiency. Gastroenterology 2018; 155(1): 130-143.e15.
[http://dx.doi.org/10.1053/j.gastro.2018.03.040] [PMID: 29604290]

[136] Liu J, Walker NM, Cook MT, Ootani A, Clarke LL. Functional Cftr in crypt epithelium of organotypic enteroid cultures from murine small intestine. Am J Physiol Cell Physiol 2012; 302(10): C1492-503.
[http://dx.doi.org/10.1152/ajpcell.00392.2011] [PMID: 22403785]

[137] Servidoni MF, Sousa M, Vinagre AM, *et al.* Rectal forceps biopsy procedure in cystic fibrosis: technical aspects and patients perspective for clinical trials feasibility. BMC Gastroenterol 2013; 13(1): 91.
[http://dx.doi.org/10.1186/1471-230X-13-91] [PMID: 23688510]

[138] Dekkers JF, Wiegerinck CL, de Jonge HR, *et al.* A functional CFTR assay using primary cystic fibrosis intestinal organoids. Nat Med 2013; 19(7): 939-45.
[http://dx.doi.org/10.1038/nm.3201] [PMID: 23727931]

[139] Boj SF, Vonk AM, Statia M, Su J, Vries RRG, Beekman JM, *et al.* Forskolin-induced Swelling in Intestinal Organoids: An In vitro Assay for Assessing Drug Response in Cystic Fibrosis Patients. J Vis Exp 2017; (120):

[140] Vonk AM, van Mourik P, Ramalho AS, *et al.* Protocol for Application, Standardization and Validation of the Forskolin-Induced Swelling Assay in Cystic Fibrosis Human Colon Organoids. STAR Protocols 2020; 1(1): 100019.
[http://dx.doi.org/10.1016/j.xpro.2020.100019] [PMID: 33111074]

[141] Dekkers JF, Berkers G, Kruisselbrink E, Vonk A, de Jonge HR, Janssens HM, *et al.* Characterizing responses to CFTR-modulating drugs using rectal organoids derived from subjects with cystic fibrosis. Science Translational Medicine 2016; 8(344): 344ra84-4.

[142] Groot KM de W. Stratifying infants with cystic fibrosis for disease severity using intestinal organoid swelling as a biomarker of CFTR function. European Respiratory Journal Available from: https://erjersjournalscom/content/52/3/1702529 2018.

[143] Noordhoek J, Gulmans V, van der Ent K, Beekman JM. Intestinal organoids and personalized medicine in cystic fibrosis. Curr Opin Pulm Med 2016; 22(6): 610-6.
[http://dx.doi.org/10.1097/MCP.0000000000000315] [PMID: 27635627]

[144] Ammon H, Müller A. Forskolin: from an ayurvedic remedy to a modern agent. Planta Med 1985; 51(6): 473-7.
[http://dx.doi.org/10.1055/s-2007-969566]

[145] Seamon KB, Daly JW. Forskolin: a unique diterpene activator of cyclic AMP-generating systems. J Cyclic Nucleotide Res 1981; 7(4): 201-24.
[PMID: 6278005]

[146] Watt FM, Huck WTS. Role of the extracellular matrix in regulating stem cell fate. Nat Rev Mol Cell Biol 2013; 14(8): 467-73.
[http://dx.doi.org/10.1038/nrm3620] [PMID: 23839578]

[147] Gjorevski N, Sachs N, Manfrin A, *et al.* Designer matrices for intestinal stem cell and organoid culture. Nature 2016; 539(7630): 560-4.
[http://dx.doi.org/10.1038/nature20168] [PMID: 27851739]

[148] Hernandez-Gordillo V, Kassis T, Lampejo A, *et al.* Fully synthetic matrices for *in vitro* culture of primary human intestinal enteroids and endometrial organoids. Biomaterials 2020; 254: 120125.
[http://dx.doi.org/10.1016/j.biomaterials.2020.120125] [PMID: 32502894]

[149] Gao Y, Bado I, Wang H, Zhang W, Rosen JM, Zhang XHF. Metastasis Organotropism: Redefining the Congenial Soil. Dev Cell 2019; 49(3): 375-91.
[http://dx.doi.org/10.1016/j.devcel.2019.04.012] [PMID: 31063756]

[150] Fujii M, Clevers H, Sato T. Modeling Human Digestive Diseases With CRISPR-Cas9–Modified Organoids. Gastroenterology 2019; 156(3): 562-76.
[http://dx.doi.org/10.1053/j.gastro.2018.11.048] [PMID: 30476497]

[151] Schwank G, Koo BK, Sasselli V, *et al.* Functional repair of CFTR by CRISPR/Cas9 in intestinal stem cell organoids of cystic fibrosis patients. Cell Stem Cell 2013; 13(6): 653-8.
[http://dx.doi.org/10.1016/j.stem.2013.11.002] [PMID: 24315439]

[152] Matano M, Date S, Shimokawa M, *et al.* Modeling colorectal cancer using CRISPR-Cas9–mediated engineering of human intestinal organoids. Nat Med 2015; 21(3): 256-62.
[http://dx.doi.org/10.1038/nm.3802] [PMID: 25706875]

[153] Jager M, Blokzijl F, Kuijk E, *et al.* Deficiency of nucleotide excision repair is associated with mutational signature observed in cancer. Genome Res 2019; 29(7): 1067-77.
[http://dx.doi.org/10.1101/gr.246223.118] [PMID: 31221724]

[154] Drost J, van Jaarsveld RH, Ponsioen B, *et al.* Sequential cancer mutations in cultured human intestinal stem cells. Nature 2015; 521(7550): 43-7.
[http://dx.doi.org/10.1038/nature14415] [PMID: 25924068]

[155] Fessler E, Drost J, van Hooff SR, *et al.* TGFβ signaling directs serrated adenomas to the mesenchymal colorectal cancer subtype. EMBO Mol Med 2016; 8(7): 745-60.
[http://dx.doi.org/10.15252/emmm.201606184] [PMID: 27221051]

[156] Kawasaki K, Fujii M, Sugimoto S, *et al.* Chromosome Engineering of Human Colon-Derived Organoids to Develop a Model of Traditional Serrated Adenoma. Gastroenterology 2020; 158(3): 638-651.e8.
[http://dx.doi.org/10.1053/j.gastro.2019.10.009] [PMID: 31622618]

[157] Shimokawa M, Ohta Y, Nishikori S, *et al.* Visualization and targeting of LGR5⁺ human colon cancer stem cells. Nature 2017; 545(7653): 187-92.
[http://dx.doi.org/10.1038/nature22081] [PMID: 28355176]

[158] Halac U, Lacaille F, Joly F, *et al.* Microvillous inclusion disease: how to improve the prognosis of a severe congenital enterocyte disorder. J Pediatr Gastroenterol Nutr 2011; 52(4): 460-5.
[http://dx.doi.org/10.1097/MPG.0b013e3181fb4559] [PMID: 21407114]

[159] Berkers G, van Mourik P, Vonk AM, *et al.* Rectal Organoids Enable Personalized Treatment of Cystic Fibrosis. Cell Rep 2019; 26(7): 1701-1708.e3.
[http://dx.doi.org/10.1016/j.celrep.2019.01.068] [PMID: 30759382]

[160] Sugimoto S, Ohta Y, Fujii M, *et al.* Reconstruction of the Human Colon Epithelium In Vivo. Cell Stem Cell 2018; 22(2): 171-176.e5.
[http://dx.doi.org/10.1016/j.stem.2017.11.012] [PMID: 29290616]

CHAPTER 4

Bone Organoids: Current Approaches, Challenges, and Potential Applications

Khushboo Dutta[1] and **Sunita Nayak**[1,*]

¹ Department of Integrative Biology, School of Bio Sciences and Technology, Vellore Institute of Technology, Vellore-632014, Tamil Nadu, India

Abstract: Organoids are complex three-dimensional microtissues formed by the self-organization of stem cells and aimed to mimic the structural and functional characteristics of human tissues. Bone comprises multiple cells with a mechanically rigid extracellular matrix (ECM). The diversity of a bone in terms of structure and complexity demands an ideal bone model with limited control of physico-chemical parameters. Potential applications of bone organoids can be seen in bone regeneration, and regulation mechanism studies and to address various bone-related disorders and defects. Approaches to creating bone organoids may include using mesenchymal stem cells (MSCs), hematopoietic stem cells (HSCs), osteoblasts, and osteoclasts in addition to endothelial cells for the vasculature generation. Moreover, in bone organoids, ECM of biological origin materials that display close resemblance with the native bone ECM are preferred or may be generated using scaffold-free methods. The mechanical load should be the primary parameter that should be considered in the model. Since bone is considered hypoxic, organoid-based bone models should include an O2 regulation mechanism to achieve a physiologic hypoxic environment. Advanced cell isolation, tissue culture, and cell differentiation techniques, along with microfluidics and tissue engineering strategies, might lead to the production of physiologically relevant bone organoids. This chapter outlines prerequisites for bone organoid development and the potential applications of bone organoid-based models in various biomedical research domains. Additionally, their limitations and future perspectives are also explored.

Keywords: Bone organoids, Bone remodeling, Biomaterial, Biocompatible, Bone tissue engineering, Extracellular matrix, Hypoxic, Hydrogels, Mesenchymal stem cell, Microfluidic devices, Microtissues, Osteoblasts, Osteoclasts, Primary cells, Polymeric beads, Scaffold-free, Self-assembly, Spheroids, 3D printing, Vascularization.

---

* **Corresponding author Sunita Nayak**: Department of Integrative Biology, School of Bio Sciences and Technology, Vellore Institute of Technology, Vellore-632014, Tamil Nadu, India;
Email: sunita013@gmail.com, sunitanayak@vit.ac.in

Manash K. Paul (Ed.)

INTRODUCTION

Currently, animal models are used as a platform to establish proof-of-concept preclinical studies in order to develop treatment options. Nonetheless, the species-specific physiological and metabolic differences hinder the extrapolation of results obtained from animal experiments to be applicable to humans. There have been reports of significant failure rates of novel promising medicines in clinical studies, even though animal studies showed the desired results [1, 2]. Thus, biomedical approaches need to be reviewed to achieve a higher degree of relevance in physiology. With the advancement in *In vitro* cell culture systems, mainly driven by tissue engineering and regenerative medicine approaches, human physiology has been reflected to a greater extent. These approaches include the isolation of primary cells and their large-scale expansion, followed by the generation of a three-dimensional (3D) tissue biomimetics scaffold using biocompatible materials. Furthermore, advanced biomaterials, 3D culture, bioreactors, and efficient microfluidic systems have also been developed as promising *In vitro* systems, and as effective alternatives to animal testing [3, 4]. The mineralized structure of bone provides support, load bearing, and protection to the internal organs. Most bone models should recapitulate the biological and functional characteristics of bone, which are crucial to the respective research question. Hence an ideal bone model should adequately depict the investigated features of bone physiology. Organoids are comprised of 3D multicellular *In vitro* tissue construct formed either of pluripotent stem cells (PSCs) or adult stem cells (ASCs) by the process of self-organization that mimics the complex *In vivo* tissues and organs. This makes them very promising models for the study of human processes, structures and cell-tissue interface in the tissue culture dish.

Several organoid models consist of various cell types that mimic their *In vivo* counterparts and are still an area of ongoing research. Bone remodeling *In vivo* necessitates the interaction between osteoblasts (OBs) and osteoclasts (OCs), and the process is directed by extrinsic stimuli like mechanical load and the effect of hormones like the parathyroid hormone. Therefore, a system designed for studying the remodeling demands the inclusion of at least OBs, OCs and external factors. Since bone extracellular matrix (ECM) resorption by the OCs is a physiological process driven by enzyme and pH changes, thus a suitable matrix is desirable for regulated degradation of the bone matrix constituents while preserving the integrity of the available scaffold [5]. Moreover, besides having differences in anatomical properties, like the cortical (or compact) and cancellous (trabecular or spongy) bone, it also possesses differences in mechanical properties due to variances in orientation and alignment of collagen fibers although they have similar material composition. Such differences, along with dissimilarities in anatomical structures and vasculature, have not been reproduced in existing methods. In addition, selecting an appropriate ECM is crucial for developing bone organoids.

Depending on culture time and biomechanics, there are several methods for the generation of bone-like ECM using various 3D cell culture systems. Utilizing 3D bioprinting to create multicellular bone organoids is a recent advancement in the field. A suitable bioink resembling the native bone ECM is needed for 3D printing to generate desired anatomical features. The parameters such as mechanical actuation, perfusion, and the exchange of gas are provided by the bioreactor for generating complex bone models. The exclusion of xenogeneic substances and materials is done in order to avoid side effects on cells. Bone organoids mimic complex human bone, and their treatment response to therapeutic targets can significantly deliver more accurate translation results to human physiology and pathology. Due to their compact size, bone organoids can be a promising platform for various potential applications depicted in Fig. (**1**).

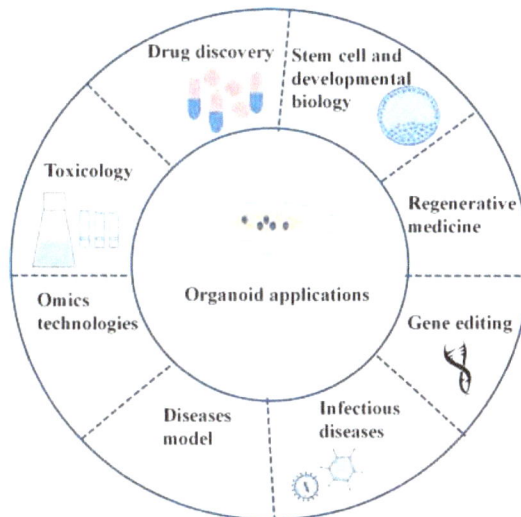

Fig. (1). Various potential applications of organoids

Physiology of Bone and Key Parameters

Bone is a mineralized, strong, and rigid connective tissue that provides structural support for the musculoskeletal system. Components of bone include ECM, a composite material with several specialized cell types. The bone ECM comprises inorganic minerals ($\sim$ 60%), organic matrix ($\sim$ 30%), water, and lipids [6]. Increased mechanical stability of the ECM is attributed to the bone mineral part known as hydroxyapatite (HA), formed largely of calcium and phosphorus. The organic component confers flexibility and elasticity and primarily consists of type I collagen. Histologically two different classes of bone tissue: namely, cortical (or compact) and cancellous (or trabecular or spongy) bone are present. The composition of the organic matrix is uniform throughout these two bone tissues [7]. The outer layer of cortical bone is referred to as the periosteum, while the endosteum provides barrier in-between the cancellous and the cortical bone. Bone marrow is localized in the cavity spaces of the cancellous bone. This includes red hematopoietic marrow (active), which is predominant during embryogenesis, and yellow fat marrow (inactive). During skeletal maturation, the red marrow is continually changed to yellow bone marrow. The presence of low levels of oxygen in the bone marrow is considered to be the critical signalling component for hematopoietic stem cells (HSCs), which are present in red bone marrow, and are also crucial for stem cell expansion [8].

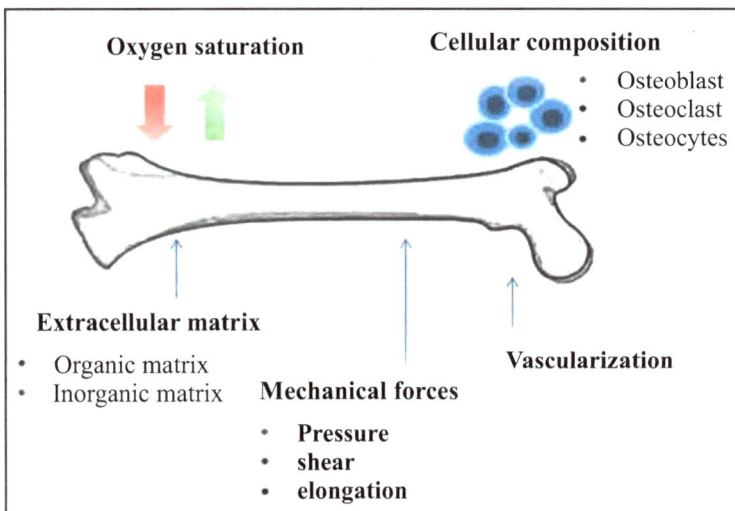

Fig. (2). Schematic illustration of key parameters in bone tissue

In bone cells, osteocytes (OCTs) constitute >90% and are involved in regulating osteoblast and osteoclast activity. The bone-forming cells, mesenchymal in origin, referred to as osteoblasts (OBs), produce osteoid, defined as the non-mineralized organic component of the bone matrix. This function is performed by the secretion of specific molecules, including collagen I, bone sialoprotein II, osteopontin, and osteocalcin. Apart from these cell types, another class of multinucleated cells derived from hematopoietic mononuclear precursor cells (monocyte-macrophage lineage), termed osteoclasts (OCs), are significant. Their primary function is bone resorption, enabling the dynamic bone remodeling phenomenon [8]. Furthermore, bone tissue is characterized by the state of hypoxia, that shows distinct impact on nearby cells. Therefore, a physiologically accurate bone model must incorporate oxygen level regulation mechanisms to produce hypoxic growth environments. Fig. (**2**) depicts the schematic summary of key bone tissue parameters that must be considered while devising a physiologic bone model.

Types and Sources of Cells for Bone Tissue Bioengineering

A bone model should incorporate diverse kinds of cells since bone is composed of different cell types. *In vitro* bone models may be created using a pure cell population and/or multiple cell types. For example, mesenchymal stromal cells (MSCs) are considered significant for investigating the soft callus generation in the process of fracture healing and ossification of endochondral bone, as observed during embryogenesis. The multicellular approach involves further incorporating other cell types, such as OBs and OCs. Additional *In vitro* bone model generation approaches include bone marrow aspirate concentrate (BMAC) and the use of micro-fragmented adipose tissue. Cell types for *In vitro* bone modeling and tissue engineering approaches are described below:

Mesenchymal Stromal Cells (MSCs)

Human MSCs are mainly harvested from adult tissue, unlike from human embryo and fetal stem cells, which are often accompanied by ethical considerations. Bone marrow-derived MSCs (BM-MSCs) are an undifferentiated population of cells and possess limited self-renewal ability. In addition, they exhibit the capability to differentiate into a multitude of cell types, such as OBs, adipocytes, and chondrocytes. These offer multiple benefits, like widespread availability, multipotency capacity, and the possibility of being utilized in allogenic applications because of their ability to evade the immune system. Depending upon the tissue source from which MSCs have been derived, functional differences exist among these cell types, which lead to site-specific phenotypes for MSCs. Other sources include adipose tissue, umbilical cord blood, skin, teeth, cartilage, pancreas, and liver. The MSCs derived from these tissues differ in properties, such as the ability to form colonies, proliferation, and ability to differentiate into multilineages.

Osteoblasts (bone-forming cells)

Although osteocytes constitute more than 90% of all bone cells, however, their complex isolation protocol involves a series of digestions and decalcification steps, preventing them from being the preferred cell type for bone tissue engineering. However, the cells undergoing terminal differentiation may be extracted from the bone tissue and further grown in 2D or 3D culture for *In vitro* experiments [9]. Peripheral blood can be a source of circulating OBs. The enzyme-based tissue digestion method for isolating bone marrow-derived OBs has been shown to have an adverse impact on the alkaline phosphatase activity (ALP) and/or the level of mineral deposition. As a result, OBs are often separated using a process known as outgrowth, in which bone fragments are placed on the plastic surface of an appropriate tissue culture plate. Donor variability causes phenotypic heterogeneity in primary cells. Using OBs from species for which particular strains are available (mouse, cow, sheep) for *In vitro* investigations minimizes variability, preventing donor-related heterogeneity.

Osteoclasts (bone-resorbing cells), Cell Lines, and Stem Cell

Investigations including remodeling, mechanism of healing, and cellular interactions that occur in the HSCs niche use HSCs and their progeny OCs to existing bone models. Human embryonic stem cells (ESCs) can generate all the bone cells but present a significant drawback due to ethical considerations. Furthermore, induced pluripotent stem cells (iPSCs), which are regarded as a connecting link between primary cells and cell lines, are pluripotent in nature derived from primary cells. Moreover, iPSCs are restricted to basic research as they may lead to inefficient differentiation, resulting in the development of malignancies upon transplantation. Additionally, cell lines like MG-63, MC3T3-E1, SaOs2, and human telomerase reverse transcriptase mesenchymal stromal cells (hTERT MSCs) present drawbacks such as malignancies, non-physiological functions, and decreased differentiation ability, that can lead to translational issues.

STRATEGIES FOR DEVELOPMENT OF BONE ORGANOIDS

Scaffold Free Self-Organization Approach

To have a better insight into osteogenic processes and mimic the *In vivo* conditions, spheroid cultures, which show a primary method for growing cells in a 3D construct, have attracted attention in recent years. This strategy might allow for the self-organization of cells [8]. Spheroids have been used as building blocks in tissue engineering to replicate the physiological phenomenon during embryonic development, such as the formation of tissues with complex 3D morphology and cell to cell interactions, which involve biological self-assembly. The concept of self-assembly serves as the foundation for the production of spheroids. The techniques involved in the fabrication of spheroids include hanging-drop spheroid cultures, microfluidic cultures, and low-adhesive substrate cultures [10]. Fabrication of 3D spheroids with precise composition and size is a major prerequisite since the cells of the spheroids undergo a diffusive supply of oxygen. Consequently, it is critical to limit the general spheroid size to a few hundred micrometers to avoid necrotic destruction to the central core cells of the spheroid [11].

Spheroids exhibit numerous properties which render them suitable for applications related to tissue engineering. These include: (i) High-throughput fabrication of large amounts of spheroids with defined sizes (ii) Enhanced regenerative potential compared to 2D cell culture system, and (iii) Complex tissue generation using co-culture, by combining different cell types in spheroid, (iv) *In vitro* preconditioning under various culture environments to enhance spheroid function, and (v) Capability of spheroids to fuse into microtissue constructs. Their 3D assembly results in enhanced biological properties like cell viability, consistent morphology, cell polarization, as well as high proliferation rate [10]. 3D spheroid cultures result in a significant increase in the differentiation potential of multipotent mesenchymal stem cells into osteogenic, adipogenic, neurogenic, chondrogenic and hepatogenic lineages when compared to 2D monolayer cultures. Furthermore, due to their high angiogenic and vasculogenic potential, spheroids can be used for efficient vascularization [10]. The generation of vascularized bone has been achieved by integrating osteogenic and endothelial cell types. Therefore, using spheroid cultures enables *In vitro* osteogenic differentiation, allowing modeling of biological phenomena, like bone formation, bone regeneration, and healing of bones.

Scaffold Based Approaches

Artificial bone can also be created using scaffolds combined with a cocktail of bone cells, later subjected to either static or dynamic culture inside a bioreactor setup. By supporting cell colonization, proliferation, migration, and stimulating cell differentiation, the scaffold acts as a primary material for the development of the desired tissue. The following characteristics should be considered while designing a scaffold for bone tissue engineering: (i) Resemblance to native bone ECM; (ii) Biocompatibility to facilitate cell adherence and cell survival; (iii) Surface characteristics conducive for cell proliferation and differentiation; (iv) Sufficient mechanical properties that can mimic those of the target tissue type; (v) A porous and permeable structure that facilitates cellular reorganization and support vascularization; and (vi) The ability to biodegrade [12]. The formation of bone organoids requires a scaffold that, ideally, supports osteoinductive as well as osteoconductive properties. Osteoinduction is the capacity of the scaffold to enhance the differentiation potential of cell types into bone-forming cells. The term osteoconduction defines the property of the scaffold to promote bone formation, cell attachment, cell proliferation, and ECM formation. In addition, the surface chemistry and structure of the scaffold determine the surface properties, and affect cell attachment and migration [13]. Examples include *In vitro* model systems using bioceramics, particularly calcium phosphate (CaP)-based materials and their associated composites, hydroxyapatite (HA), bicalcium phosphates (BCPs), and β-tricalcium phosphate (β-TCP).

Bone Models Based on Hydrogels

Hydrogels are networks of hydrophilic polymers that are cross-linked to one another and possess structural similarities to components based on macromolecules. Hydrogels can be synthesized from any hydrophilic polymer, nevertheless there are multiple chemical compositions and physical properties [14]. Hydrogels with excellent biocompatibility can be used for a variety of applications in the body, for example, as matrix material for the immobilization of cells, applications in tissue engineering and regenerative medicine. Hydrogels can be classified as:

Naturally Derived Polymer Hydrogels

These are naturally occurring and can be obtained from fibrin, alginate, hyaluronic acid, silk fibroin, gelatin, and collagen. Scaffolds, beads, or injectable hydrogels are often coated with natural protein in combination with other biomaterials. Proteins like bone morphogenetic protein (like BMP-2) can be added to hydrogels to facilitate bone formation *in vitro*. Similarly, osteogenesis can be enhanced by synthesizing hyaluronic acid-based crosslinked hydrogels [15]. Natural polymers are biocompatibile, and with enhanced degradation by enzymes present *in vivo*. However, they suffer from certain limitations, like inconsistent hydration and elastic properties.

Synthetic Polymer Hydrogels

These include polyethylene glycol (PEG), methacrylate, and synthetic peptides. PEG polymers lack the ability to facilitate cell adhesion, and therefore, their modification is done by the incorporation of short peptides (*e.g.*, RGD sequence, a cell adhesion peptide) or the molecular structure of PEG is altered. Nanocomposite hydrogels can be prepared by doping hydrogels with nanodiamonds (NDs), and can serve as a 3D scaffold for the purpose of drug delivery, promoting osteogenic differentiation of mesenchymal stem cells [16]. The advantages include enhanced consistency and the capacity to change properties, such as cell binding and degradation. Synthetic hydrogels suffer from certain limitations, such as weak mechanical strength, and are unable to sequester growth factors, resulting in burst release.

Bead-Based Approaches for Bone Tissue Engineering

This approach involves using polymeric-based beads to construct a 3D scaffold having a well-distinct microarchitecture, which facilitates enhanced migration of cells. Beads have the potential to serve as a medium to mediate the controlled/sustained release of substances, such as drugs inside of the scaffold material. The release is time-dependent and also depends on the size of the pore [17]. Calcium alginate beads have been used in which the microstructure thus developed enabled cell migration owing to pore size, resulting in the differentiation of human MSCs towards the osteogenic lineage. Bioactive ceramics, such as Hydroxyapatite (HA) and tricalcium phosphate (TCP), may also be used in the production of beads. The beads thus generated will possess the properties of both the materials, such as high mechanical strength and adhesive features of HA and bioadsorbable properties of TCP. Moreover, alginate hydrogels containing osteoinductive and osteoconductive materials have shown promise as potential cell carriers in the field of bone tissue engineering.

3D Printing

The use of 3D printing technology enables the production of scaffolds that possess chemical and biomechanical characteristics similar to that of a bone. Earlier, the stability of the constructs was achieved by sintering of the deposited material. Therefore, these approaches were not suitable for incorporating cells in the printing process. The use of three-dimensional (3D) printing technology in the production of cell-free scaffolds tailored for implantation has significant potential as a viable strategy for bone reconstructive surgery. Bioprinting techniques utilized for tissue engineering include inkjet writing (IW), laser-assisted forward transfer (LIFT), extrusion printing (EP), and stereolithography (SLA). These techniques allow the integration of living cells in the scaffolds. Furthermore, organoids can be generated using bioprinting technology since the technique offers the fabrication of precise structural features of bone and results in efficient and exact cell deposition [18, 19]. Table **1** shows the various 3D printing techniques for the fabrication of bone scaffolds. Vascularization can be introduced in the organoid from the beginning, which will eventually result in an efficient oxygen exchange, cell nutrients, and essential metabolites. Extrusion printing (EP) is the most commonly used method for bone bioprinting due to its ability to accommodate hydrogels of diverse viscosities and high cell densities. However, EP suffers from a limitation. The mechanical bioink extrusion *via* a nozzle makes the deposition process easier in EP, which in turn leads to high shear forces that may affect cell survival, especially in the context of stem cells [20].

Table 1. 3D printing techniques for fabrication of bone scaffolds.

Technique	Procedure	Polymers Used	Pros and Cons	Refs.
3D plotting/ direct ink writing	The liquid phase ink is dispensed out of nozzles under controlled flow rates, and deposition takes place through layer-by-layer.	Polycaprolactone PCL, Hydroxyapatite (HA), Bioactive glass, polylactic acid (PLA), polyethylene glycol (PEG), Poly(hydroxymethyl glycolide-co-ε-caprolactone), PLA/(PEG)/G5 glass, alginate composite, Bioactive 6P53B glass.	Mild conditioning of process results in the plotting of drugs and biomolecules. Heating/ post-processing required can affect incorporated biomolecules.	[21]
Extrusion printing (EP) Fused deposition modeling (FDM) Low-temperature deposition manufacturing (LDM)	Thermoplastic materials (filaments) are used. Filaments are pushed into a liquefier and are extruded from a computer-controlled nozzle. Non-heating liquefying processing of materials.	(PCL), (HA), tricalcium phosphate (TCP), Polypropylene (PP), (PLA), gelatine, alginate. PLA, Multi-walled carbon nanotubes (MWCNTs), beta-TCP	It can deposit various polymers, hydrogels, and encapsulate cells. Difficulty in stacking hierarchical constructs. Long production time Material restriction due to the need for molten phase.	[22]
Selective laser sintering (SLS)	A powder bed is prepared, followed by a layer-by-layer addition of powder,	PCL, calcium phosphate (CaP), nano HA, β-TCP,	Support and post-processing are not needed.	[23]
-	and then each layer is sintered using a laser source.	poly(hydroxybutyrate-co-hydroxyvalerate) (PHBV), carbonated nhydroxyapatite (CHAp), poly(L-lactic acid) (PLLA).	Feature resolution is contingent upon the diameter of the laser beam.	-

(Table 1) cont.....

Technique	Procedure	Polymers Used	Pros and Cons	Refs.
Stereolithography (SLA)	The print platform is immersed in a photopolymer solution and is subjected to focused light exposure.	Poly(propylene fumarate) (PPF), PPF, DEF-HA, Diethyl fumarate (DEF), PDLLA, β-TCP, HA.	Intricate internal features can be generated, with the possibility of patterning proteins, growth factors, and cells in a specific orientation in space. Only relevant to photopolymers.	[24]
Laser-assisted forward transfer (LIFT)	The material is coated on a transparent quartz disk. Deposition is controlled by laser pulse energy.	HA, Zirconia, HA/MG63 osteoblast-like cell, nano HA, human osteoprogenitor cell, human umbilical vein endothelial cell.	Condition is ambient and quantitatively controlled. Relevant for inorganic materials, organic materials, and cells.	[25]
Robotic-assisted deposition	This process entails the precise deposition of liquid material *via* a nozzle, followed by consolidation achieved by a change from a liquid to a gel state.	HA/PLA, HA/PCL,6P53B glass/PCL	3D nozzle movement is independent. Thickness can be precisely controlled and platform/support not needed. Material restriction.	[26]

Biocomposite Inks for 3D Printing Techniques

Printable biocomposite inks can be classified as natural, synthetic, and functional polymers. Fig. (**3**) lists the various printable biocomposite inks.

Natural Polymers

These polymers have been extensively utilized within the domain of tissue engineering. These materials are regarded as prospective biomaterials due to their resemblance with the components of the body's native tissue or organs. Natural polymers that are protein-based, such as gelatin, collagen, and ECM-based bioink, are capable of regenerating the epithelial layer, which is an important factor in creating functional tubular tissue. In contrast, it may be shown that alginate bioink has a comparatively lower level of biological functionality in comparison to natural polymers derived from proteins. However, it shows properties such as controllable printability and outstanding biocompatibility [27].

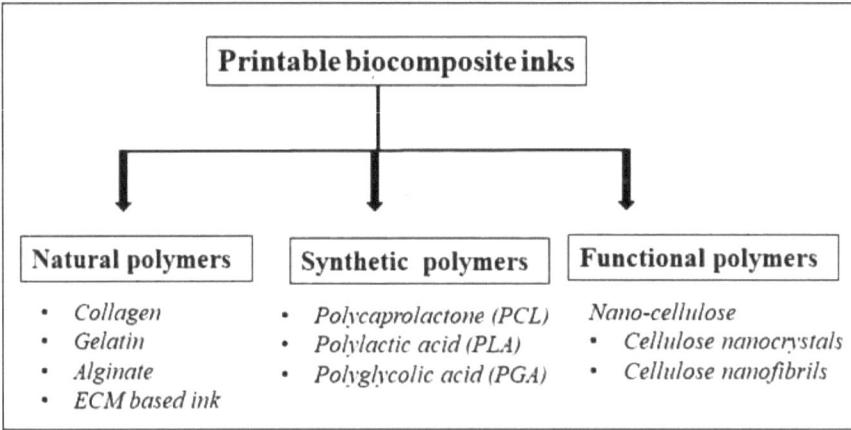

Fig. (3). The various printable biocomposite inks used for 3D printing techniques

Collagen: It consists of proline and glycine, with the polypeptides arranged in a triple-helical fashion. Collagen I is the most abundant in the human body and is most commonly used in 3D bioprinting. It is considered to be low in toxicity, though shows little cross-species immunological reaction. It permits the incorporation of viable cells along with ECM components and biochemical molecules. Nevertheless, the incorporation of a cross-linker or the implementation of a gelation process facilitated by temperature is required in order to facilitate the formation of 3D structures. However, the mechanical strength and bioprinting properties are dependent on viscosity.

Gelatin: It is a partly hydrolyzed form of collagen that is biodegradable and biocompatible. Gelatin is generally used with different polymer materials, for instance, with PCL, hyaluronic acid, chitosan hydrogel, alginate, and silk. Gelatin methacrylate (GelMA) has long been used to modify photocrosslinkable polymers [28]. The promotion of cell attachment and proliferation is facilitated by the gelatin-based bioink, which contains an abundance of integrin-binding motifs, namely the RGD sequence. Furthermore, gelatin has the thermosensitive characteristic of transitioning into a low-viscosity soluble state when dissolved in water, hence retaining its ability to respond to changes in temperature at physiological levels. Consequently, rigid 3D tissue structures cannot be built solely by using gelatin-based bioink.

Decellularized extracellular matrix (dECM) Ink: The fabrication of dECM-based ink, a compelling material for 3D bioprinting, involves the process of decellularization of the desired tissue. The chemical composition resembles the tissue microenvironment. The components include glycoprotein, proteoglycans, and collagenous protein. dECM-based inks have been successfully used for various tissues, for instance, skin, bone, blood vessels, liver, and kidney. These polymers have the intrinsic property of temperature-sensitive gelation under physiological conditions [17].

Alginate: Alginic acid, referred to as alginate, is an anionic polymer that is naturally present in the cellular structures of brown algae. The substance exhibits hydrophilicity and, when hydrated, undergoes a transformation into a gelatinous state. This substance has been employed as a material for wound dressings wounds, as well as an ink for the fabrication of 3D structures for tissue engineering applications. Alginate shows excellent biocompatibility, with structural similarity to natural ECM, and is bioinert in nature. It is able to form a hydrogel by a mechanism involving polymerization using multivalent cations (*e.g.*, Ca^{2+}, Ba^{2+}). Alginate has no cell-adhesive site, therefore peptide sequence, such as RGD is often required for attaining cell viability and differentiation. However, it displays inferior biological activity as compared to protein-based natural polymers.

Synthetic Polymers

Synthetic polymers closely mimic the structure of the target tissue. In addition to this, these polymers are amenable to complete degradation after implantation without any significant side effects.

Polycaprolactone (PCL): An aliphatic polyester has been prevalently used as a biomaterial for 3D bioprinting. It is biocompatible and offers superior printability properties owing to its low melting point and glass-transition temperature. The regulation of the degradation rate may be achieved by the strategic mixing of various ratios of polymers and copolymers. Degradation involves bulk erosion through hydrolysis.

Polylactic acid (PLA): PLA is also one of the prevalently used polymers for generating tissue engineered structures. Owing to its thermoplastic properties, PLA has been used for 3D bioprinting. Despite differences depending on molecular weight (MW), PLA exhibits significantly high mechanical strength. However, the molecular weight affects biodegradability. In addition to this, high-MW PLA may cause inflammation and infection *In vivo* [29]. Thus, molecular weight properties must be taken into consideration before 3D bioprinting, in order to mimic the mechanical characteristics of the target tissues.

Polyglycolic acid (PGA): PGA is a thermoplastic material having a high melting point and glass transition temperature. Compared to PLA, it is more acidic and hydrophilic. The properties include high mechanical strength and also better biocompatibility. In tissue engineering, the porous scaffolds based on PGA are generated by solvent casting and compression molding [30]. However, PGA suffers from certain limitations. It is extremely vulnerable to degradation and therefore, needs to be precisely controlled. Moreover, an increase in the acid concentrations in the surrounding tissues, due to glycolic acid release during the degradation of PGA, may cause tissue damage.

Functional Polymers

Functional polymers are used to develop complex structures, which are developed in an effort to enhance the mechanical properties and positive biochemical factors [27]. Biocomposite inks are categorized as functional polymers, which include hydrogel-type inks, providing an optimized microenvironment for a viable cell. These materials should have adequate rheological properties in order to maintain an appropriate shape during bioprinting, and also the significant cross-linking capabilities, which will further allow to retain the fidelity of 3D structure after the bioprinting. Overall, functional polymers enhance the stability of bioprinting and fidelity, particularly when mixed with biomaterials like nanocellulose. Nanocellulose includes cellulosic nanomaterials, such as cellulose nanocrystals (CNC), as well as cellulose nanofibrils (CNF) [31]. The nanocellulose-based bioink allows bioprinting of both 2D structures as well as 3D constructs with high precision.

Dynamic Culture Systems for Culturing Bone Models under Controlled Conditions

Bioreactors

The biological characteristics of bone include its highly vascular nature, with ECM being a composite material, and hypoxic condition with oxygen gradients governing the stem cell-specific characteristic. In addition, it bears a constant mechanical load as a result of the body's locomotion. In a tissue-engineered construct or organoid model, the biological properties of bone may be replicated. However, suitable bioreactor systems must be devised in order to simulate the physical characteristics such as a hypoxic environment and mechanical load. In the case of bioreactors for culturing bone, instead of seeding the different cell types, all at once, the preferred strategy is to mimic the cellular composition of bone by host ingrowth. Furthermore, the rotating wall reactors or spinner flasks allow the generation of bone cells in a high volume, and mechanical forces by the perturbation of the culture medium in the form of shear stress [32]. A substantial limitation in the *In vitro* culture of larger organoids or scaffold-based grafts is the absence of relevant vasculature, which leads to the attenuated diffusion of nutrients, oxygen, and metabolites through the constructs. Therefore, in order to avert the formation of necrotic regions, perfusion is required for larger organoids or grafts. Systems capable of perfusion culture have been incorporated in tissue-engineered grafts employing critical-size bone defect treatment, *e.g.*, a tube-like bioreactor with unidirectional flow. In order to generate the mechanical forces that are constantly present in the bone as a result of locomotion, shear stress is applied by perfusing the organoid, and compression due to mechanical load leads to direct strain and hydrostatic pressure. Furthermore, the determination of Young's modulus of the organoids over time needs to be done, if biological activity leading to changes in stiffness is to be investigated. In addition to this, the applied mechanical load should be adapted to the shifts in Young's modulus due to cell-cell interaction to retain a consistent rate of compressive strain on bone organoids throughout a period. A recent approach involves using magnetic fields in the bioreactor that induce the transduction of mechanical forces, which eradicates the contamination risk generated *via* the typically necessary moveable components for mechanical load application [33].

Microfluidics

Organ-on-a-chip refers to the bioreactor technology encompassing microfluidic systems for simulating the functions of organs or certain physiological parameters present. These can also be referred to as miniaturized bioreactor systems or microphysiological systems. The tissue-specific properties, including mechanical strains and oxygen levels, can be attained, and the co-culturing of various cell types in 2D or 3D is also allowed in these systems. The role of microfluidics in the context of bone involves simulating the bone marrow niche, mimicking the 3D network of osteoclasts (OCTs), modeling vasculogenesis, and studying the MSCs differentiation to the osteogenic lineage under the given conditions of mechanical strain [34, 35]. Table **2** lists the various microfluidic systems designed to model a particular aspect of bone biology.

Table 2. The various microfluidic designs based on the aspect of bone biology.

Design of Microfluidic System	Aspect Being Modeled	Ref.
Cage made of polymethylmethacrylate (PMMA), which is filled with a hydrogel	Critical parameters governing angiogenesis in bone.	[36]
System based on polydimethylsiloxane (PDMS)	Angiogenesis	[15]
Organoid positioned within two layers of PDMS/incorporation in an existing multi-organ chip.	Achieving bone organoid microfluidic perfusion culture.	[3, 34]

(Table 2) cont.....

Design of Microfluidic System	Aspect Being Modeled	Ref.
Microfluidic perfusion of several layers consisting of HA microbeads interspersed with osteoclasts.	Recreating the OCT network found in native bone.	[3]
Microbeads and cells placed in a PDMS device followed by mechanical stimulation in a cyclic manner by the use of bidirectional alteration in perfusion rate.	Simulating shearing stress and hydrostatic pressures in the canaliculi of bone.	[35)
Microfluidic chip allowing the simultaneous pneumatic actuation of several culture chambers. Three layers of PMMA form the main body of the device and incorporating a pliable PDMS layer, application of cyclic pneumatic force.	The impact of mechanical strain on the process of osteogenic development in MSCs derived from various sources.	[37]

The ECMs for the above-mentioned microfluidic systems are generated by materials including demineralized bone powder used for ectopic bone formation, ceramic scaffolds, HA microbeads, and fibrin- or collagen-based hydrogels [34].

APPLICATIONS OF BONE ORGANOIDS

The rigid, mineralized structure of bone and inaccessibility of the cellular components located within the tissue makes it difficult to study the events *in vivo*, such as the continuous adaptation of the bone to the environment in order to meet the metabolic demands. Organoids and organotypic cultures, being unique pertaining to properties such as self-organization and development of facets that might be used to comprehend aberrations in molecular pathways leading to a particular disease, are being increasingly utilized as complex platforms for replicating anatomical structures. Their reduced size, which in turn allows for the mass applications and multiple permutations of the subjected environment or conditions, offers significant advantages for other applications, such as drug screening and personalized medicine. However, the developmental stages, when it comes to the bone, have been studied using spheroid morphology, and organoid models are still not fully existent. Unlike soft tissues, the formation of bone tissues needs the careful orchestration of various phases, including soft matrix deposition by progenitor cells, and also the inorganic phase in which a correct physical alignment has to be developed, as well as a biologically active mineral phase [38]. The establishment of bona fide bone organoids has not yet been achieved. An approach to form spheroids using human adult bone precursor cells, which are heterogeneous populations of bone progenitors including osteoblasts, has been developed, and these can further promote the development of tiny pieces of crystalline bone termed microspicules [39].

Disease Modelling

Human organoids are mainly used in modeling human diseases to establish platforms for high throughput drug screening, genotype and phenotype testing, disease-specific bio-banking, and personalized therapies. Patient-derived xenografts (PDXs) have been established as a feasible alternative for the purpose of developing improved models that may effectively mimic tissue architecture, genetics, and therapeutic response. *In vitro* models of cancer have been created using organoids obtained from tissue resections, tissue biopsies, and even circulating tumor cells. These cancer-derived organoids retain the tumor's genetic and behavioral characteristics. They are similar to patient-derived xenografts (PDXs), but have certain advantages over them, such as they can be expanded *In vitro* and can be used for drug screening [40]. Recently, a microscale bone organoid prototype has been developed, which can allow the study of bone processes at the cell-tissue interface. A combination of primary osteoblastic and osteoclastic cells was sown onto the femoral head micro-trabeculae, simulating the appropriate phenotypes and functions. Subsequently, a pathological state of reduced mechanical stimulation was modeled by the insertion of the constructs into a simulated microgravity bioreactor. Osteoclastic bone resorption sites were subsequently detected in these constructs [38].

Study of Tissue Development

The mechanism of bone formation is comparatively challenging to study in both *In vivo* and *Ex vivo* conditions when compared to other organs due to its highly mineralized nature, which is difficult to penetrate, and also it binds many chromogenic agents and therefore, needs destruction to identify internal histological features. Thus, the processing, which is usually invasive in nature in case of bone, leads to the destruction of the embedded cellular network, which is crucial in order to understand the tissue hierarchy and the mechanism of conversion of mechanical stimuli into biochemical signals by the cells. The microenvironment involving bone formation at the cellular level is temporally and spatially coupled and relies on the delicate balance between bone effector cells, such as osteoblasts, osteocytes, *etc.*, which is difficult to replicate in a laboratory setting [38]. Therefore, *ex-vivo* models which are capable of recapitulating the events that occur during bone remodeling are needed. Organoids and organotypic cultures are capable of functioning as a complete unit, and they also enable to study and detect the early-stage cellular processes.

Drug Development, Screening, and Precision Medicine

Since organoids are near-physiological architectures, therefore, they can efficiently recapitulate the primary tumors and also drug responses. They can even help in optimizing therapeutic strategies for each patient. The scalability of organoids is particularly important in drug testing applications. Organoids can also be combined with already existing 2D and bioengineered disease models.

Moreover, organoid-derived cells may be carefully isolated and subsequently subjected to 2D cell culture. Furthermore, applying organ-on-a-chip platforms to organoids or cells isolated from organoids can prove to be a better alternative. Furthermore, the use of microfluidics may be employed in the context of organoids to introduce controlled fluid flow or constrain their development within certain spatial configurations [41].

CONCLUDING REMARKS

Organoid technology holds great potential in accurately modeling the physiological tissue environment as well as human diseases. Bone tissue engineering remains challenging due to its complex roles in hematopoiesis, locomotion, organ protection, and mineral homeostasis. Therefore, the majority of bone models, instead of recapitulating the entire bone, cover the aspects of bone biology and function that are essential for the respective research query. An ideal bone model should be highly complex to suitably represent the specific aspects of bone physiology that are under investigation. Bone remodeling involves interactions between osteoblasts (OBs) and osteoclasts (OCTs) and is directed by external signaling cues such as mechanical load and parathyroid hormones *in vivo*. Therefore, the system designed for investigating the remodeling phenomenon would need at least these two cell types, *i.e.*, OBs and OCs, and also an external cue that is driving the process, along with a suitable matrix. The proper selection of an extracellular matrix is necessary to produce a functional bone organoid. The generation of multicellular bone organoids by the bioprinting approach needs to be explored. The characterization and processing of multiple higher-resolution bioinks resembling the ECM of native bone would be an advantage. In addition, 3D printing enables the manufacture of standardized organoids with the necessary anatomical features, which, unlike other bone organoid generation methods, permits the repeatable generation of accurate anatomical characteristics.

Furthermore, bioreactor systems are able to provide perfusion, thus facilitating vascularization, mechanical actuation, and gaseous exchange for the generation of complex bone models. However, an ideal bone model would be a physiologic bone itself. Certain key parameters need to be considered while designing an ideal bone model. The multiple aspects of bone biology can be accepted in the model of choice, depending on the research question. In addition to this, limited control of the physiochemical parameters should be allowed, with costs as low as possible. The model should contain a minimal set of cells, such as MSCs, HSCs, OBs, and OCTs, as well as endothelial cells for vasculature. Moreover, ECM should preferably be composed of components of biological origin and closely mimic the original bone ECM. Mechanical stress, which is present in bone throughout an organism's lifespan, must be taken into account while creating a bone model since it tremendously influences the physiology of bone cells. A physiologic bone model or organoid should include a system regulating oxygen levels to maintain hypoxic growth conditions. Advanced bone models with the aforementioned aspects can be generated with the developments in cell isolation techniques, 3D culture methods, and differentiation protocols, combined with advances in tissue engineering and microfluidics.

REFERENCES

[1]   Knight A. Systematic reviews of animal experiments demonstrate poor human clinical and toxicological utility. Altern Lab Anim 2007; 35(6): 641-59.
[http://dx.doi.org/10.1177/026119290703500610]

[2]   Weidner C, Steinfath M, Opitz E, Oelgeschläger M, Schönfelder G. Defining the optimal animal model for translational research using gene set enrichment analysis. EMBO Mol Med 2016; 8(8): 831-8.
[http://dx.doi.org/10.15252/emmm.201506025] [PMID: 27311961]

[3]   Sieber S, Wirth L, Cavak N, *et al.* Bone marrow-on-a-chip: Long-term culture of human haematopoietic stem cells in a three-dimensional microfluidic environment. J Tissue Eng Regen Med 2018; 12(2): 479-89.
[http://dx.doi.org/10.1002/term.2507] [PMID: 28658717]

[4]   Zheng F, Fu F, Cheng Y, Wang C, Zhao Y, Gu Z. Organ-on-a-chip systems: microengineering to biomimic living systems. Small 2016; 12(17): 2253-82.
[http://dx.doi.org/10.1002/smll.201503208] [PMID: 26901595]

[5]   Väänänen HK, Laitala-Leinonen T. Osteoclast lineage and function. Arch Biochem Biophys 2008; 473(2): 132-8.
[http://dx.doi.org/10.1016/j.abb.2008.03.037] [PMID: 18424258]

[6]   Clarke B. Normal bone anatomy and physiology. Clin J Am Soc Nephrol 2008; Suppl 3(Suppl 3): S131-9.
[http://dx.doi.org/10.2215/CJN.04151206]

[7]   Buckwalter JA, Glimcher MJ, Cooper RR, Recker R. Bone biology. I: Structure, blood supply, cells, matrix, and mineralization. Instr Course Lect 1996; 45: 371-86.
[PMID: 8727757]

[8]   Scheinpflug J, Pfeiffenberger M, Damerau A, *et al.* Journey into bone models: a review. Genes (Basel) 2018; 9(5): 247.
[http://dx.doi.org/10.3390/genes9050247] [PMID: 29748516]

[9]   Jonsson KB, Frost A, Nilsson O, Ljunghall S, Ljunggren Ö. Three isolation techniques for primary culture of human osteoblast-like cells: A comparison. Acta Orthop Scand 1999; 70(4): 365-73.
[http://dx.doi.org/10.3109/17453679908997826] [PMID: 10569267]

[10]   Laschke MW, Menger MD. Life is 3D: Boosting spheroid function for tissue engineering. Trends Biotechnol 2017; 35(2): 133-44.
[http://dx.doi.org/10.1016/j.tibtech.2016.08.004] [PMID: 27634310]

[11]   Groebe K, Mueller-Klieser W. On the relation between size of necrosis and diameter of tumor spheroids. Int J Radiat Oncol Biol Phys 1996; 34(2): 395-401.
[http://dx.doi.org/10.1016/0360-3016(95)02065-9] [PMID: 8567341]

[12]   Stevens B, Yang Y, Mohandas A, Stucker B, Nguyen KT. A review of materials, fabrication methods, and strategies used to enhance bone regeneration in engineered bone tissues. J Biomed Mater Res B Appl Biomater 2008; 85B(2): 573-82.

[http://dx.doi.org/10.1002/jbm.b.30962] [PMID: 17937408]

[13]   Li Z, Gong Y, Sun S, *et al.* Differential regulation of stiffness, topography, and dimension of substrates in rat mesenchymal stem cells. Biomaterials 2013; 34(31): 7616-25.
[http://dx.doi.org/10.1016/j.biomaterials.2013.06.059] [PMID: 23863454]

[14]   Dhivya S, Saravanan S, Sastry TP, Selvamurugan N. Nanohydroxyapatite-reinforced chitosan composite hydrogel for bone tissue repair *in vitro* and in vivo. J Nanobiotechnology 2015; 13(1): 40.
[http://dx.doi.org/10.1186/s12951-015-0099-z] [PMID: 26065678]

[15]   Jusoh N, Oh S, Kim S, Kim J, Jeon NL. Microfluidic vascularized bone tissue model with hydroxyapatite-incorporated extracellular matrix. Lab Chip 2015; 15(20): 3984-8.
[http://dx.doi.org/10.1039/C5LC00698H] [PMID: 26288174]

[16]   Pacelli S, Maloney R, Chakravarti AR, *et al.* Controlling adult stem cell behavior using nanodiamond-reinforced hydrogel: implication in bone regeneration therapy. Sci Rep 2017; 7(1): 6577.
[http://dx.doi.org/10.1038/s41598-017-06028-y] [PMID: 28747768]

[17]   La WG, Jang J, Kim BS, Lee MS, Cho DW, Yang HS. Systemically replicated organic and inorganic bony microenvironment for new bone formation generated by a 3D printing technology. RSC Advances 2016; 6(14): 11546-53.
[http://dx.doi.org/10.1039/C5RA20218C]

[18]   Ji S, Guvendiren M. Recent advances in bioink design for 3d bioprinting of tissues and organs. Front Bioeng Biotechnol 2017; 5: 23.
[http://dx.doi.org/10.3389/fbioe.2017.00023] [PMID: 28424770]

[19]   Genova T, Roato I, Carossa M, Motta C, Cavagnetto D, Mussano F. advances on bone substitutes through 3d bioprinting. Int J Mol Sci 2020; 21(19): 7012.
[http://dx.doi.org/10.3390/ijms21197012] [PMID: 32977633]

[20]   Park JY, Shim JH, Choi SA, *et al.* 3D printing technology to control BMP-2 and VEGF delivery spatially and temporally to promote large-volume bone regeneration. J Mater Chem B Mater Biol Med 2015; 3(27): 5415-25.
[http://dx.doi.org/10.1039/C5TB00637F] [PMID: 32262513]

[21]   Luo Y, Wu C, Lode A, Gelinsky M. Hierarchical mesoporous bioactive glass/alginate composite scaffolds fabricated by three-dimensional plotting for bone tissue engineering. Biofabrication 2012; 5(1)015005
[http://dx.doi.org/10.1088/1758-5082/5/1/015005] [PMID: 23228963]

[22]   Xiong Z, Yan Y, Zhang R, Sun L. Fabrication of porous poly(l-lactic acid) scaffolds for bone tissue engineering *via* precise extrusion. Scr Mater 2001; 45(7): 773-9.
[http://dx.doi.org/10.1016/S1359-6462(01)01094-6]

[23]   Shuai C, Gao C, Nie Y, Hu H, Zhou Y, Peng S. Structure and properties of nano-hydroxypatite scaffolds for bone tissue engineering with a selective laser sintering system. Nanotechnology 2011; 22(28)285703
[http://dx.doi.org/10.1088/0957-4484/22/28/285703] [PMID: 21642759]

[24]   Ronca A, Ambrosio L, Grijpma DW. Preparation of designed poly(d,l-lactide)/nanosized hydroxyapatite composite structures by stereolithography. Acta Biomater 2013; 9(4): 5989-96.
[http://dx.doi.org/10.1016/j.actbio.2012.12.004] [PMID: 23232210]

[25]   Yusupov V, Churbanov S, Churbanova E, *et al.* Laser-induced forward transfer hydrogel printing: A defined route for highly controlled process. International Journal of Bioprinting 2020; 6(3): 77-92.
[http://dx.doi.org/10.18063/ijb.v6i3.271] [PMID: 33094193]

[26]   Russias J, Saiz E, Deville S, *et al.* Fabrication and *in vitro* characterization of three-dimensional organic/inorganic scaffolds by robocasting. J Biomed Mater Res A 2007; 83A(2): 434-45.
[http://dx.doi.org/10.1002/jbm.a.31237] [PMID: 17465019]

[27]   Gao G, Lee JH, Jang J, *et al.* Tissue Engineered Bio-Blood-Vessels Constructed Using a Tissue-

Specific Bioink and 3D Coaxial Cell Printing Technique: A Novel Therapy for Ischemic Disease. Adv Funct Mater 2017; 27(33)1700798
[http://dx.doi.org/10.1002/adfm.201700798]

[28]   Gauvin R, Chen YC, Lee JW, *et al.* Microfabrication of complex porous tissue engineering scaffolds using 3D projection stereolithography. Biomaterials 2012; 33(15): 3824-34.
[http://dx.doi.org/10.1016/j.biomaterials.2012.01.048] [PMID: 22365811]

[29]   Andreopoulos AG, Hatzi EC, Doxastakis M. Controlled release systems based on poly(lactic acid). An *in vitro* and *in vivo* study. J Mater Sci Mater Med 2000; 11(6): 393-7.
[http://dx.doi.org/10.1023/A:1008990109419] [PMID: 15348021]

[30]   Sabir MI, Xu X, Li L. A review on biodegradable polymeric materials for bone tissue engineering applications. J Mater Sci 2009; 44(21): 5713-24.
[http://dx.doi.org/10.1007/s10853-009-3770-7]

[31]   Markstedt K, Mantas A, Tournier I, Martínez Ávila H, Hägg D, Gatenholm P. 3D bioprinting human chondrocytes with nanocellulose-alginate bioink for cartilage tissue engineering applications. Biomacromolecules 2015; 16(5): 1489-96.
[http://dx.doi.org/10.1021/acs.biomac.5b00188] [PMID: 25806996]

[32]   Rauh J, Milan F, Günther KP, Stiehler M. Bioreactor systems for bone tissue engineering. Tissue Eng Part B Rev 2011; 17(4): 263-80.
[http://dx.doi.org/10.1089/ten.teb.2010.0612] [PMID: 21495897]

[33]   Dobson J, Cartmell SH, Keramane A, El Haj AJ. Principles and design of a novel magnetic force mechanical conditioning bioreactor for tissue engineering, stem cell conditioning, and dynamic *in vitro* screening. IEEE Trans Nanobiosci 2006; 5(3): 173-7.
[http://dx.doi.org/10.1109/TNB.2006.880823] [PMID: 16999242]

[34]   Torisawa Y, Spina CS, Mammoto T, *et al.* Bone marrow–on–a–chip replicates hematopoietic niche physiology in vitro. Nat Methods 2014; 11(6): 663-9.
[http://dx.doi.org/10.1038/nmeth.2938] [PMID: 24793454]

[35]   Sun Q, Choudhary S, Mannion C, Kissin Y, Zilberberg J, Lee WY. *Ex vivo* replication of phenotypic functions of osteocytes through biomimetic 3D bone tissue construction. Bone 2018; 106: 148-55.
[http://dx.doi.org/10.1016/j.bone.2017.10.019] [PMID: 29066313]

[36]   Bersini S, Gilardi M, Arrigoni C, *et al.* Human *in vitro* 3D co-culture model to engineer vascularized bone-mimicking tissues combining computational tools and statistical experimental approach. Biomaterials 2016; 76: 157-72.
[http://dx.doi.org/10.1016/j.biomaterials.2015.10.057] [PMID: 26524536]

[37]   Park SH, Sim WY, Min BH, Yang SS, Khademhosseini A, Kaplan DL. Chip-based comparison of the osteogenesis of human bone marrow- and adipose tissue-derived mesenchymal stem cells under mechanical stimulation. PLoS One 2012; 7(9)e46689
[http://dx.doi.org/10.1371/journal.pone.0046689] [PMID: 23029565]

[38]   Iordachescu A, Hughes EAB, Joseph S, Hill EJ, Grover LM, Metcalfe AD. Trabecular bone organoids: a micron-scale 'humanised' prototype designed to study the effects of microgravity and degeneration. npj Microgravity 2021; 7(1): 17.

[39]   Kale S, Biermann S, Edwards C, Tarnowski C, Morris M, Long MW. Three-dimensional cellular development is essential for *ex vivo* formation of human bone. Nat Biotechnol 2000; 18(9): 954-8.
[http://dx.doi.org/10.1038/79439] [PMID: 10973215]

[40]   Weeber F, Ooft SN, Dijkstra KK, Voest EE. Tumor organoids as a pre-clinical cancer model for drug discovery. Cell Chem Biol 2017; 24(9): 1092-100.
[http://dx.doi.org/10.1016/j.chembiol.2017.06.012] [PMID: 28757181]

[41]   Karzbrun E, Kshirsagar A, Cohen SR, Hanna JH, Reiner O. Human brain organoids on a chip reveal the physics of folding. Nat Phys 2018; 14(5): 515-22.
[http://dx.doi.org/10.1038/s41567-018-0046-7] [PMID: 29760764]

CHAPTER 5

Cardiac Organoids: Promises and Future Challenges

Malay Chaklader[1,*] and **Beverly Rothermel**[1,2,*]

[1] *Department of Internal Medicine, Division of Cardiology, University of Texas Southwestern Medical Center, 5323 Harry Hines Blvd, Dallas, TX 75390, Texas*

[2] *Department of Molecular Biology, University of Texas Southwestern Medical Center, 5323 Harry Hines Blvd, Dallas, TX 75390, Texas*

Abstract: The development of state-of-the-art, *In vitro* three-dimensional organoid culture methodologies, represents a quantum leap in stem cell technology and tissue engineering. In contrast to traditional two-dimensional cell culture and rodent models, organoids generated from patient-derived cells can dramatically increase the precision and relevance of *In vitro* approaches to model human development and disease. Currently, the most well-established organoid systems are those for the intestine, brain, bone, kidney, and eye. Surprisingly, research using cardiac organoids, or "cardioids, is still in a nascent phase, lagging significantly behind these other, more mature, tissue-specific organoid platforms. Consequently, there is an ongoing need to develop more robust and reproducible protocols capable of yielding self-organizing cardiac organoids that assemble following valid cardiogenic principles, recapitulating the microanatomy and cellular hierarchy observed during *In vivo* cardiac development. Cardiovascular disease is currently the primary cause of death in developed countries, and its prevalence is growing worldwide. Improved cardiac organoid technologies have the potential to facilitate and enhance the application of the new wave of personalized medicine aimed at addressing cardiovascular disease. This book chapter will discuss the development of cardiac organoid research, its present state, and future challenges in detail.

Keywords: Cardiac organoids, Cardiovascular disease, Embryogenesis, Personalized medicine.

INTRODUCTION

The journey from a single totipotent cell to a complex, multicellular adult organism is an elegantly coordinated process of self-renewal, proliferation,

---

[*] **Corresponding authors Malay Chaklader and Beverly Rothermel:** Department of Internal Medicine, Division of Cardiology, University of Texas Southwestern Medical Center, 5323 Harry Hines Blvd, Dallas, TX 75390, Texas and Department of Molecular Biology, University of Texas Southwestern Medical Center, 5323 Harry Hines Blvd, Dallas, TX 75390, Texas; E-mails: malay.chaklader@utsouthwestern.edu, beverly.rothermel@utsouthwestern.edu

Manash K. Paul (Ed.)

differentiation, self-organization, and self-assembly. Provided with the proper microenvironment, dissociated cells can reassemble to regenerate differentiated structures, including a variety of embryonic organs, adult tissues, and even entire organisms. Taking advantage of these properties, researchers have developed "organoids" or mini organs. An organoid is an *In vitro* 3D micro-tissue structure that recapitulates fundamental, *In vivo* properties of the corresponding parental organ. As such, organoids can provide a flexible *In vitro* platform for experimental manipulation of specific aspects of tissue and organ biology in a controlled environment isolated from the influence and complexity of the intact organism [1, 2]. Although growing cultured cells in a 3D matrix is an age-old technique, the term "organoid" is relatively new to the scientific community and is used to denote cellular assemblies derived from either pluripotent stem cells (PSCs) or adult stem cells (ASCs). The PSCs used to generate organoids can be either embryonic stem cells (ES) or induced pluripotent stem cells (iPSCs). In contrast, ASCs can come from a variety of organs or tissues and may therefore already be partially committed toward specific cell or tissue lineages. Although the self-organization and self-assembly processes that govern organoid formation *In vitro* are not identical to those that occur during *In vivo* embryonic organ development, they can accurately model key phenotypic features. It is important to note that tissue morphology is not simply a direct manifestation of the genetic code but is the outcome of numerous interactive dialogues and iterative readjustments between the underlying genetic template and the biophysical processes that ultimately shape the complex tissue architecture [3]. During organoid development, the regenerative potential of the starting pool of stem cells, their interactions with one another, and the experimental environment, ultimately determine the characteristics of the growing structure.

For more than a decade, starting with either PSCs or ASCs, it has been possible to derive complex tissue-like structures that recapitulate essential properties of the brain, retina, intestine, and other organs [4]. However, the development of cardiac organoid technology to model heart development has lagged behind those of other tissue-specific organoids. This chapter highlights cardiac organoid research, current development, application, and future challenges. We also highlight and compare various tools and methodologies available for human organoid research. We also provide useful information encouraging basic biologists to consider using this emerging platform to study human cardiac pathophysiology.

BACKGROUND OF ORGANOIDS

3D Culture Models

3D cell culture initially evolved from techniques culturing suspension cell cultures, with or without the introduction of a scaffold to bypass direct contact with plastic culture surfaces. In this context, a scaffold is a bio-mimetic extracellular matrix (ECM) used to model an *In vivo* cellular niche. Feeder cells have also been used as a scaffold to support cell growth. Feeder cells are a layer of live cells that have been rendered growth-arrested but provide cell-cell contacts and secrete growth factors that help the cells of interest to proliferate. Feeders differ from a coculture system because only the cell type of interest is capable of proliferating. In terms of cell-free scaffolds, Matrigel™, produced by Corning Life Sciences, is one of the most often used. It is derived from a complex mixture of proteins secreted from Engelbreth-Holm-Swarm (EHS) mouse sarcoma cells [5]. Many adhesive proteins in Matrigel™ mimic the extracellular environment, providing mechano-transduction and ECM-like signals to the cells. In contrast, scaffold-free 3D cell culture techniques primarily rely on hanging drop culture methods using a combination of a well-defined culture medium, gravity, and surface tension [6]. Culturing of organoids at an "air-liquid interface" has also been used successfully by initially growing cells, or layers of cells and feeder cells submerged in a defined medium, then allowing the media to gradually evaporate or be removed, exposing cells to the air in a way that can stimulate cell polarization and differentiation [7, 8].

Self-organization is an essential developmental characteristic during embryogenesis, and this same property is fundamental to organoid technology. In 1907, Henry Peters Wilson first demonstrated the capacity for self-organization by regenerating an entire, mature organism from dissociated sponge cells [9]. Numerous research groups built on these initial dissociation-reaggregation studies by demonstrating the ability of dissociated amphibian pronephros [10] and chick embryos [11] to regenerate various organs. Organoid research and the concept of self-organization gained further momentum following significant advances in the field of pluripotent stem cell biology because of the potential for modeling human disease and personalized medicine [12 - 15]. In 1987, Li *et al.* [16], for the first time, demonstrated that ECM secreted from EHS (now known as Matrigel™) induced mammary epithelial cells to self-organize into 3D duct-like structures complete with lumens and the ability to secrete milk proteins, a stark contrast to their behavior in a 2D cell culture system. Another seminal experiment demonstrated that alveolar type II epithelial cells sustain a differentiated form in a 3D ECM environment in contrast to their undifferentiated state when cultured in a 2D system [17]. These findings are consistent with the concept of cell-matrix interactions being indispensable for tissue homeostasis and differentiation. Eiraku *et al.* [18] revolutionized the field of organoid research by using a 3D aggregation cell culture method to grow cerebral cortex tissue from ESCs. From then onwards, organoid research shifted from primarily 2D approaches to 3D technologies. In 2009, Hans Clever's group showed that adult intestinal cells expressing the stem cell marker Leucine-Rich Repeat Containing G Protein-Coupled Receptor 5 (*LRG5*+) form three-dimensional crypt-villus organoids or "mini-guts" on Matrigel™, progressing through a process of self-organization and differentiation outside the context of an *In vivo* mesenchymal niche [19]. This seminal work was the first example of adult stem cell-based organoid culture. It opened the door widely for advances in other cells and tissue lineages, including the generation of both mesendoderm (*e.g.*, liver, lung, kidney) and neuroectoderm (brain and retina) derived organs from both ASCs and PSCs.

Organoids vs. Spheroid

Organoids and spheroids are products of advanced 3D cell culture methods and provide a conceptual bridge between cell culture and animal models. In general, spheroids and organoids are both 3D microstructures composed of multiple cells and cell types. Although the terms organoid and spheroid are often used interchangeably in the literature, several characteristic features distinguish them from one another. Lancaster and Knoblich defined an organoid as a "collection of organ-specific cell types that develops from stem cells or organ progenitors and self-organizes through cell sorting and spatially restricted lineage commitment like *in vivo*" [1]. Thus, organoids aim to model the properties of healthy organs. In contrast, spheroids are multicellular tumor models that were first described in the early 70s and were generated by culturing cancer cell lines under non-adherent conditions [2]. Tumor spheres (TS) are used to model cancer stem cell expansion. The starting material for tissue-derived tumor spheres and multicellular spheroids is typically obtained by biopsy and mechanical dissociation of tumor tissue [20]. Furthermore, an organoid is generally the product of a higher order of 3D self-assembly like a "mini-organ," often capable of independently carrying out key processes such as lactation or insulin secretion. In contrast, the generation and characteristics of spheroid cultures are more dependent on matrix support rather than the phenocopying properties of a specific organ or tissue. Organoids resemble the originating tissue histologically and are genetically stable. They can be cultured from a minimal amount of starting tissue, are amenable to genetic manipulations, and can be expanded for long-term culture or cryopreservation [4, 21].

CARDIAC ORGANOIDS

As previously noted, research involving cardiac organoids is in its infancy compared to many other organoid systems, perhaps only truly coming of age in 2020. As an innovative biological tool, the goal of human cardiac organoids (hCOs) is to model *In vitro* specific functions and developmental features consistent with the human heart *in vivo*. Compared to conventional cell culture/2D systems, cells within organoids tend to obtain a higher level of maturation, allowing them to more accurately model *In vivo* properties. The term hCO has been used to describe various *In vitro* culture techniques, ranging from tissue explants to organoid-chip systems [3]. The primary objective for hCOs is to provide a high-fidelity *In vitro* 3D cellular model with specialized functions similar to those of the human heart that can be used to study aspects of human disease and screen for new therapeutic approaches.

Empirical Approaches to Generating Human Cardiac Organoids

In vitro production of hCO primarily depends on the self-organization of human-derived iPSCs [22 - 24]. Groups focused on biomaterial science have designed and tested a variety of biocompatible scaffolds, including cell hydrogel matrices [25, 26], biomaterial-based micromolds [27, 28], and 3D-printed biomaterials [29, 30], for their ability to mimic the complex *In vivo* microenvironment and promote *In vitro* production of hCO. Approaches to initiate the generation of hCOs primarily depend on initially dispersing cellular aggregates in a defined cell culture medium or immersing them in a 3D matrix or scaffold. Hydrogels are a three-dimensional network of hydrophilic polymers that can swell in the water while maintaining their structure. Unlike suspension culture, collagen-conjugated hydrogels are more effective in generating human iPSC-derived embryoid bodies (EBs) able to differentiate into a myocardium-like tissue with nascent blood vessels [30]. The relative stiffness of the collagen-conjugated hydrogel matrix can impact results by better modeling the environment of myocardial tissue, thereby encouraging proliferation and differentiation of cardiomyocytes. The optimal rigidity of the 3D collagen-hydrogel matrix expedites the formation of contracting hCOs that can then be used in translational drug screening. Matrices with inconsistent rigidity fail to provide proper mechanical support for hCO formation. Therefore, stringent procedures for hydrogel production must be followed for reliable production of hCO.

Surface topology and electrical stimulation are necessary input signals that guide the development of cardiomyocytes *in vivo*. Recently, Mills *et al.* introduced the "Heart-Dyno" system for high throughput fabrication of hCOs. For this system, they constructed a 96-well-format device using an epoxy-based negative photoresist (SU-8 photolithography) combined with polydimethylsiloxane (PDMS) casting. This device provides a universal and rapid format for performing drug screening on hCOs. A mixture of hPSC-CMs, type I collagen, and Matrigel™ was prepared on ice and then pipetted into each well of the device for high throughput screenings that can be used to investigate the underlying mechanisms of cell cycle arrest [26] or assay compounds for drug discovery [31]. Their data illustrate the potential power of using a contractile hCO model as an *In vitro* substitute for the native human heart.

Spatiotemporal patterning during heart development involves a complex input of biomolecular, structural, and functional signals. To recapitulate this *in vitro*, researchers use geometrically patterned micromolds to create well-defined topological and signaling niches. The constraints of these defined niches can be used to drive the differentiation of PSCs that reflects the appropriate temporal and chamber-specific organization. These models, incorporating classical organoids and micropatterning approaches, have produced three-dimensional microcavities with appropriate spatial and cellular fate specification characteristics [27, 28]. Indeed, cardiac organoids assembled with biomaterial-based microchambers are emerging as flexible, alternative platforms for investigating cardiac contractility and repair.

Despite the advances in strategies to produce lab-grown "mini-hearts" or cardiac organoids, most protocols are neither user-friendly nor easily reproduced in a standard cell biology laboratory, even one equipped for stem cell research. Many protocols depend on special devices or technologies only available in a sophisticated tissue engineering laboratory. To address these issues, Hofbauer*et al.* 2021 from the Austrian Academy of Science developed an organoid protocol that is straightforward and easily replicated in a standard stem cell culture facility [32]. The unique methodologies of this protocol are summarised in Table **1** and compared side-by-side with other published protocols.

Table 1. Methodological comparison of the hCO production protocols used successfully by research groups worldwide.

Research Groups/Parameters	Hofbauer *et al.* 2021 [32]	Richards *et al.* 2020 [35]	Filippo Buono *et al.* 2020 [41]	Mills *et al.* 2017, 2019 [40, 46] Voges *et al.* 2017 [23]	Keung *et al.* 2019 [22]	Israeli *et al.* [42]
Source of cardiomyocytes	1. H9 hESC 2. H7 hESC 3. Wild-Type iPSC	hiPSC-CMs	1. Cor.4U-CMs 2.WTSIi020-A 3.UKKi025-A	1. hESC 2. hES3 female donor	hES2	1. iPSC 2. H9 hESC

(Table 1) cont.....

Research Groups/Parameters	Hofbauer *et al.* 2021 [32]	Richards *et al.* 2020 [35]	Filippo Buono *et al.* 2020 [41]	Mills *et al.* 2017, 2019 [40, 46] Voges *et al.* 2017 [23]	Keung *et al.* 2019 [22]	Israeli *et al.*[42]
Source of endothelial cells and Fibroblasts or stromal cells	hPSC	**1. HCvFB** (Human cardiac ventricular Fibroblasts). **2. HUVEC** (Human umbilical vein endothelial cells) **3. HADSCs** (Human adipose-derived stem cells,)	**1. HCFs** (Human Cardiac Fibroblasts). **2. HCMECs** (Human Cardiac Microvascular Endothelial Cells)	**1.hPSCs** **2. HCMEC** (Human cardiac microvascular endothelial Cells,))	**HDFs** (human dermal fibroblasts)	hiPSCs
CM:EC: FB	Not defined.	**1:1** [hiPSC-CMs and non-myocytes]	**3:5:2 (CM: HCMEC: HCF)**	Not defined	**10:1** (CM: HDFs)	Not defined
Cell seeding density and Plate type	5000 cells/well of ultra-low attachment 96-well plates.	~150,000 cells/agarose hydrogel mold and maintained in 12 wells plate.	100,000 cells/well of Greiner HLA Terasaki 60-well plates.	50000 cardiac cells were seeded in Heart-dyno culture system (HdCS).	11,000,000 cells in Ultra-low-attachment cell culture dishes.	10,000 cells/well in round bottom ultra-low attachment 96-well plates.
Media for stem cell and cardiomyocyte maintenance	E8 + ROCK inhibitor CDM medium containing FGF2 LY294002, Activin A, BMP4, and CHIR99021.	iCell Cardiomyocyte Plating Medium (CDI)	**1.** mTeSR1 Medium. **2.** STEM diff Cardiomyocyte Differentiation Kit. **3.** STEM diff Maintenance Kit.	**1**. mTeSR-1 **2.** CTRL medium for organoid (α-MEM GlutaMAX, 10% (FBS), 200 mM L-ascorbic acid 2-phosphate sesquimagnesium salt hydrate.	**1.** mTeSR-1 **2.** StemPro 34 for the organoid	E8 Flex Media
Media for EC and FB maintenance	Following differentiation, ECs were maintained in CDM media supplemented with VEGF.	FGM-2 medium for HCvFB. EGM-3 medium for HUVEC. Low-glucose DMEM with glutamine for HADSCs.	**1.** Endothelial Cell Medium (ECM) for HCMEC. **2.** HCF Growth Medium.	Not defined	Not defined	None

(Table 1) cont.....

Research Groups/Parameters	Hofbauer *et al.* 2021 [32]	Richards *et al.* 2020 [35]	Filippo Buono *et al.* 2020 [41]	Mills *et al.* 2017, 2019 [40, 46] Voges *et al.* 2017 [23]	Keung *et al.* 2019 [22]	Israeli *et al.*[42]
Matrix	Vitronectin Laminin-511 E8 fragment and Laminin-521	0.1% gelatin	1. 10 μg/mL VitronectinXF in CellAdhere dilution buffer for stem cell. 2. Poly---Lysine coating for organoid.	Matrigel-coated	2 mg/mL collagen, 0.9 mg/mL Matrigel	Growth factor reduced Matrigel for stem cell culture

Assembloids

Assembloids are a new generation of organoids, usually composed of multiple regions and/or cell lineages of the same organ in 3D culture. They are helpful for modeling interactions between different regions of the same organ as they provide a wider diversity of cell-cell interactions to better model *In vivo* biology. This approach has been useful for complex organs such as the brain [33]. The complex, compartmentalized nature of the heart makes it an excellent candidate for assembloid technology. In the future, it may be possible to produce region-specific cardiac organoids and assemble them to yield more complex cardiac organoids with appropriate electrophysical coupling and chamber specification that would greatly expand their utility for drug screening and the study of heart disease.

APPLICATIONS OF CARDIAC ORGANOIDS

Organoids to Model Cardiovascular Disease and Screen for new Drug Therapies

Developing animal models of human disease often relies on chemical and surgical manipulations or requires time-consuming gene modifications to replicate the symptoms and pathologies associated with a specific human disease. Despite the many advantages of animal models, there are substantial limitations, including differences in genetic synteny, metabolism, and endurance between the animal model and humans. These differences can lead to substantial bottlenecks in the translation of animal data to application in humans. Because human organoids are composed of human-derived cells, they can be superior for modeling specific aspects of human biology. The techniques for hCOs research are improving daily [34], and in the following sections, we highlight their application in the study of specific, complex disease conditions.

Myocardial Infarction Model

In vitro, modeling of myocardial infarction is an indispensable tool for studying the biology of human heart attacks in the laboratory. In 2017, Voges *et al.* developed an hCO model involving a circular mold seeded with a collagen matrix and hPSC-CMs. They modeled an acute myocardial infarction in these hCOs using dry ice to deliver a localized cryoinjury. The hCOs developed by Voges *et al.* retained attributes of fetal cardiac tissue and therefore possessed the regeneration capacity of neonatal human hearts. Consequently, there was a complete functional recovery two weeks after acute injury due to the proliferation and migration of cardiomyocytes in the uninjured "remote" regions of the hCOs [37].

In a concurrent study, Richards *et al.* created hCO in non-adhesive agarose hydrogel molds using a mixture of 50% hPSC-CM and 50% non-myocytes (at a 4:2:1 ratio of human cardiac fibroblasts: human umbilical vein endothelial cells: human adipose-derived stem cells). Following the generation of hCOs, they simulated a myocardial infarction by maintaining the hCOs under hypoxia (1% O_2) and treating them with 1 μM noradrenaline. This organoid-based model of myocardial infarction recapitulated appropriate pathological changes, including a shift in metabolism, increased fibrosis, and altered calcium handling when assessed at transcriptional, structural, and functional levels [35]. It is important to note, however, that both of these hCO-based myocardial infarction models share the limitation of cellular diversity, notably, the lack of immune/inflammatory cells that are usually present in a live heart *In vivo* as the immune system plays a vital role in response to myocardial infarction, particularly about the process of fibrosis [36].

Heart Failure Model

Tiburcy *et al.* developed hCOs that resemble postnatal myocardium in terms of key structural and functional aspects [24]. They confirmed the maturation of their hCOs based on morphology, systolic twitch forces, a positive force-frequency response, their inotropic responses to β-adrenergic stimulation, and their transcriptional profile. They used neurohumoral stimulation of their hCOs to create a human heart failure-like condition. Of major importance, their hCO model responded to chronic catecholamine toxicity by displaying contractile dysfunction, cellular hypertrophy, and apoptosis accompanied by loss of positive force-frequency responses, adrenergic signal desensitization, and release of classical biomarkers of heart failure, such as the N-terminal pro-B-type natriuretic peptide. In addition, treatment with metoprolol (a β1-adrenergic receptor blocker) and phenoxy benzylamine (an α1-adrenergic receptor blocker) prevented the progression of the hCOs to heart failure phenotype. These outcomes provide proof of the principle that *In vitro* hCO technology can be successfully applied to model human heart failure and used as a screening platform for new and repurposed drugs [24]. A significant limitation is that the hCOs developed by Tiburcy *et al.* resemble a 13-week old human fetal heart in terms of histomorphometry and functionality. Developing hCOs with complete adult characteristics remains a significant challenge.

Congenital Cardiac Disease Model

Similar to the case of myocardial infarction and heart failure, congenital heart disease has also been successfully modeled in hCOs. Recently, Yang *et al.* studied novel familial cardiomyopathy of impaired contractility due to a mutation (E848G) in the human beta myosin heavy-chain (*MYH7*) using patient-specific iPSC-CMs. In humans, *MYH7* is expressed both in the embryonic and adult hearts. To generate hCOs, they use a cell suspension of hiPSC-CMs and human marrow stromal cells in a customized polymethyl siloxane (PDMS) mold. They subjected the hCOs to a force transducer and a length controller to assess the tissue construct force. Sarcomere alignment was significantly compromised in hCOs made with E848G mutant cells compared to those generated from wild-type cells. Furthermore, mutant hCOs significantly reduced systolic function with a minimal change in diastolic function. The ability to assess myofibril contraction in 3D hCO systems substantially increases their physiological relevance and potential future application in personalized medicine [37].

In another recent study, Long *et al.* used iPSC-derived hCOs to demonstrate successful CRISPR-mediated restoration of dystrophin expression in cells from patients with Duchenne muscular dystrophy (DMD). Furthermore, they showed that only 50% of the cardiomyocyte pool required gene-edit corrections to restore normal contractile function. The outcomes of these studies are particularly significant as they validate the use of hCOs as a platform for testing the functional effects of genome editing strategies and, therefore, their utility as preclinical tools for assessing gene-editing therapies [38].

Arrhythmia Model

hCOs can generate spontaneous and autonomous action potentials and respond to induced pacing. They display higher conduction velocities than two-dimensional cell culture models and therefore are improved platforms for studying acquired and congenital cardiac arrhythmias. Initially, research into familial or congenital arrhythmias focused on the cellular electrophysiological and ionic phenotypes of hiPSC-CMs. Consequently, the field stands poised to move toward hCO technology as an ideal platform for investigating more critical electrophysiological events, such as conduction and re-entry, in arrhythmogenic syndromes, including but not limited to short QT syndrome (SQTS), long QT syndrome (LQTS), *etc.* Recently, Shinnawi *et al.* used hCOs composed of hiPSC from SQTS patients to compare parameters in organoids with and without CRISPR-based genome editing of the mutated gene. Using hCOs technology, they could mimic mutant arrhythmogenicity *in vitro* and demonstrate corrective genome editing. They used this model as a platform to test the efficacy of newly developed candidate antiarrhythmic molecules. Their screening and functional tests demonstrated that quinidine and disopyramide could prolong action potential duration and suppress arrhythmogenicity. However, sotalol did not exhibit any antiarrhythmic activity. This study provides proof of principle regarding the use of hCOs for modeling arrhythmia syndromes *In vitro* and further supports the use of hCOs for screening drug candidates tailored to the genetic background of a specific patient as well as reducing the need for animal studies [39].

Goldfracht *et al.* generated hCOs from hPSC-CMs with chamber-specific characteristics (ventricular and atrial) with the aid of chamber-specific small molecules (*e.g.*, bone morphogenetic protein 4, activin A, and retinoic acid). They initiated the formation of their hCOs in circular casting molds and subsequently transferred them into a passive silicon stretcher to form chamber-specific, ring-shaped hCOs. Chamber-specific hCOs exhibited unique atrial versus ventricular phenotypes, including morphology, gene expression, electrophysiological, and contractile properties. Serendipitously, they observed that the atrial hCOs displayed a variety of arrhythmias at baseline and were amenable to intervention by electrical shock. These results demonstrate the potential utility of atrial hCOs as models to study recurrent arrhythmias, such as atrial fibrillation. Consequently, arrhythmic atrial hCOs are also likely to be excellent platforms for screening antiarrhythmic pharmacological candidates [40].

Limitations of Cardiac Organoids

Whether one anticipates using cardiac organoids in their research or simply needs to evaluate findings reported by other groups, it is essential to understand both the strengths and limitations of this model system. Like cardiomyocytes derived from iPSCs in 2D culture, the properties of cardiomyocytes in hCOs tend to resemble those of immature cardiomyocytes rather than adults. This can include immaturity of sarcomere structure and isoform content, excitation-contraction coupling, energy metabolism, calcium handling, mechanotransduction, and modes of cell-cell interaction. Methods that impose conditions of increased load, such as pacing or adrenergic stimulation, can shift engineered tissue toward a more mature phenotype. Indeed, this approach has proven effective in the technology of cardiac patches. Although the term "cardiac organoid" has sometimes been applied to describe engineered cardiac patches, we posit that cardiac patches engineered under mechanostress provide a model of cardiac remodeling in response to changes in demand. In contrast, hCOs, generated under the guidance of self-directed assembly and organization principles, provide a better model of innate embryonic patterning. Each approach is best used to address different questions. Furthermore, combining approaches may be valuable in yielding new experimental models tailored to address specific questions.

CONCLUSION

Cardiac organoids have come of age. Coupled with iPSC technology, they provide versatile tools for both exploring fundamental biological concepts and improving therapeutic outcomes. Their ability to replicate elemental cardiac properties, such as genetically determined contractile or conduction defects, on a microscale is propelling them toward application as platforms for drug discovery. Such screens have the benefit of being carried out in the genetic milieu of a specific individual and are thus also ideally suited for application in personalized medicine approaches aimed at guiding the best potential drug therapies for a specific patient. We envision significant advancements in hCO technologies that will allow for greater complexity, improved vascularization, and co-culture to model immune infiltration.

ABBREVIATION

PSCs   Pluripotent stem cells

ASCs   Adult stem cells

ES   Embryonic stem cells

iPSCs   Induced pluripotent stem cells

ECM   Extracellular matrix

EHS   Engelbreth-Holm-Swarm cells

LGR5   Leucine-rich repeat-containing G protein-coupled receptor

TS   Tumor spheres

hCOs   Human cardiac organoids

Ebs   Embryoid bodies

MYH7   Myosin heavy-chain

DMD   Duchenne muscular dystrophy

SQTS   Short QT syndrome

LQTS   Long QT syndrome

REFERENCES

[1]   Lancaster MA, Knoblich JA. Organogenesis in a dish: Modeling development and disease using organoid technologies. Science 2014; 345(6194)1247125
[http://dx.doi.org/10.1126/science.1247125] [PMID: 25035496]

[2]   Sutherland RM, McCredie JA, Inch WR. Growth of multicell spheroids in tissue culture as a model of nodular carcinomas. J Natl Cancer Inst 1971; 46(1): 113-20.
[PMID: 5101993]

[3]   Rossi G, Manfrin A, Lutolf MP. Progress and potential in organoid research. Nat Rev Genet 2018; 19(11): 671-87.
[http://dx.doi.org/10.1038/s41576-018-0051-9] [PMID: 30228295]

[4]   Schutgens F, Clevers H. Human Organoids: Tools for Understanding Biology and Treating Diseases. Annu Rev Pathol 2020; 15(1): 211-34.
[http://dx.doi.org/10.1146/annurev-pathmechdis-012419-032611] [PMID: 31550983]

[5]   Orkin RW, Gehron P, McGoodwin EB, Martin GR, Valentine T, Swarm R. A murine tumor producing a matrix of basement membrane. J Exp Med 1977; 145(1): 204-20.
[http://dx.doi.org/10.1084/jem.145.1.204] [PMID: 830788]

[6]   Timmins NE, Nielsen LK. Generation of multicellular tumor spheroids by the hanging-drop method. Methods Mol Med 2007; 140: 141-51.
[http://dx.doi.org/10.1007/978-1-59745-443-8_8] [PMID: 18085207]

[7]   Turner DA, Baillie-Johnson P, Martinez Arias A. Organoids and the genetically encoded self-assembly of embryonic stem cells. BioEssays 2016; 38(2): 181-91.
[http://dx.doi.org/10.1002/bies.201500111] [PMID: 26666846]

[8]   Kalabis J, Wong GS, Vega ME, *et al.* Isolation and characterization of mouse and human esophageal epithelial cells in 3D organotypic culture. Nat Protoc 2012; 7(2): 235-46.
[http://dx.doi.org/10.1038/nprot.2011.437] [PMID: 22240585]

[9]   Wilson HV. A new method by which sponges may be artificially reared. Science 1907; 25(649): 912-5.
[http://dx.doi.org/10.1126/science.25.649.912] [PMID: 17842577]

[10]   Holtfreter J. Experimental studies on the development of the pronephros. Rev Can Biol 1944; 3: 220-50.

[11]   Weiss P, Taylor AC. Reconstitution of complete organs from single-cell suspensions of chick embryos in advanced stages of differentiation. Proc Natl Acad Sci USA 1960; 46(9): 1177-85.
[http://dx.doi.org/10.1073/pnas.46.9.1177] [PMID: 16590731]

[12]   Takahashi K, Tanabe K, Ohnuki M, *et al.* Induction of pluripotent stem cells from adult human fibroblasts by defined factors. Cell 2007; 131(5): 861-72.
[http://dx.doi.org/10.1016/j.cell.2007.11.019] [PMID: 18035408]

[13]   Takahashi K, Yamanaka S. Induction of pluripotent stem cells from mouse embryonic and adult fibroblast cultures by defined factors. Cell 2006; 126(4): 663-76.
[http://dx.doi.org/10.1016/j.cell.2006.07.024] [PMID: 16904174]

[14]   Thomson JA, Itskovitz-Eldor J, Shapiro SS, *et al.* Embryonic stem cell lines derived from human blastocysts. Science 1998; 282(5391): 1145-7.
[http://dx.doi.org/10.1126/science.282.5391.1145] [PMID: 9804556]

[15]   Yu J, Vodyanik MA, Smuga-Otto K, *et al.* Induced pluripotent stem cell lines derived from human somatic cells. Science 2007; 318(5858): 1917-20.
[http://dx.doi.org/10.1126/science.1151526] [PMID: 18029452]

[16]   Li ML, Aggeler J, Farson DA, Hatier C, Hassell J, Bissell MJ. Influence of a reconstituted basement membrane and its components on casein gene expression and secretion in mouse mammary epithelial

cells. Proc Natl Acad Sci USA 1987; 84(1): 136-40.
[http://dx.doi.org/10.1073/pnas.84.1.136] [PMID: 3467345]

[17]   Shannon JM, Mason RJ, Jennings SD. Functional differentiation of alveolar type II epithelial cells in vitro: Effects of cell shape, cell-matrix interactions and cell-cell interactions. Biochim Biophys Acta Mol Cell Res 1987; 931(2): 143-56.
[http://dx.doi.org/10.1016/0167-4889(87)90200-X] [PMID: 3663713]

[18]   Eiraku M, Watanabe K, Matsuo-Takasaki M, *et al.* Self-organized formation of polarized cortical tissues from ESCs and its active manipulation by extrinsic signals. Cell Stem Cell 2008; 3(5): 519-32.
[http://dx.doi.org/10.1016/j.stem.2008.09.002] [PMID: 18983967]

[19]   Weiswald LB, Bellet D, Dangles-Marie V. Spherical cancer models in tumor biology. Neoplasia 2015; 17(1): 1-15.
[http://dx.doi.org/10.1016/j.neo.2014.12.004] [PMID: 25622895]

[20]   Hendriks D, Artegiani B, Hu H, Chuva de Sousa Lopes S, Clevers H. Establishment of human fetal hepatocyte organoids and CRISPR–Cas9-based gene knockin and knockout in organoid cultures from human liver. Nat Protoc 2021; 16(1): 182-217.
[http://dx.doi.org/10.1038/s41596-020-00411-2] [PMID: 33247284]

[21]   Keung W, Chan PKW, Backeris PC, *et al.* Human cardiac ventricular-like organoid chambers and tissue strips from pluripotent stem cells as a two-tiered assay for inotropic responses. Clin Pharmacol Ther 2019; 106(2): 402-14.
[http://dx.doi.org/10.1002/cpt.1385] [PMID: 30723889]

[22]   Voges HK, Mills RJ, Elliott DA, Parton RG, Porrello ER, Hudson JE. Development of a human cardiac organoid injury model reveals innate regenerative potential. Development 2017; 144(6)dev.143966
[http://dx.doi.org/10.1242/dev.143966] [PMID: 28174241]

[23]   Tiburcy M, Hudson JE, Balfanz P, *et al.* Defined engineered human myocardium with advanced maturation for applications in heart failure modeling and repair. Circulation 2017; 135(19): 1832-47.
[http://dx.doi.org/10.1161/CIRCULATIONAHA.116.024145] [PMID: 28167635]

[24]   Shkumatov A, Baek K, Kong H. Matrix rigidity-modulated cardiovascular organoid formation from embryoid bodies. PLoS One 2014; 9(4)e94764
[http://dx.doi.org/10.1371/journal.pone.0094764] [PMID: 24732893]

[25]   Mills RJ, Titmarsh DM, Koenig X, *et al.* Functional screening in human cardiac organoids reveals a metabolic mechanism for cardiomyocyte cell cycle arrest. Proc Natl Acad Sci USA 2017; 114(40): E8372-81.
[http://dx.doi.org/10.1073/pnas.1707316114] [PMID: 28916735]

[26]   Hoang P, Wang J, Conklin BR, Healy KE, Ma Z. Generation of spatial-patterned early-developing cardiac organoids using human pluripotent stem cells. Nat Protoc 2018; 13(4): 723-37.
[http://dx.doi.org/10.1038/nprot.2018.006] [PMID: 29543795]

[27]   Ma Z, Wang J, Loskill P, *et al.* Self-organizing human cardiac microchambers mediated by geometric confinement. Nat Commun 2015; 6(1): 7413.
[http://dx.doi.org/10.1038/ncomms8413] [PMID: 26172574]

[28]   Noor N, Shapira A, Edri R, Gal I, Wertheim L, Dvir T. 3D printing of personalized thick and perfusable cardiac patches and hearts. Adv Sci (Weinh) 2019; 6(11)1900344
[http://dx.doi.org/10.1002/advs.201900344] [PMID: 31179230]

[29]   Lee A, Hudson AR, Shiwarski DJ, *et al.* 3D bioprinting of collagen to rebuild components of the human heart. Science 2019; 365(6452): 482-7.
[http://dx.doi.org/10.1126/science.aav9051] [PMID: 31371612]

[30]   Shkumatov A, Baek K, Kong H. Matrix rigidity-modulated cardiovascular organoid formation from embryoid bodies. PLoS One 2014; 9(4)e94764

[http://dx.doi.org/10.1371/journal.pone.0094764] [PMID: 24732893]

[31] Mills RJ, Parker BL, Quaife-Ryan GA, *et al.* Drug screening in human PSC-cardiac organoids identifies pro-proliferative compounds acting *via* the mevalonate pathway. Cell Stem Cell 2019; 24(6): 895-907.e6.
[http://dx.doi.org/10.1016/j.stem.2019.03.009] [PMID: 30930147]

[32] Hofbauer P, Jahnel SM, Papai N, *et al.* Cardioids reveal self-organizing principles of human cardiogenesis. Cell 2021; 184(12): 3299-3317.e22.
[http://dx.doi.org/10.1016/j.cell.2021.04.034] [PMID: 34019794]

[33] Makrygianni EA, Chrousos GP. From Brain Organoids to Networking Assembloids: Implications for Neuroendocrinology and Stress Medicine. Front Physiol 2021; 12621970
[http://dx.doi.org/10.3389/fphys.2021.621970] [PMID: 34177605]

[34] Nugraha B, Buono MF, Emmert MY. Modelling human cardiac diseases with 3D organoid. Eur Heart J 2018; 39(48): 4234-7.
[http://dx.doi.org/10.1093/eurheartj/ehy765] [PMID: 30576473]

[35] Richards DJ, Li Y, Kerr CM, *et al.* Human cardiac organoids for the modelling of myocardial infarction and drug cardiotoxicity. Nat Biomed Eng 2020; 4(4): 446-62.
[http://dx.doi.org/10.1038/s41551-020-0539-4] [PMID: 32284552]

[36] Forte E, Furtado MB, Rosenthal N. The interstitium in cardiac repair: role of the immune–stromal cell interplay. Nat Rev Cardiol 2018; 15(10): 601-16.
[http://dx.doi.org/10.1038/s41569-018-0077-x] [PMID: 30181596]

[37] Yang KC, Breitbart A, De Lange WJ, *et al.* Novel adult-onset systolic cardiomyopathy due to MYH7 E848G mutation in patient-derived induced pluripotent stem cells. JACC Basic Transl Sci 2018; 3(6): 728-40.
[http://dx.doi.org/10.1016/j.jacbts.2018.08.008] [PMID: 30623132]

[38] Long C, Li H, Tiburcy M, *et al.* Correction of diverse muscular dystrophy mutations in human engineered heart muscle by single-site genome editing. Sci Adv 2018; 4(1)eaap9004
[http://dx.doi.org/10.1126/sciadv.aap9004] [PMID: 29404407]

[39] Shinnawi R, Shaheen N, Huber I, *et al.* Modeling reentry in the short QT syndrome with human-induced pluripotent stem cell-derived cardiac cell sheets. J Am Coll Cardiol 2019; 73(18): 2310-24.
[http://dx.doi.org/10.1016/j.jacc.2019.02.055] [PMID: 31072576]

[40] Goldfracht I, Protze S, Shiti A, *et al.* Generating ring-shaped engineered heart tissues from ventricular and atrial human pluripotent stem cell-derived cardiomyocytes. Nat Commun 2020; 11(1): 75.
[http://dx.doi.org/10.1038/s41467-019-13868-x] [PMID: 31911598]

[41] Filippo Buono M, von Boehmer L, Strang J, Hoerstrup SP, Emmert MY, Nugraha B. Human Cardiac Organoids for Modeling Genetic Cardiomyopathy. Cells 2020; 9(7): 1733.
[http://dx.doi.org/10.3390/cells9071733] [PMID: 32698471]

[42] Lewis-Israeli YR, Wasserman AH, Gabalski MA, *et al.* Self-assembling human heart organoids for the modeling of cardiac development and congenital heart disease. Nat Commun 12
[http://dx.doi.org/10.1038/s41467-021-25329-5]

CHAPTER 6

Respiratory System-Based *In Vitro* Antiviral Drug Repurposing Strategies for Sars-Cov-2

Dilara Genc[1], Ahmet Katı[2,3], Amit Kumar Mandal[4], Suvankar Ghorai[5], Hanen Salami[6], Sare Nur Kanari ElHefnawi[3] and Sevde Altuntas[3,7,*]

[1] *Kadir Has University, Undergraduate Program of Bioinformatics and Genetics, Fatih, 34230, Istanbul, Turkey*

[2] *University of Health Sciences Turkey, Biotechnology Department, Uskudar, 34662, Istanbul, Turkey*

[3] *University of Health Sciences Turkey, Validebag Research Park, Experimental Medicine Research and Application Center, Uskudar, 34662, Istanbul, Turkey*

[4] *Raiganj University, Centre for Nanotechnology Sciences (CeNS) & Chemical Biology Laboratory, Department of Sericulture, North Dinajpur, West Bengal-733134, India*

[5] *Virology Laboratory, Department of Microbiology, North Dinajpur, West Bengal-733134, India*

[6] *Laboratory of Treatment and Valorization of Water Rejects (LTVRH), Water Researches and Technologies Center (CERTE), Borj-Cedria Technopark, University of Carthage, 8020, Soliman, Tunisia*

[7] *University of Health Sciences Turkey, Tissue Engineering Department, Uskudar, 34662, Istanbul, Turkey*

Abstract: To date, no known drug therapy is available for COVID-19. Further, the complicated vaccination processes like limited infrastructure, insufficient know-how, and regulatory restrictions on vaccines caused this pandemic episode more badly. Due to the lack of ready-to-use vaccination, millions of people have been severely infected by SARS-CoV-2. Additionally, the increasing contagion risk of the SARS-CoV-2 variants makes drug repurposing studies more critical. Conventionally, antiviral drug repurposing has been conducted on two-dimensional (2D) cell culture systems or *in vivo*-based experimental setups. Recently, *In vitro* three-dimensional (3D) cell culture techniques have proven more coherent in mimicking host-pathogen interactions and exploring or repurposing drugs than other 2D cell culture methods. 3D culture techniques like organoids, bioprinting, and microfluidics/organ-on-a-chip have just been started to mimic the natural microenvironment respiratory system infected with SARS-CoV-2. These techniques avoid the need for animals in agreement with the 3R principles (Replacement, Reduction, and Refinement) to enhance animal welfare. Herein, SARS-CoV-2-host interaction and 3D cell culture techniques have been

* Corresponding Author Sevde Altuntaş: University of Health Sciences Turkey, Validebag Research Park, Experimental Medicine Research and Application Center, Uskudar, 34662, Istanbul, Turkey; E-mail: sevde.altuntas@sbu.edu.tr\

Manash K. Paul (Ed.)

proposed for drug screening and repurposing models through representative examples. This study will frame tissue engineering strategies for studying SARS-CoV-2 infection and enlightening host-virus interactions.

Keywords: SARS-CoV-2, 3D cell culture model, Antiviral, Alveolar tissue, Drug repurposing, Drug screening, Drug screening, Infection.

INTRODUCTION

Every once a year, new infectious pathogens may appear and are only detected when an epidemic or pandemic occurs. Early detection and drug repurposing studies may reduce the number of deaths related to these contagious diseases. Notably, *in vitro* conditions for modeling respiratory system diseases still requires more attention due to the lack of fundamental knowledge on the prognosis of such diseases. Globally, researchers are showing a great effort to find the treatments to bring the COVID-19 pandemic under control. COVID-19 patients develop Acute Respiratory Distress Syndrome (ARDS) with renal failure, pericarditis, and disseminated intravascular coagulation (DIC), including death. Several clinical trials and computer-aided drug design direction studies have focused on detecting the hidden benefits of any existing drugs to improve survival or control COVID-19 [1].

The drug repurposing strategies were found to be more effective in accelerating infectious control modalities compared to the long and complex process of vaccine or drug development [1, 2]. Drug repurposing strategies are of great importance to accelerate the R&D process without any preclinical steps compared with conventional drug discovery approaches. Additionally, this process has the potential to reduce the total development costs and discovery time involved in drug development [3]. Nevertheless, the platforms like computational and experimental approaches can be adapted to vaccine efficacy studies as they are intended for human-specific design [4, 5]. Such repurposing studies may aid in optimizing the diagnosis process for COVID-19 and support healthcare systems [6].

For several decades, many 2D and 3D *In vitro* cell culture techniques have been developed for drug repurposing. To check the alarming spread of COVID-19, both Food and Drug Administration (FDA) approved and unapproved drugs have been tested to cure this disease [7]. The studies have reported that 3D models show more consistency with real scenarios due to physical, chemical, mechanical, and biological similarities with the native tissue than 2D models [8 - 10].

Table **1** shows 2D and 3D cell culture techniques and their properties. The air-liquid interface (ALI) models significantly contribute while mimicking the natural environment of organs such as skin, lungs, and brain [11 - 13]. However, organotypic models have drawbacks, such as limited cellular diversity and life terms compared to spherical culture techniques [14]. Compared to 2D models, 3D cell culturing strategies such as spherical cultures (organoids), organ-on-a-chip systems, and bioprinting technologies can combine tissue engineering and virology studies to construct ideal platforms to model drug repurposing systems [15].

Table 1. Comparison of cell culture techniques for modeling the respiratory system based on COVID-19.

Techniques	Advantages	Disadvantages
Air-Liquid Interface	- Mimics in vivo systems better than 2D cultures as it consists of ECM* - Ability to tune the mechanical &biochemical properties of the gel/scaffold - Mechanical and biochemical signaling support for cells - High reproducibility - Co-culture ability - Ease of using pathogen transfection studies	- Additional prep-time needed - Relatively high-degree of cell-matrix vs. cell-cell interactions - Difficulties of imaging and integrating to HCS** systems
Spharical Cultures Cell cluster 3D	- Compact, multicellular aggregates - Well-known and established technology for spheroids based on the traditional Harrison's hanging drop technique - Ease of adapting to HTS*** and HCS platforms - Sensitive size control for spherical models - Not need to expertise and high quality equipments - Suitable for 3D tissue models - Co-culture ability - Reproducibility	- Expertise for imaging process - Require sensitive culture conditions - Susceptibility to mass transport limitations - Non-automated culture techniques - Labor intensive culture protocols at times
Microfluidics / Organ-on-chip	- In order to mimic physiological conditions, combines microfabrication and tissue engineering approaches - Ability to manipulate and stimulate the cells, tissues, and ECM by using physicochemical, electrical, and mechanical factors - Provide interconnection among different types of tissues - Reducing reagents and supplies consumption	- Technology readiness level 4 or 5 and limited commercialized products - Required highly-specialized techniques and expertise - Hard to adapt to HTS - High-cost consumables - Lack of standardization/validation

Abbreviations; ECM: Extracellular matrix; HTS: High Throughput System; HCS: High Content System.

Organ-on-a-chip systems and bioprinting technologies are frequently studied to create an appropriate organ model environment [16, 17]. However, they need specialized equipment, expertise, and complex fabrication protocols. The major drawbacks of these techniques include difficulty in adapting to high throughput screening. The spherical culture techniques may provide an appropriate platform for host-pathogen interaction and drug repurposing studies. It is thought that organoids are among the best contributions of stem cell studies to the scientific world. The term organoid is defined as three-dimensional (3D) mini-tissue cultures produced *In vitro* from primary tissues, induced pluripotent stem cells (iPSCs), embryonic stem cells (ESCs), or adult stem cells to mimic specific organ components and exhibit organ-specific behaviors [18]. Therefore, human-sourced organoids can be modeled to discover COVID-19 mechanisms and progressions. For instance, it was currently shown that organoids could be used as an alternative experimental platform to understand SARS-CoV-2-host interaction with the lung and other organs [19]. Additionally, patient-specific drug therapy pathways can be defined for the infection using organoids [20].

Herein, we focus on the drug repurposing studies for COVID-19 based on *In vitro* models, which can mimic the branched and alveolar structures of human lungs, offering a rapid and reliable solution. In particular, this review aims to guide cell culture experts in the innovations and improvements of the current drug screening strategies for the disease. We categorized certain cell culture techniques for drug screening strategies based on COVID-19 and provided representative examples of their applications. The review concludes with perspectives on the possible drug screening models, such as a hybrid form of the ALI and spherical culture techniques.

CELL CULTURE TECHNIQUES FOR DRUG REPURPOSING APPLICATIONS IN COVID-19 MODELS

During the emergence of the SARS-CoV-2 pandemic, the demand for antiviral drug repurposing has been incremented. Even though there are clinical trials of the disease, the aggressive progression of it is needed in real tissue culture models. To properly model the human lung *In vitro* conditions, ALI models, spherical cultures (organoids), and organ-on-a-chip systems have been developed by different research groups [21 - 24].

Although the cell culture models can apply to drug screening technologies, they are generally used to understand disease mechanisms or sensing technologies [25, 26]. Here, the potential of techniques was stressed by supporting the examples. Table **2** explains studies with various potential antiviral drugs screened on different cell culture techniques.

Air Liquid Interface (ALI) Models

ALI has been typically selected as a respiratory system model [12, 14]. In the ALI models, the cells are grown until they reach confluency on a permeable filter. The medium, which is added to the apical side of the filter is detracted from the well in ALI models. Thus, the cells can contact the air, which triggers their differentiation and tissue formation [24]. Such techniques have been used to investigate COVID-19 pathogenesis. It was reported that immune molecules such as type I and type III interferon (IFN) could inhibit SARS-CoV-2 infection in the ALI alveolar model [27]. The transcriptional profiling analysis also showed the induction of pro-inflammatory cytokines and chemokines in the infected models. Zhu *et al.* reported that SARS-CoV-2 exhibited a similar morphogenic process like other coronaviruses except for plaque-like cytopathic effects [28]. The immunohistological studies suggest that SARS-CoV-2 can directly transmit from cell to cell, making ALI models an ideal candidate for understanding the mechanism involved in infectious diseases [29].

Alternatively, ALI models can provide real-like information in inhalation toxicology and are generally considered between conventional *In vitro* models and animal-based strategies. Using the infected models in drug repurosing fields can develop this perspective, and the models can be used to find the best or rapid solution during pandemics. Remdesivir, a potential antiviral drug, has been reported to diminish the recovery time in patients hospitalized with COVID-19 [30]. The results of *In vitro* analysis of the cell culture models are expected to correlate with the clinical study. The ALI models were currently employed to demonstrate Remdesivir effects on SARS-CoV-2 with rapid real-life information. The earlier study showed that the drug possesses negligible toxic impacts on the ALI models and inhibits infectious virus production in a dose-dependent manner [31]. Similar results were confirmed by different research groups who reported that the antiviral activity of Remdesivir is significantly superior to hydroxychloroquine and interferons [32]. However, although the ALI culture technique is a beneficial model for proximal airway cells, alveolar stem cell niches have a more complex structure and may require stromal cell support. Moreover, the drawbacks of the technique prevent applying the model to high content/throughput screening systems, crucial for analyzing the high number of samples and, when required, more detail in a short time span.

Table 2. Screened drugs for COVID-19 using various cell culture techniques based on the respiratory system.

Cell Culture Methods	Screened Drugs	Results	Ref
ALI	IFN I IFN III	Pre-treatment with type I and type III IFN inhibited SARS-CoV-2 replication.	[Vanderheiden *et al.*, 2020]
	Remdesivir	Remdesivir potency depends on the dose and shows no toxicity on the ALI model.	[Mulay *et al.*, 2021; Pruijssers *et al.*, 2020]
	Hydroxychloroquine	Hydroxychloroquine reduced viral replication potency, but it is limited compared to the Remdesivir effect.	[Mulay *et al.*, 2021]
Spherical Models	Imatinib MPA QNHC	MPA and QNHC prevent SARS-CoV-2-entry virus infection in the organoids rather than hindering luciferase activity in a dose-dependent study.	[Han *et al.*, 2020]
	IFN λ_1	SARS-CoV-2 viral titers are responsive to low amount IFN λ_1 treatment for lung organoids grown in bronchoalveolar ALI culture.	[Lamers *et al.*, 2021]
	Anti-ACE2	Anti-ACE2 treatment could prohibit virus entry into cells.	[Huang *et al.*, 2020; Pei *et al.*, 2020]
	Remdesivir	Remdesivir potently inhibited SARS-CoV-2 replication in lung organoids.	[Pei *et al.*, 2020]
Organ-on-a-chip	Amodiaquine DesethylamodiaquineToremifene Clomiphene Chloroquine Hydroxychloroquine Arbidol	Amodiaquine, desethylamodiaquin, toremiphene, and clomiphene inhibited viral entry and showed no toxicity. Hydroxychloroquine, chloroquine, and arbidol showed no effect on pseudo-typed SARS-CoV-2.	[Si *et al.*]
	Remdesivir	Treatment with remdesivir can inhibit viral replication and lessen barrier disruption on a chip.	[Zhang *et al.*, 2021]

ALI: Air-liquid interface, IFN: Interferon, MPA: Mycophenolic acid, QNHC: Quinacrinedihydrochloride

Spherical Cultures

These techniques employ embryonic stem cells, induced pluripotent stem cells, and patient-derived tissues [33 - 37]. They show similarities with native tissue but have limitations like requirements of domain expertise, is labor-intensive, and require automated fabrication strategies. Unlike ALI techniques, the technologies can quickly adapt to high throughput/content screening systems, making them ideal for drug screening studies with large sample groups (Table **1**) [34].

Developing antiviral drug screening platforms requires a deep understanding of the developmental biology of the organs and the applicability of disease models on cell culture strategies. For COVID-19-targeted alveolar models, the knowledge of developmental biology becomes more significant since stem cells may have a prominent role in the COVID-19 prognosis. It was currently reported that the epithelial stem cells (H2-K1, BASCs, basal cells) in respiratory tissue express angiotensin-converting enzyme marker (ACE2) and other SARS-CoV-2 entry factors like furin [38, 39]. Stem cell-based lung organoid generation strategies can provide an environment for various cell lines in the same niche through physic-chemical manipulations, helping in mimicking disease conditions [18]. Additionally, the lung epithelium contains different types of stem cell populations [40]. For instance, the basal stem cells, capable of self-renewal and mucociliary differentiation, can cause secretion and produce ciliary cells during abrasion, rupture, or injury located on the airway epithelium. There are two typical cells within the alveolar epithelium: Type I and Type II. Type I alveolar epithelial cells (AEC1s) establish a selective air-blood barrier for diffusion. Type II alveolar epithelial cells (AEC2s) are responsible for repairing the alveolar epithelium in any alveolar injuries and surfactant system functions that inhibit alveolar collapse by reducing surface tension [41]. AEC2s can also act as stem cells with their self-renewal and differentiation into AEC1s, making them crucial for the disease models [42]. Any functional disability, infections, or acute injuries on AEC2s can lead to respiratory system problems, including respiratory syncytial virus (RSV) [43]. Hence, ACE2-rich AEC2-specific cell culture models can give more accurate results when evaluated for drug repurposing.

Mainly, to mimic ACE2$^+$ tissues, which are abundant in the organs such as the brain, lung, intestine, and kidney, the technique can present a multicellular and complex environment, very similar to the native tissue [44]. Virus uptaking in ACE2$^+$ capillary alveolar organoids has been significantly reduced after anti-ACE2 treatment [45]. Additionally, Lamer *et al.* designed SARS-CoV-2 infected-ACE2$^+$ small intestinal organoids. The study also indicated that intestinal enterocytes are the virus's target cell and intestinal organoids could be a significant model for understanding infection biology [46]. Contradictorily, it has been recently shown that ciliated cells express ACE2 marker more constitutively compared to AEC2s, with increasing expression toward the proximal part of the respiratory system [47]. The results indicate that multicellular strategies can provide a wider perspective on conducting virology studies instead of single cell-focused *In vitro* analyses.

The organoid-based disease models have demonstrated that viral penetration could occur around the $ACE2^+$ cell population in organoids; thus, anti-ACE2 treatment could check virus entry into the cells, reducing disease symptoms [48]. Han *et al.* suggested that the alveolar organoids can provide an ergonomic platform for rapid screening of FDA-approved COVID-19 drugs. This study proposed that imatinib and mycophenolic acid could reduce viral infection [49]. Pei *et al.* also reported that human lung organoids might serve as a model for pathophysiological studies to elucidate the mechanism involved in SARS-CoV-2 infection [50]. The study also explored the COVID-19 treatment potential of remdesivir and anti-ACE2 on alveolar organoids *via* inhibiting SARS-CoV-2 replication.

Additionally, the RNA-seq analysis of the infected organoids indicated that the lipid metabolism was surprisingly down-regulated. Although the studies focused on a few medicines or in-silico based assumptions, a potential research area has emerged to generate high-throughput drug screening systems to expedite the development of therapies for COVID-19. It was reported that 21 drugs, including Remdesivir, have inhibited the replication of SARS-CoV-2 virus using tissue explant models and infected Vero E6 cell lines that exhibit dose-related responses [51]. Weiwei *et al.* also reported a 2D high-throughput drug screening methodology to evaluate replication kinetics and drug responses of SARS-CoV-2 in Vero E6 cells [52]. The histopathology results of infected organoid models showed that the effects of COVID19 differ from organ to organ [53]. For instance, metabolic downregulation has been observed in hepatocytes grown in liver organoids [36], while altered Tau protein distribution has been detected in brain organoids after infection [54]. Therefore, the spherical culture techniques may provide reliable, fast, and high-throughput drug screening strategies to find appropriate drugs [55]. However, high-throughput screening strategies for the SARS-CoV-2 infected-alveolar organoids have not been reported to the best of our knowledge.

Directly targeting the virus and/or indirectly targeting the virus through host modulation strategies are currently employed strategies. To date, approximately 500 clinical trials have been registered on ClinicalTrials.gov [56]. The compounds exhibiting antiviral properties against CoVs like chloroquine, remdesivir, ribavirin, and favipiravir were tested with conventional techniques under *In vitro* conditions and referred to clinical trials. However, the limited cell culture techniques are insufficient for initiating phase 1 studies. Even with vast and robust *In vivo* experiments designed, patient and disease-specific conditions can alter results. To illustrate clinical outcomes, chloroquine has acute unwanted effects on patients, although the drug was supposed to be the best COVID-19 treatment at the beginning of the pandemic [57, 58]. This situation also demonstrates that organ-like systems may be more helpful in understanding disease mechanisms and pharmaceutical dynamics. Hence, advancing the development of respiratory organoid culture models has contributed to understanding prognosis and advancing drug design and drug repurposing for SARS-CoV-2 [59].

Organ-on-a-chip

Another culture technique pioneered in lung development is the organ-on-a-chip, which can also determine toxicological and pharmacological responses for new drug designs [60]. The current COVID-19 pandemic challenge has made lung airway chips a powerful platform to repurpose the approved drugs for SARS-CoV-2. It has been recently reported that a microfluidic system, which has two main parts as air channel and cell carrier surface, is designed to model human infection by SARS-CoV-2 [61]. In the study, the virus-like particles carrying key entry factors of the virus were introduced to the air channel, increasing the expression levels of ACE2 and transmembrane protease serine 2 (TMPRSS2), which is responsible for viral uptake into alveolar cells. For example, hydroxychloroquine has been one of the most debatable drugs for thetreatment of SARS-CoV-2. Boulware *et al.* reported no significant difference in the rate of occurrence of new COVID-19 cases between patients getting hydroxychloroquine and placebo [62]. This suggests that there is no clinical evidence of using hydroxychloroquine for the disease.

Moreover, Longlong *et al.* reported that hydroxychloroquine showed a reduction in SARS-CoV-2 S protein entry in Huh-7 cells in a dose-dependent manner without diminishing the viral infection in human airway chips [63]. In addition, Zhang *et al.* reported that the treatment with Remdesivir could inhibit viral replication and alleviate barrier disruption on the human alveolar chip [64]. In another study, amodiaquine and toremifene were repurposed as potential inhibitors for SARS-CoV-2 entrance to the cells *via* the system [64]. Overall, the study proposes that lung-on-a-chip technology could provide a similar environment to model the infection, which may expedite drug repurposing for the virus.

Furthermore, the organoids-on-a-chip techniques can help mimic the native organ environment. This method can make measurable organ-to-organ variabilities of side effects of the drug under *In vitro* conditions and open a gate for personalized medicine applications. However, they need more validation and cost-friendly fabrication protocols. Accordingly, spherical culture techniques may be prominent compared to other methods due to their applicability to high content/throughput screening systems [65]. They can be utilized for large sample groups for drug screening studies like we need under pandemic conditions.

CONCLUSION

The review showed that drug screening historical pathways could be examined through three different branches. The first generation of drug repurposing studies was started on animal test platforms. But most animal model applications do not reach the clinical phase and have many pros and cons. The second generation is 2D and 3D cell culture testing platforms. They might help mimic human tissue more precisely compared to the animal model. Lastly, the third-generation studies concentrate on combining the all-testing platforms such as organoids-in-a-chip or ALI-supported chip models. It has been a few decades since 3D cell culture techniques started to be applied for host-pathogen applications as third-generation *In vitro* studies. An enormous number of unique 3D cell culture models differ in fabrication strategies, operation, and extracellular matrix environment for conducting 3D culture of various cell types. These models have enabled endless versatility in the drug repurosing field by providing universal or patient-specific platforms for various cell lines and biological applications. In particular, the models can be used to understand host-pathogen interaction to find a cure for the COVID-19 pandemic and its drug repurposing studies.

Although ALI, spherical, and microfluidics/organ-on-a-chip models have separately shown significantly promising outcomes for developing useful drug screening platforms for COVID-19, several improvements are still required to succeed in more complex *In vitro* systems. The physicochemical conditions, such as the continuous supply of nutrients, gas exchange, cell-cell or cell-environment interactions, and vascularization, are critical for closer recapitulation of the native tissue environment. Comparison of the models used in COVID-19 drug repurposing studies shows various drawbacks of the models individually, but combining them may provide new insight to get more robust results. In our opinion, the combination of the organ-on-a-chip and spherical culture model serves as an alternative technique to repurpose the drugs. Controlling the fluid's viscosity around the spherical structure can present an optimal host-pathogen interaction since the mucosal environment has a significant role in virus retention [66, 67]. The model can also be applied to high throughput/content screening systems to collect more information from the cellular bodies or rapidly complete serial sample analysis [68].

Additionally, combining ALI-spherical culture techniques may provide the most suitable environment for cross-talking among the cells. The hybrid model can be utilized to mimic the cytokine storm for infections. We believe that future cell biology and tissue engineering techniques will provide complex and tissue-mimic *In vitro* systems for drug screening applications, convenient for predicting patient-specific drugs and possible therapeutic effects.

ACKNOWLEDGEMENT

The authors acknowledge funding from the Health Institutes of Turkey awards 7913/8795.

REFERENCES

[1]   Ulm JW, Nelson SF. COVID-19 drug repurposing: Summary statistics on current clinical trials and promising untested candidates. Transbound Emerg Dis 2020; 00(00): 5.
[PMID: 32619318]

[2]   Rumlová M, Ruml T. *In vitro* methods for testing antiviral drugs. Biotechnol Adv 2018; 36(3): 557-76.
[http://dx.doi.org/10.1016/j.biotechadv.2017.12.016] [PMID: 29292156]

[3]   Parvathaneni V, Kulkarni NS, Muth A, Gupta V. Drug repurposing: a promising tool to accelerate the drug discovery process. Drug Discov Today 2019; 24(10): 2076-85.
[http://dx.doi.org/10.1016/j.drudis.2019.06.014] [PMID: 31238113]

[4]   Pushpakom S, Iorio F, Eyers PA, *et al.* Drug repurposing: progress, challenges and recommendations. Nat Rev Drug Discov 2019; 18(1): 41-58.
[http://dx.doi.org/10.1038/nrd.2018.168] [PMID: 30310233]

[5]   Hong KJ, Seo SH. Organoid as a culture system for viral vaccine strains. Clin Exp Vaccine Res 2018; 7(2): 145-8.
[http://dx.doi.org/10.7774/cevr.2018.7.2.145] [PMID: 30112354]

[6]   Wang YC, Lee YT, Yang T, Sun JR, Shen CF, Cheng CM. Current diagnostic tools for coronaviruses–From laboratory diagnosis to POC diagnosis for COVID -19. Bioeng Transl Med 2020; 5(3)e10177
[http://dx.doi.org/10.1002/btm2.10177] [PMID: 32838038]

[7]   Chang R, Sun WZ. Repositioning chloroquine as antiviral prophylaxis against COVID-19: potential and challenges. Drug Discov Today 2020; 25(10): 1786-92.
[http://dx.doi.org/10.1016/j.drudis.2020.06.030] [PMID: 32629169]

[8]   Gupta N, Liu JR, Patel B, Solomon DE, Vaidya B, Gupta V. Microfluidics-based 3D cell culture models: Utility in novel drug discovery and delivery research. Bioeng Transl Med 2016; 1(1): 63-81.
[http://dx.doi.org/10.1002/btm2.10013] [PMID: 29313007]

[9]   Shuai HH, Yang CY, Harn HIC, *et al.* Using surfaces to modulate the morphology and structure of attached cells – a case of cancer cells on chitosan membranes. Chem Sci (Camb) 2013; 4(8): 3058-67.
[http://dx.doi.org/10.1039/c3sc50533b]

[10]   Yang WY, Chen LC, Jhuang YT, *et al.* Injection of hybrid 3D spheroids composed of podocytes, mesenchymal stem cells, and vascular endothelial cells into the renal cortex improves kidney function and replenishes glomerular podocytes. Bioeng Transl Med 2021; 6(2)e10212
[http://dx.doi.org/10.1002/btm2.10212] [PMID: 34027096]

[11]   Giandomenico SL, Mierau SB, Gibbons GM, *et al.* Cerebral organoids at the air liquid interface generate diverse nerve tracts with functional output. Nat Neurosci 2019; 22(4): 669-79.

[http://dx.doi.org/10.1038/s41593-019-0350-2] [PMID: 30886407]

[12]   van Riet S, Ninaber DK, Mikkers HMM, *et al. In vitro* modelling of alveolar repair at the air-liquid interface using alveolar epithelial cells derived from human induced pluripotent stem cells. Sci Rep 2020; 10(1): 5499.
[http://dx.doi.org/10.1038/s41598-020-62226-1] [PMID: 32218519]

[13]   Todd C, Hewitt SD, Kempenaar J, Noz K, Thody AJ, Ponec M. Co-culture of human melanocytes and keratinocytes in a skin equivalent model: effect of ultraviolet radiation. Arch Dermatol Res 1993; 285(8): 455-9.
[http://dx.doi.org/10.1007/BF00376817] [PMID: 8274033]

[14]   Cao XF, Coyle JP, Xiong R, Wang YY, Heflich RH, Ren BP, *et al.* Invited review: human air-liqui--interface organotypic airway tissue models derived from primary tracheobronchial epithelial cells-overview and perspectives. In Vitro Cell Dev Biol Anim 2020; 11(1): 29.
[PMID: 33175307]

[15]   Shpichka A, Bikmulina P, Peshkova M, *et al.* Engineering a Model to Study Viral Infections: Bioprinting, Microfluidics, and Organoids to Defeat Coronavirus Disease 2019 (COVID-19). International Journal of Bioprinting 2020; 6(4): 10-29.
[http://dx.doi.org/10.18063/ijb.v6i4.302] [PMID: 33089000]

[16]   Beg S, Almalki WH, Malik A, *et al.* 3D printing for drug delivery and biomedical applications. Drug Discov Today 2020; 25(9): 1668-81.
[http://dx.doi.org/10.1016/j.drudis.2020.07.007] [PMID: 32687871]

[17]   Yu F, Choudhury D. Microfluidic bioprinting for organ-on-a-chip models. Drug Discov Today 2019; 24(6): 1248-57.
[http://dx.doi.org/10.1016/j.drudis.2019.03.025] [PMID: 30940562]

[18]   Kim J, Koo BK, Knoblich JA. Human organoids: model systems for human biology and medicine. Nat Rev Mol Cell Biol 2020; 21(10): 571-84.
[http://dx.doi.org/10.1038/s41580-020-0259-3] [PMID: 32636524]

[19]   Mahalingam R, Dharmalingam P, Santhanam A, Kotla S, Davuluri G, Karmouty-Quintana H, *et al.* Single-cell RNA sequencing analysis of SARS-CoV-2 entry receptors in human organoids. J Cell Physiol 2020; 9.
[PMID: 32944935]

[20]   Ramos-Lopez O, Daimiel L, Ramírez de Molina A, Martínez-Urbistondo D, Vargas JA, Martínez JA. Exploring Host Genetic Polymorphisms Involved in SARS-CoV Infection Outcomes: Implications for Personalized Medicine in COVID-19. Int J Genomics 2020; 2020: 1-8.
[http://dx.doi.org/10.1155/2020/6901217] [PMID: 33110916]

[21]   Randell SH, Fulcher ML, O'Neal W, Olsen JC. Primary Epithelial Cell Models for Cystic Fibrosis Research.Cystic Fibrosis: Diagnosis and Protocols, Vol Ii: Methods and Resources to Understand Cystic Fibrosis Methods in Molecular Biology 742. Totowa: Humana Press Inc 2011; pp. 285-310.
[http://dx.doi.org/10.1007/978-1-61779-120-8_18]

[22]   Miller AJ, Dye BR, Ferrer-Torres D, *et al.* Generation of lung organoids from human pluripotent stem cells in vitro. Nat Protoc 2019; 14(2): 518-40.
[http://dx.doi.org/10.1038/s41596-018-0104-8] [PMID: 30664680]

[23]   Sachs N, Papaspyropoulos A, Zomer-van Ommen DD, *et al.* Long-term expanding human airway organoids for disease modeling. EMBO J 2019; 38(4)e100300
[http://dx.doi.org/10.15252/embj.2018100300] [PMID: 30643021]

[24]   Choi KYG, Wu BC, Lee AHY, Baquir B, Hancock REW. Utilizing Organoid and Air-Liquid Interface Models as a Screening Method in the Development of New Host Defense Peptides. Front Cell Infect Microbiol 2020; 10: 228.
[http://dx.doi.org/10.3389/fcimb.2020.00228] [PMID: 32509598]

[25] Brandenberger C, Rothen-Rutishauser B, Mühlfeld C, *et al.* Effects and uptake of gold nanoparticles deposited at the air–liquid interface of a human epithelial airway model. Toxicol Appl Pharmacol 2010; 242(1): 56-65.
[http://dx.doi.org/10.1016/j.taap.2009.09.014] [PMID: 19796648]

[26] Strikoudis A, Cieślak A, Loffredo L, *et al.* Modeling of Fibrotic Lung Disease Using 3D Organoids Derived from Human Pluripotent Stem Cells. Cell Rep 2019; 27(12): 3709-3723.e5.
[http://dx.doi.org/10.1016/j.celrep.2019.05.077] [PMID: 31216486]

[27] Vanderheiden A, Ralfs P, Chirkova T, *et al.* Type I and Type III Interferons Restrict SARS-CoV-2 Infection of Human Airway Epithelial Cultures. J Virol 2020; 94(19)e00985-20
[http://dx.doi.org/10.1128/JVI.00985-20] [PMID: 32699094]

[28] Zhu N, Wang W, Liu Z, *et al.* Morphogenesis and cytopathic effect of SARS-CoV-2 infection in human airway epithelial cells. Nat Commun 2020; 11(1): 3910.
[http://dx.doi.org/10.1038/s41467-020-17796-z] [PMID: 32764693]

[29] Mulay A, Konda B, Garcia G Jr, *et al.* SARS-CoV-2 infection of primary human lung epithelium for COVID-19 modeling and drug discovery. Cell Rep 2021; 35(5)109055
[http://dx.doi.org/10.1016/j.celrep.2021.109055] [PMID: 33905739]

[30] Beigel JH, Tomashek KM, Dodd LE, *et al.* Remdesivir for the Treatment of Covid-19 — Final Report. N Engl J Med 2020; 383(19): 1813-26.
[http://dx.doi.org/10.1056/NEJMoa2007764] [PMID: 32445440]

[31] Pruijssers AJ, George AS, Schäfer A, *et al.* Remdesivir Inhibits SARS-CoV-2 in Human Lung Cells and Chimeric SARS-CoV Expressing the SARS-CoV-2 RNA Polymerase in Mice. Cell Rep 2020; 32(3)107940
[http://dx.doi.org/10.1016/j.celrep.2020.107940] [PMID: 32668216]

[32] Uzunova K, Filipova E, Pavlova V, Vekov T. Insights into antiviral mechanisms of remdesivir, lopinavir/ritonavir and chloroquine/hydroxychloroquine affecting the new SARS-CoV-2. Biomed Pharmacother 2020; 131110668
[http://dx.doi.org/10.1016/j.biopha.2020.110668] [PMID: 32861965]

[33] van der Vaart J, Clevers H. Airway organoids as models of human disease. J Intern Med 2021; 289(5): 604-13.
[http://dx.doi.org/10.1111/joim.13075] [PMID: 32350962]

[34] Liu C, Qin T, Huang Y, Li Y, Chen G, Sun C. Drug screening model meets cancer organoid technology. Transl Oncol 2020; 13(11)100840
[http://dx.doi.org/10.1016/j.tranon.2020.100840] [PMID: 32822897]

[35] Li Z, Qian Y, Li W, *et al.* Human Lung Adenocarcinoma-Derived Organoid Models for Drug Screening. iScience 2020; 23(8)101411
[http://dx.doi.org/10.1016/j.isci.2020.101411] [PMID: 32771979]

[36] Zhao B, Ni C, Gao R, *et al.* Recapitulation of SARS-CoV-2 infection and cholangiocyte damage with human liver ductal organoids. Protein Cell 2020; 11(10): 771-5.
[http://dx.doi.org/10.1007/s13238-020-00718-6] [PMID: 32303993]

[37] Thomas H. Organoid modelling of NAFLD. Nat Rev Gastroenterol Hepatol 2019; 16(8): 454-5.
[http://dx.doi.org/10.1038/s41575-019-0181-3] [PMID: 31267040]

[38] Valyaeva AA, Zharikova AA, Kasianov AS, Vassetzky YS, Sheval EV. Expression of SARS-CoV-2 entry factors in lung epithelial stem cells and its potential implications for COVID-19. Sci Rep 2020; 10(1): 17772.
[http://dx.doi.org/10.1038/s41598-020-74598-5] [PMID: 33082395]

[39] Mathilde L, Serge C, Stanislas V, René L, Pierre JD. Developing new drugs that activate the protective arm of the renin–angiotensin system as a potential treatment for respiratory failure in COVID-19 patients. Drug Discov Today In Press

[40]   Li F, He J, Wei J, Cho WC, Liu X. Diversity of epithelial stem cell types in adult lung. Stem Cells Int 2015; 2015: 1-11.
[http://dx.doi.org/10.1155/2015/728307] [PMID: 25810726]

[41]   Fehrenbach H. Alveolar epithelial type II cell: defender of the alveolus revisited. Respir Res 2001; 2(1): 33-46.
[http://dx.doi.org/10.1186/rr36] [PMID: 11686863]

[42]   Barkauskas CE, Cronce MJ, Rackley CR, *et al.* Type 2 alveolar cells are stem cells in adult lung. J Clin Invest 2013; 123(7): 3025-36.
[http://dx.doi.org/10.1172/JCI68782] [PMID: 23921127]

[43]   Gu H, Xie Z, Li T, *et al.* Angiotensin-converting enzyme 2 inhibits lung injury induced by respiratory syncytial virus. Sci Rep 2016; 6(1): 19840.
[http://dx.doi.org/10.1038/srep19840] [PMID: 26813885]

[44]   Yang WY, Chen LC, Jhuang YT, Lin YJ, Hung PY, Ko YC, *et al.* Injection of hybrid 3D spheroids composed of podocytes, mesenchymal stem cells, and vascular endothelial cells into the renal cortex improves kidney function and replenishes glomerular podocytes. Bioengineering Translational Medicine

[45]   Monteil V, Kwon H, Prado P, *et al.* Inhibition of SARS-CoV-2 Infections in Engineered Human Tissues Using Clinical-Grade Soluble Human ACE2. Cell 2020; 181(4): 905-913.e7.
[http://dx.doi.org/10.1016/j.cell.2020.04.004] [PMID: 32333836]

[46]   Lamers MM, Beumer J, van der Vaart J, *et al.* SARS-CoV-2 productively infects human gut enterocytes. Science 2020; 369(6499): 50-4.
[http://dx.doi.org/10.1126/science.abc1669] [PMID: 32358202]

[47]   Lamers MM, van der Vaart J, Knoops K, *et al.* An organoid-derived bronchioalveolar model for SARS-CoV-2 infection of human alveolar type II-like cells. EMBO J 2021; 40(5)e105912
[http://dx.doi.org/10.15252/embj.2020105912] [PMID: 33283287]

[48]   Huang KY, Lin MS, Kuo TC, *et al.* Humanized COVID-19 decoy antibody effectively blocks viral entry and prevents SARS-CoV-2 infection. EMBO Mol Med 2021; 13(1)e12828
[http://dx.doi.org/10.15252/emmm.202012828] [PMID: 33159417]

[49]   Han Y, Yang L, Duan X, Duan F, Nilsson-Payant BE, Yaron TM, *et al.* Identification of Candidate COVID-19 Therapeutics using hPSC-derived Lung Organoids. bioRxiv 2020.
[http://dx.doi.org/10.1101/2020.05.05.079095]

[50]   Pei RJ, Feng JQ, Zhang YC, Sun H, Li L, Yang XJ, *et al.* Host metabolism dysregulation and cell tropism identification in human airway and alveolar organoids upon SARS-CoV-2 infection. Protein Cell 2020; 17.
[PMID: 33314005]

[51]   Riva L, Yuan S, Yin X, *et al.* Discovery of SARS-CoV-2 antiviral drugs through large-scale compound repurposing. Nature 2020; 586(7827): 113-9.
[http://dx.doi.org/10.1038/s41586-020-2577-1] [PMID: 32707573]

[52]   Wan W, Zhu S, Li S, *et al.* High-Throughput Screening of an FDA-Approved Drug Library Identifies Inhibitors against Arenaviruses and SARS-CoV-2. ACS Infect Dis 2021; 7(6): 1409-22.
[http://dx.doi.org/10.1021/acsinfecdis.0c00486] [PMID: 33183004]

[53]   Simoneau CR, Ott M. Modeling Multi-organ Infection by SARS-CoV-2 Using Stem Cell Technology. Cell Stem Cell 2020; 27(6): 859-68.
[http://dx.doi.org/10.1016/j.stem.2020.11.012] [PMID: 33275899]

[54]   Ramani A, Müller L, Ostermann PN, *et al.* SARS -CoV-2 targets neurons of 3D human brain organoids. EMBO J 2020; 39(20)e106230
[http://dx.doi.org/10.15252/embj.2020106230] [PMID: 32876341]

[55] Phan N, Hong JJ, Tofig B, *et al.* A simple high-throughput approach identifies actionable drug sensitivities in patient-derived tumor organoids. Commun Biol 2019; 2(1): 78.
[http://dx.doi.org/10.1038/s42003-019-0305-x] [PMID: 30820473]

[56] Ahidjo BA, Loe MWC, Ng YL, Mok CK, Chu JJH. Current Perspective of Antiviral Strategies against COVID-19. ACS Infect Dis 2020; 6(7): 1624-34.
[http://dx.doi.org/10.1021/acsinfecdis.0c00236] [PMID: 32485102]

[57] Gautret P, Lagier JC, Honoré S, Hoang VT, Colson P, Raoult D. Hydroxychloroquine and azithromycin as a treatment of COVID-19: results of an open label non-randomized clinical trial revisited. Int J Antimicrob Agents 2021; 57(1)106243
[http://dx.doi.org/10.1016/j.ijantimicag.2020.106243] [PMID: 33408014]

[58] Bessière F, Roccia H, Delinière A, *et al.* Assessment of QT Intervals in a Case Series of Patients With Coronavirus Disease 2019 (COVID-19) Infection Treated With Hydroxychloroquine Alone or in Combination With Azithromycin in an Intensive Care Unit. JAMA Cardiol 2020; 5(9): 1067-9.
[http://dx.doi.org/10.1001/jamacardio.2020.1787] [PMID: 32936266]

[59] Jelte van der Vaart MML, Bart L. Haagmans, Hans Clevers. Advancing lung organoids for COVID-19 research. Dis Model Mech 2021; 14(6): 1-6.

[60] Cong Y, Han X, Wang Y, *et al.* Drug Toxicity Evaluation Based on Organ-on-a-chip Technology: A Review. Micromachines (Basel) 2020; 11(4): 381.
[http://dx.doi.org/10.3390/mi11040381] [PMID: 32260191]

[61] Tang H, Abouleila Y, Si L, *et al.* Human Organs-on-Chips for Virology. Trends Microbiol 2020; 28(11): 934-46.
[http://dx.doi.org/10.1016/j.tim.2020.06.005] [PMID: 32674988]

[62] Boulware DR, Pullen MF, Bangdiwala AS, *et al.* A Randomized Trial of Hydroxychloroquine as Postexposure Prophylaxis for Covid-19. N Engl J Med 2020; 383(6): 517-25.
[http://dx.doi.org/10.1056/NEJMoa2016638] [PMID: 32492293]

[63] Si LL, Bai HQ, Rodas M, Cao WJ, Oh CY, Jiang AD, *et al.* A human-airway-on-a-chip for the rapid identification of candidate antiviral therapeutics and prophylactics. Nature Biomedical Engineering 18
[http://dx.doi.org/10.1038/s41551-021-00718-9]

[64] Zhang M, Wang P, Luo R, *et al.* Biomimetic Human Disease Model of SARS-CoV-2-Induced Lung Injury and Immune Responses on Organ Chip System. Adv Sci (Weinh) 2021; 8(3)2002928
[http://dx.doi.org/10.1002/advs.202002928] [PMID: 33173719]

[65] Kang SM, Kim D, Lee JH, Takayama S, Park JY. Engineered Microsystems for Spheroid and Organoid Studies. Adv Healthc Mater 2020; 18.
[PMID: 33185040]

[66] Lu WJ, Liu XQ, Wang T, Liu F, Zhu AR, Lin YP, *et al.* Elevated MUC1 and MUC5AC mucin protein levels in airway mucus of critical ill COVID-19 patients. J Med Virol 2020.
[PMID: 32776556]

[67] Liu L, Liu Z, Chen H, *et al.* Subunit Nanovaccine with Potent Cellular and Mucosal Immunity for COVID-19. ACS Appl Bio Mater 2020; 3(9): 5633-8.
[http://dx.doi.org/10.1021/acsabm.0c00668] [PMID: 35021794]

[68] Sedláková V, Kloučková M, Garlíková Z, *et al.* Options for modeling the respiratory system: inserts, scaffolds and microfluidic chips. Drug Discov Today 2019; 24(4): 971-82.
[http://dx.doi.org/10.1016/j.drudis.2019.03.006] [PMID: 30877077]

CHAPTER 7

Organoids: New Research Tool in Cancer Diagnostics and Therapeutics

Pravin D. Potdar[1,2,*]

[1] *Former Head, Department of Molecular Medicine and Biology, Jaslok Hospital & Research Centre, Mumbai, 400053, Maharashtra, India*

[2] *Chairman, Institutional Ethics Committee, Dr. A. P. J. Kalam Educational and Research Centre, Mumbai, Maharashtra, India*

Abstract: Cancer remains the leading cause of mortality in the world, despite several cutting-edge technologies and established therapeutic regimens for cancer treatment. Therefore, the key to developing accurate and effective therapeutics is having a comprehensive knowledge of these complex molecular events. Patient-derived organoids (PDOs) represent a perfect model for studying cancer drug resistance and therapy. These cancer organoid models are cheaper alternatives to xenograft models and traditional two-dimensional (2D) cell culture model systems. All cancer organoid models are developed using iPSC-derived spheroids and tumor cells from different sources, which are then processed on a matrigel scaffold to get cancer organoids. The major advantage of these model systems is that they can recapitulate many functional and genetic characteristics of the same tumor tissues "*in vitro*". These cancer organoids can be passaged, frozen, and preserved for further high-throughput screening analysis. PDOs are powerful tools for evaluating mutational profiles and testing cancer drugs for personalized therapy. Cancer organoids can also be used to study tumor microenvironment cell types by co-culturing the required cell types involved in the process of transformation, which allows us to study tumor microenvironment and tissue-tissue interactions in the tumor development and metastasis process. This leads to more accurate predictions of the process of tumor development and evaluation of responses of cancer drug-resistance in a particular patient to develop personalized therapies for cancer. However, several limitations to these cancer organoid models must be addressed and resolved to get a perfect system for cancer drug evaluation. Several scientists are working on it by developing standardized protocols and reagents to generate individual tissue organoids. It is hoped that major developments in technologies, such as organoids-on-chips, 3D bio-printing, and advanced imaging techniques, will improve the handling of these organoids more precisely. Further CRISPR-Cas9-based gene editing technology allows us to bioengineer normal organoids by introducing any combination of cancer gene alterations to derive cancer

---

[*] **Corresponding Author Pravin D. Potdar**: Former Head, Department of Molecular Medicine and Biology, Jaslok Hospital & Research Centre, Mumbai, 400053, Maharashtra, India., USA.; Chairman, Institutional Ethics Committee, Dr. A. P. J. Kalam Educational and Research Centre, Mumbai, Maharashtra, India., E-Mail: ppravin012@gmail.com

Manash K. Paul (Ed.)

organoids. In this review, we focused on the development and improvement of various normal and cancer organoids for targeted tissues such as the lung, breast, colon, liver, and kidney and their use as model systems for cancer drug discovery and personalized therapy. We have also highlighted some of the uses of the latest technologies, such as microfluidics chips and 3D bioprinting, for deriving better cancer organoids-based *in vitro* models for future research on cancer therapeutics.

Keywords: Breast cancer, Colon cancer, Cancer organoids, Lung cancer, Personalised therapy, Tumour Microenvironment (TME), 3D organoid model system.

INTRODUCTION

GLOBOCAN 2020 has reported approximately 20 million new cancer cases and almost 10 million people dying from cancer in 2020 globally [1]. Thus, cancer has become one of the leading causes of death worldwide, even though there are many innovative technologies as well as therapeutics protocols established for the cure of cancer [2]. However, there are still many issues that must be addressed in order to improve cancer therapy. So far, cancer has been treated as a global and homogeneous disease, and tumors are considered a uniform population of cells. Thus, the exact understanding of these complex phenomena is the most critical fact in designing precise and efficient therapies. During cancer progression, tumors become highly heterogeneous, having a mixed population of cancer cells with different molecular features with diverse responsiveness to various cancer therapies. The heterogeneity of tumor cells is the key factor responsible for the generation of resistant phenotypes and a critical challenge in managing cancer therapies. Therefore most cancer research programs focus on finding new and efficient therapies that can target cancer stem cells and related drug-resistant cells to reduce drug resistance in these cancer therapeutics. Thus, several different technologies are currently under evaluation to find specific targeted cancer drugs to cure cancer.

For many decades, two-dimensional (2D) cell cultures were considered as the best model system for drug screening and biological assays [3]. These culture systems mainly involve monolayer culture, which grows in flat layers on plastic surfaces. These model systems do not mimic the *in vivo* cell organization due to the lack of cell-cell/cell-matrix interactions in the original tissue architecture and the microenvironment structure around it. Therefore three-dimensional (3D) cell cultures have recently been introduced as a pioneering platform, allowing cell growth and interaction with each other in 3D. Recent advances in 3D culture-based organoid technology have opened new avenues for developing novel *in vitro* model systems resembling human cancer models [4]. These organoid cancer models play an important role in the development of cancer therapeutics to get precise treatment for metastatic cancer patients. Normal organoids can be developed from embryonic stem cells (ESCs) and adult stem cells (ASCs), which recapitulate the original organ architecture of the same organ. These normal organoids can be genetically modified by various innovative gene editing technologies to get disease-specific organoids for therapeutic drug screening and the development of personalized treatment regimes. In this book chapter, we highlight the development of cancer organoids and their use in monitoring cancer therapeutics for metastatic cancer patients. It also explores recent advances in the development of 3D cell culture model systems for developing various organoid cultures for particular cancers on their application in a pharmacological analysis. Finally, special emphasis is paid to organoid bioengineering concepts, 3D constructs that mimic the 3D architecture of intact organs, and the associated technologies to study the exact mechanism of action of therapeutic drugs on the inhibition of cancer growth for a better cure for this disease.

STEM CELLS AND THEIR APPLICATION IN ORGANOIDS RESEARCH

Rheinwald and Green *et al.*, 1975 first cultured human diploid epidermal cells in serial culture using 3T3 lethally irradiated cells as a feeder layer. As epidermal cells required the presence of fibroblast cells for their growth and to initiate colony formation, they have used lethally irradiated 3T3 cells with appropriate density for this culture [5]. Rheinwald and Green have further shown that each colony consists of keratinocytes which ultimately form a stratified squamous epithelium, where dividing cells are confined at the lowest layers. Hydrocortisone was added to the growth medium to make the colony morphology distinctive and maintain proliferation at a slightly greater rate. These epidermal cells have a finite culture lifetime ranging from 20-50 cell generations. Thereafter several studies have developed human cell lines and used them as 2D culture model systems for various drug discovery programs and in various research programs in the field of biomedical sciences [6]. Organoids are the latest innovation in the field of tissue culture technology and stem cell research. Organoids are 3-dimensional tissue culture systems that originate from specific stem cells and recapitulate the structural and functional properties of human organs. These organoids can be established from Adult Stem Cells (ASCs), Embryonic Stem Cells (ESC), and induced Pluripotency Stem Cells (iPSCs). Human organoids have become an excellent model system for studying the exact cellular and molecular mechanism of normal and human diseases and screening drugs for their precise therapeutic use to better cure diseases [7]. Thus, organoid technology has played an important role in the fields of oncology, neurology, and infectious diseases. However, there are many challenges regarding protocol development, cost, and the total time required to set up this culture system for better effectiveness and precise treatment.

Induced pluripotency stem cells are being used for organoid development. Yamanaka and Takanashi, 2006 established a method to develop iPSCs from mouse fibroblasts. They have reprogrammed these fibroblasts by transfecting four pluripotency-inducing transcription factors namely, Oct3/4, Sox2, Klf4, and c-Myc to get iPSCs cells [8]. Dr. Yamanaka received the Nobel Prize for this outstanding work in the year 2012 [9]. Thus, human iPSC technology is currently used in regenerative medicine and various drug discovery programs [10]. With recent developments in CRISPR-Cas9 gene editing technology, we can establish human iPSC cells from any normal or disease cell type to study cellular and molecular mechanisms of this disease to know the cause of the disease [11, 12]. Thus, human iPSC technology is vital in stem cell-based therapy and drug development.

ESCs are present in the early development of human embryos. They have two major properties, self-renewal capacity and pluripotency. These cells have great potential to give rise to most of the human body's tissues. Therefore, ESCs can be used to establish all tissue organoids by supplementing various growth factors in the medium, including the liver, lung, stomach, intestine, kidney, and brain [12 - 14]. Adult Stem Cells (ASCs) are the undifferentiated cells in a tissue or organ located in well-specified microenvironments. These cells have a limited differentiation potential and play an important role in maintaining tissue homeostasis [15]. Several studies have shown that the ASCs having leucine-ric--repeat-containing G-protein-coupled receptor 5 (Lgr5) gene can develop epithelial organoids from normal and diseased organs within one week after culturing these cells. As ASC-derived epithelial organoids are mainly involved in tissue repair, these organoids can only be derived from tissue having regenerative capacity. Thus, ASCs can be used to derive epithelial organoids from the lung, liver, intestine, and pancreas having a regenerative capacity [15, 16].

ORGANOIDS AND THEIR APPLICATIONS IN BIOMEDICAL SCIENCES

Recent developments in stem cell research, enable us to generate self-organizing, stem cell-derived organoids in 3-dimension [17]. Human stem cells and patient-derived iPSCs may be used to create organoids for toxicity, drug discovery, and regenerative medicine [17]. Thus 3-dimensional organoid technology has opened up new avenues for various research and development programs in the field of cancer therapeutics and drug discovery (Fig. **1**). Different aspects of tumor initiation and progression are widely studied using organoid cultures derived from healthy tissues, including the role of pathogens or specific cancer genes and many other diseases. Organoids are considered to be miniature versions of organs, and they often display a very accurate microanatomy of that tissue *in vitro*. This makes them invaluable for *in vitro* modeling to study various drug adverse effects and to understand the susceptibility to drug-induced toxicities. In this review, I have outlined some historical advances in this field and describe some of the major recent developments in 3D human organoid formation from the intestine, liver, brain, lung, kidney, *etc.*

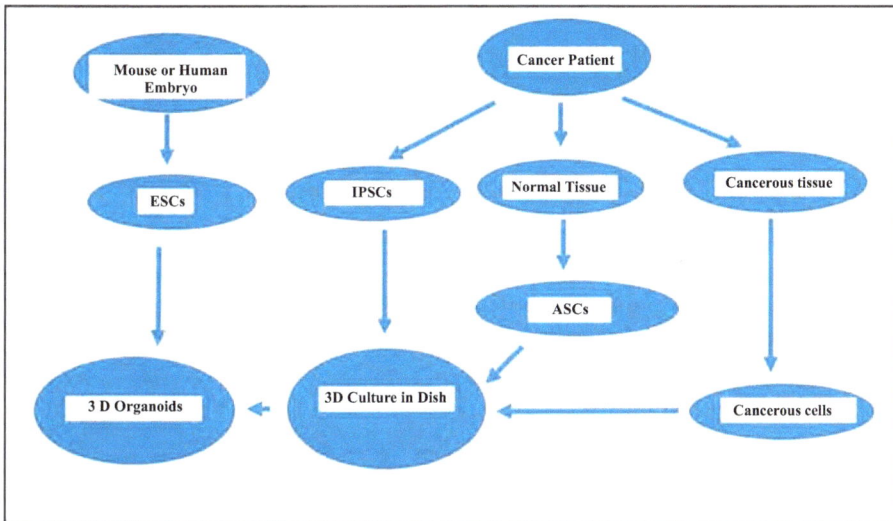

Fig. (1). Schematic diagram showing the basic developmental protocol of 3D normal and cancer organoids from human embryos and cancer patient tissue biopsy.

Hans Clever in 2009 for the first time showed that in adult mammals, the most rapidly self-renewing tissue is the intestinal epithelium, having 4 to 6 crypt stem cells residing above paneth cells in the small intestine [4]. They have further suggested that the Lgr5 gene is one of the intestinal Wnt target genes responsible for its restricted crypt expression. Following this, more studies have revealed the abundant expression of Lgr5 in the cycling columnar cells at the base of the crypt. The Lgr5-positive crypt base columnar cells gave rise to all epithelial lineages over a period of 60-days, suggesting that these cells represent the stem cells of the small intestine and colon. The expression pattern of Lgr5 suggests that it can mark stem cells in multiple adult tissues and cancers. Sasai *et al.* 2013 are the first to introduce an improved 3D aggregate cell culture protocol that permits highly efficient ESC-differentiation into cortical progenitors and functional projection neurons [18]. Then, using this ESC culture, they demonstrated the recapitulation of embryonic corticogenesis and its manipulation *in vitro* in the context of layer-specific neurogenesis and regional specification. Finally, they have demonstrated that cortical progenitors produced in this culture by using both mouse and human ESCs spontaneously formed patterned structures that mimicked the early aspect of corticogenesis. They have discussed the remarkable ability of ESC-derived cortical neuroepithelium with regard to self-organized tissue formation. Lancaster *et al.* 2013 in their Nature publication, first time suggested that due to the complexity of the human brain, there are many difficulties to be faced by clinicians in studying and treating brain disorders. They have further highlighted that for this reason, specific *in vitro* model systems can be developed to study human brain development [19]. In this regard, they have developed human pluripotent stem cell-derived 3D human brain organoids also called cerebral organoids. They have used RNA interference and patient-specific iPSCs to develop a model for microcephaly, a neurological disorder, and it is challenging to study this condition in experimental mice. In this study, they have demonstrated that there is a premature neuronal differentiation in patient organoids developed by them and suggested that this model system can help them explain the disease phenotype [19]. So overall, it shows that there is an excellent possibility of generating any 3-dimensional tissue organoids from ESCs, ASCs, and iPSCs, which can recapitulate with the original tissue architecture in culture and are very useful for future programs in understanding the cellular and molecular mechanism of various diseases. Next, I will review some of the works on normal intestinal organoids, normal liver organoids, normal lung organoids, and normal brain organoids to understand their role in cancer therapeutics.

Normal Intestinal Organoids

The intestines are vital organs subdivided into small and large intestines. The small intestine is mainly divided into the duodenum, jejunum, and ilium. The intestinal crypt contains LGR5+ stem cells, which are slow-growing cells but give rise to highly proliferative and multipotent progenitor transit-amplifying (TA) cells. Mainly digestion occurs in the small intestine, whereas the large intestine's primary function is to absorb water and ions, along with vitamins and short-chain fatty acids (SCFA) synthesized by the gut commensal bacteria. These bacteria help maintain normal gut health and protect the small intestine from pathogenic microbes by enhancing mucus production. The intestinal epithelium comprises several types of cells, which include enterocytes, goblet cells, paneth cells, and enteroendocrine cells mainly derived from LGR5+ stem cells. Self-renewal and differentiation of LGR5+ stem cells are mainly regulated through the various signaling pathways, including Wnt3a, epidermal growth factor (EGF), transforming growth factor α (TGFα), and Notch ligand D114 [20, 21]. Dysregulation of these pathways is responsible for the development of colon cancers [22]. The identification of LGR5+ stem cells and the subsequent characterization of the intestinal stem-cell niche have paved the way for the development of *in vitro* intestinal 3D organoid cultures and the ability to amplify intestinal epithelium [23 - 25].

Hans Clever group has significantly contributed towards developing intestinal organoids since they first developed the same in 2009 [4, 26 - 29]. They have shown that in adult mammals, the intestinal epithelium is the most rapidly self-renewing tissue and Lgr5 gene is one of the responsible intestinal Wnt target genes for its crypt expression [4]. Thus they have shown that Lgr5 expressing stem cells generate all epithelial lineages for over 60 days in multiple adult intestinal tissues and even in cancers. Spence *et al.* 2011 have established 3-dimensional intestine organoids from the differentiation of human iPSCs into intestinal tissue by manipulating various growth factors to mimic embryonic intestinal development [30]. They have shown that these intestinal organoids have polarized columnar epithelium with the presence of goblet cells, paneth cells, and enteroendocrine cells. They have further identified that the combined effect of WNT3A and FGF4 activity permits hindgut morphogenesis [30]. Hannan *et al.* 2013 have developed differentiated human foregut stem cells (hFSCs) from human iPSCs (hiPSCs) and suggested that the expansion of a multipotent foregut progenitor cell may be of great interest with regard to future clinical applications in biomedical research in this field [31]. McCracken *et al.* 2014 have first time reported the 3-dimensional human gastric organoids by using differentiated hESCs and hiPSCs in culture [32]. In their experiment, McCracken *et al.* used iPSCs from gastric cells to form endoderm spheroids in the presence of retinoic acid. When transferred into retinoic acid-supplemented matrigel culture, these spheroid cells formed gastric mucous, endocrine cells, and LGR5-expressing stem cells. Thus, they have made an upcoming 3D organoid model for investigating human gastric diseases [32].

In general, the intestinal mucosa forms the first line of defense against enteric pathogens salmonellae. Forbester *et al.* 2015 used intestinal organoids (iHOs) derived from hiPSCs and explored the interaction of *Salmonella enterica* serovar Typhimurium with iHOs [33]. Forbes *et al.* 2015 further suggested that hiPSC-derived intestine organoids are a promising model system for assessment of the interactions between enteric pathogens such as *S. Typhimurium* [33]. Forbester group in the year 2019 further showed that iHOs had demonstrated differentiation of intestinal epithelium into goblet cells, enteroendocrine cells, paneth cells, and enterocytes, which were confirmed by immunostaining, transmission electron microscopy (TEM), and quantitative PCR (qPCR) [34]. Hill *et al.* 2017 have established a method for the microinjection of hiPSC-derived HIOs and tissue-derived intestinal organoids. They have further demonstrated an approach for analyzing and interpreting the data generated using these methods [35]. The development of human iHO has emerged as a new innovative technology for studying intestinal functions and cellular processes involved in normal and disease conditions [36]. Human intestinal organoid culture has become an essential assay for therapeutic research and drug development programs for intestinal diseases. The three-dimensional self-renewal model of human intestinal organoids showed the presence of an intestinal variety of stem cells and their differentiated epithelial progeny, which allows us for a more robust exploration of cellular activity happening in the small and large intestine *in vivo* than the traditional 2D cell lines used for these studies in gastrointestinal pathology [37]. Thus human intestinal organoids remain a promising 3D model for disease modeling, drug and toxicity testing, and host-pathogen interaction studies in gastrointestinal pathology.

With the use of CRISPR-Cas9 technology, human intestinal organoids hold great potential for human-specific applications like gene therapy, treating chronic inflammatory disorders, and drug discovery programs [38]. Intestinal 3D organoids are being used to study gastrointestinal diseases. However, there are great challenges in modeling organoids for immune cells and microbiota interaction studies. Almeqdadi *et al.* 2019 have recently described the development process of intestinal organoids to study the pathophysiology of intestinal diseases [39]. The microenvironment of the intestine is a very complex system and has continuous cross-talk between the intestinal epithelium, immune cells, enteric microbiota, and intestinal metabolites. Sara *et al.* 2019 have considered these factors and developed intestinal organoids by using a co-culture system and exploring the advantages of organoid technology and their applications in various fields of gastrointestinal pathology [40]. The use of intestinal organoids in biomedical research has proven to be an interesting and useful model system for understanding the cellular and molecular mechanisms involved in gastrointestinal pathology.

Normal Liver Organoids

Significant advances in the field of gastroenterology allow us to understand the pathophysiology of liver diseases. It has been seen that there is a high incidence of liver problems and more than 20 million people die per year from liver diseases. Therefore, there is a great need to investigate the possibility of providing effective therapeutic options to liver patients [41]. Liver diseases mainly include chronic liver viral obesity-related fatty liver degeneration, Wilson's disease, drug-related liver diseases, and liver malignancies [42]. Recently, liver organoid technology has been established for studying the human biology of normal and liver diseases. Three-dimensional mini-organ structures of the liver are grown in a 3D matrix where the structure, functions, and signaling pathways of the liver can be efficiently studied [43]. The liver organoids can be derived from either hiPSCs or liver-specific stem cells [43]. Similarly, the normal liver organoid can be modified further using genome editing and bioengineering technologies to develop patient-specific liver organoids for clinical applications. Thus, liver organoids have opened up a new era for understanding the pathophysiology of liver diseases and can be useful in drug discovery programs. So far, 2D cell culture and animal models have been used to comprehend the molecular mechanisms of liver development and pathogenesis. However, when grown in 2D cultures, primary hepatocyte cultures fail to replicate, have limited division capacity, lose liver-specific gene expression, and cannot maintain stable cytochrome P450 after a few passages [44, 45]. These cultures also lack cell-cell and cell-extracellular matrix (ECM) interactions, which are essential for maintaining biological functions [46].

The liver is the target organ and produces bile, which helps in cholesterol metabolism, promotes blood detoxification, and regulates blood homeostasis. The liver lobule is mainly composed of heterogeneous cell types, with approximately 70% being hepatocytes and 30% cholangiocytes. The main principle of liver organoid technology is to grow liver cells *in vitro* so as to mimic *in vivo* hepatogenesis. Generally, an endoderm is formed during gastrulation, and the liver buds are formed from the foregut endoderm by the induction of various signaling molecules. Once the liver buds are formed, the hepatoblasts undergo expansion and differentiate into hepatocytes and biliary epithelium, and the adjacent mesoderm differentiates into liver fibroblasts and stellate cells. Thus, the maturation of the hepatocytes and cholangiocytes completes the normal structure of liver organoids [47].

iPSC-derived 3D Liver Organoid Models

Takebe *et al.* 2013 were the first to develop hepatic organoids from iPSCs using human mesenchymal stem cell co-culture model systems [48]. They have taken hepatic progenitor stem cells from differentiated iPSCs in a 2D cell culture, and then co-cultured them with human mesenchymal stem cells (MSCs) in Matrigel-embedded culture to get spontaneously iPSC-liver buds (LB). After transplantation of iPSCs LB, human vasculature structures in iPSC-LB became functional, and found that from day 10 until day 45 after transplantation, the engrafted LBs have started secreting albumin into the circulation of the recipient mouse. This demonstrated the ability of LBs to regenerate and rescue drug-induced lethal liver failure [48]. These findings are confirmed further by Camp *et al.* 2017 by single-cell RNA sequencing of 3D liver bud organoids. They found a good similarity between the 3D liver bud and fetal liver cells and confirmed further that the vascular endothelial growth factor (VEGF) crosstalk potentiates endothelial network formation and hepatoblast differentiation by regulating human liver organoid development [49].

Further to these studies, Gaun *et al.* 2017 have developed 3D human hepatic organoids (HOs) model from iPSCs differentiation resembling the human liver during its embryonic development and having both cell types of liver, *i.e.*, hepatocytes and cholangiocytes. This new 3-D liver organoid model can be modified using genome editing technologies to characterize the pathogenesis of human liver hereditary diseases [50]. Wu *et al.* 2019 established functional hepatobiliary organoids without exogenous cells or genetic manipulation from human iPSCs. They suggested that this model will be useful in several aspects of hepatobiliary organogenesis and provide great promise for drug development study and liver transplantation [51]. Akabari *et al.* 2019 have developed and characterized the endoderm-derived hepatic organoids (eHEPO) culture system using human iPSC-derived EpCAM-positive endodermal cells and shown that this system can be developed within 2 weeks and expanded for more than 16 months without any loss of differentiation capacity to mature hepatocytes. Thus, they have suggested that eHEPO is an excellent model system for unlimited cell sources to generate functional hepatic organoids [52].

Collin de l'Hortet *et al.* 2019 have developed fatty acid liver organoids by differentiating edited human iPSCs with a controllable expression of SIRT1 into hepatocytes. The SIRT1 gene plays an essential role in the development of fatty liver in inbred animals. They have further confirmed their finding by knocking down the SIRT1 gene with gene editing technology. They found an increase in fatty acid biosynthesis that exacerbates fat accumulation in liver tissue [53]. Thus this genetically edited human liver organoid may become an important model for investigating human liver biology and disease. Recently Ouchi *et al.* 2019 have used iPSCs from healthy and diseased liver cells to develop reproducible multi-cellular human liver organoids consisting of hepatocyte, stellate, and Kupffer-like cells. These preliminary results of the study by Ouchi *et al.* on developing liver organoids have offered a new avenue for studying liver inflammation and fibrosis in humans at the personalized level and also facilitate the drug discovery program for several liver diseases [54]. Ramali *et al.* 2020 have developed an excellent hepatic organoid model by using human ESCs and human iPSCs in serum-free media that comprise hepatocytes and cholangiocytes and have structural features of the liver. They have confirmed these findings by studying the secretion of albumin, apolipoprotein B, and cytochrome P450 activity by hepatocytes and the secretion of gamma-glutamyl transferase and alkaline phosphatase activity by cholangiocytes. Ramali *et al.* have used immunostaining assay, flow cytometry, and live imaging to confirm these findings [55, 56]. Thus, using various strategies, most of these investigators have used iPSCs to form intermediate endoderm stages for liver organoids giving rise to a functional final liver organoid resembling human liver tissue.

Adult Stem Cell-Derived 3D Liver Organoids

The liver is one of the most regenerative tissues in the human body, which can regenerate fully to recover mass and function after several injuries. It generally regenerates *via* replication of existing cells under certain conditions and *via* differentiation from specialized resident stem cells. Furugama *et al.* 2011 first time reported that Sox9 is expressed throughout the biliary ductal epithelial hepatocyte cells and generated physiologically from Sox9-expressing progenitor stem cells [57]. They have further shown through hepatic injury experiments the involvement of Sox9-positive precursors in liver regeneration [57]. Dorrell *et al.* 2011 reported that SOX9 was exclusively expressed by the subpopulation of normal liver and was highly enriched in injured liver cells. Their study further suggested that liver cells that proliferate during liver regeneration are nothing but the progeny of Sox9-expressing precursor cells [58]. Whereas, Suner *et al.* 2012 reported that liver progenitor cells or biliary cells are terminally differentiated into liver hepatocytes only after chronic liver injury [59]. Kim *et al.* 2005 have already shown that after injection of R-spondin1, there is an extensive proliferative effect on intestinal epithelium *via* Wnt-mediated signaling pathways [60]. It has been shown that the Wnt/β-catenin signaling pathways play essential roles in the self-renewal and maintenance of ASCs. Carmon *et al.* 2011 have shown that R-spondins (RSPOs) are ligands for orphan receptors LGR4 and LGR5 stem cells to regulate *via* the Wnt/β-catenin signaling pathway and thus help in stem cell growth in normal and cancerous organoids [61].

Huch *et al.* 2013 have used the same concept as Carmon *et al.* 2013 to use R-spondins to regulate Wnt/ β-catenin pathways for the expansion of LGR5 cells in 3-dimensional culture systems. They have shown that Lgr5-LacZ cells are not expressed in normal adult liver, but small Lgr5-LacZ(+) cells appeared upon damage indicating the Wnt-signaling activation. They have further shown that these LGR5 cells can be differentiated into functional hepatocytes *in vivo* and *in vitro* [62]. Huch *et al.* 2015 have extended their study and demonstrated the long-term expansion of adult bile duct-derived progenitor cells generated from human liver cells, which can easily be differentiated into functional hepatocytes *in vitro* and *in vivo* and also have stable and normal karyotypes [63]. Whereas, Raven *et al.* 2017 have shown that during liver injury, hepatocyte regeneration can occur from a non-hepatocyte origin [64], and Wnt signaling can influence the homeostatic self-renewal of these hepatocytes [65, 66]. Wang *et al.* 2015 have identified self-renewing diploid Axin2 + liver progenitor cells which express the Tbx3 marker and are different from mature hepatocytes [67]. Hu *et al.* 2018 have established the long-term 3D liver organoid culturing system for mouse and human primary hepatocytes and grown for many months by retaining its morphological and gene expression properties. They have also grown organoids from cholangiocytes [68]. Generally, we have seen that liver organoids are generated *in vitro* from bile-duct epithelial cells and not from hepatocytes. However, Peng *et al.* 2018 used TNFα to promote and expand these hepatocytes in long-term 3D culture for more than 6 months. These results were confirmed by single-cell RNA sequencing for hepatocyte markers [69]. The 3D cell culture systems currently available for liver organoids do not resemble the exact human liver development. Therefore, Vyas *et al.* 2018 have successfully used acellular liver ECM scaffolds to recapitulate the hepatobiliary organogenesis in these liver organoids. Thus Vyas *et al.* have concluded that their study will be used for further studying the mechanisms of hepatic and biliary development and drug discovery programs [70].

Cholangiocyte Organoids

Cholangiocytes are biliary epithelial cells and mainly originate from hepatoblasts. Dainat *et al.* 2014 have developed functional cholangiocytes-like cells by differentiation of human embryonic stem cells (hESCs) and from HepaRG-derived hepatoblasts using specific supplemented growth medium. These cells expressed cytokeratin 7, osteopontin, SOX9, and hepatocyte nuclear factor 6 as makers for cholangiocytes confirming these results. They have suggested that this *in vitro* model system will be useful for studying the molecular mechanisms of bile duct development [71]. Ogawa *et al.* 2015 developed a protocol which developed cholangiocytes by differentiation of human pluripotent stem cells (hPSCs) and showed that this 3-dimensional model gave rise to similar ductal structures. They have further demonstrated that these cholangiocytes possess all epithelial functions and can be used to study biliary development and liver diseases [72]. Sampaziotis *et al.* 2015 developed cholangiocytes-like cells from human induced pluripotent stem cells by using a serum-free growth medium and characterized for various cellular and molecular biomarkers. They have suggested that this model system can be used for drug discovery programs for various liver disorders [73].

Normal Lung Organoids

The lung organoids recapitulate the 3D organizational structure and function of the lung *in vitro* [74 - 76]. It simulates the developmental process of the lung, such as the alveolus, airways, and lung buds. Lung organoids are commonly used to study pulmonary diseases, including Covid 19 infection [77]. These organoids can be used as a model system to study human lung development, processes in regenerative medicine, and drug discovery programs [78]. Shannon *et al.* 1987 have first time grown rat type II cells on Engelbreth-Holm-Swarm (EHS) tumor basement membrane matrix within 4 days in culture and confirmed cell cuboidal morphology [79]. In 1991, Kopf Maier *et al.* cultured lung cancer cells by creating gas-liquid interfaces that differentiate to give rise to lung organoid structures similar to the original lung tissues [80]. The air-liquid interface (ALI) 3D culture system also supports the proliferation and differentiation of airway stem cells and recapitulates a pseudostratified mucociliary epithelial structure of airways. Nettesheim group, in 1993, developed rat tracheal epithelial cell differentiation protocol *in vitro* by using an ALI culture system [81]. Thereafter, Paul Nettesheim and Potdar in 1999, cultured human tracheobronchial epithelial cells using an ALI culture system and demonstrated the presence of ciliated cells and mucous-secreting cells *in vitro* within 14 days of culture using a specific medium having growth factors. Further, they cloned 3 novel genes, GGT-rel, KPL1, and KPL2, during mucociliary differentiation of tracheobronchial epithelial cells [82 - 85].

Rock *et al.* 2009 isolated mouse basal cells by FACS analysis and further described a simple *in vitro* clonal sphere-forming assay where they found self-renewal basal cells that differentiated into ciliated cells. They have suggested that these basal cells are stem cells and have 2 cell-surface markers, ITGA6 and nerve growth factor receptor (NGFR) [86]. Barkauskas *et al.* 2013 have shown that Type 2 alveolar cells (AEC2s) are stem cells with self-renewal and differentiated properties. They have stated that when AEC2s were placed into a 3D culture by co-culturing with primary PDGFRα$^+$ lung stromal cells, it gave rise to self-renewing alveolospheres [87]. Gotoh *et al.* have developed alveolar epithelial spheroids from the differentiation of hPSCs. They have identified carboxypeptidase M (CPM) as a surface marker and shown that CPM+ cells were differentiated into alveolar epithelial cells [88]. Dye *et al.* 2015 first developed normal lung organoids by differentiating human iPSCs and ESCs by manipulating various signaling pathways. Human Lung Organoids (HLOs) consist of well-organized structural features of native lungs. Furthermore, by RNA-sequencing, they have shown that HLOs are similar to human fetal lungs, and therefore, it is an excellent human lung organoids model system to study human lung development and lung diseases [89].

The major factor in the development of lung organoids depends upon the final application of lung organoids. The other factor is the microenvironment in which these cells grow and form organoids, one is Matrigel which differentiates lung cells to form 3D lung organoids [74, 87, 89, 90]. Some investigators have also used various strategies to form microenvironments to produce lung organoids. Wilkinson *et al.* bioengineered lung organoids using multiple cell types and the cell-cell self-assembly property. In a bioreactor, they added iPSCs-derived mesenchymal cells, small airway epithelial cells, human umbilical vein endothelial cells, and collagen-coated alginate beads (scaffold) to build 3D patterns resembling the human lung [91, 92]. The beads offered a niche for the laboratory-grown lung-like tissue and may be utilized to analyze lung problems that are difficult to assess using conventional 2D models [91]. The ALI culture system has also been used by Fulcher *et al.* in 2013, which supports the proliferation and differentiation of airway stem cells and recapitulates only the pseudostratified mucociliary epithelial structure of airways [93]. Konishi *et al.* 2016 used Multi-ciliated airway cells (MCACs) generated by the differentiation of human pluripotent stem cells and formed airway epithelial progenitor cell spheroids. Then with the induction of MCACs, they observed the motile ciliated cells with microtubule arrangement and dynein arms. Thus these lung organoids can be used to study human lung diseases [94]. These types of lung organoids have provided a powerful platform for studying various pulmonary diseases [91, 92, 95].

Normal Brain Organoids

The brain is a complex organ in the human body. Therefore, studying the biological functions of brain development and the mechanism of neurological disorders is challenging. Currently, human brain studies are mainly based on the availability of post-mortem brain tissue; therefore, most studies are available on the use of animal models and nonhuman primates. Thus, there is a great need to establish a brain organoid model system to recapitulate human brain developmental processes and molecular mechanisms underlying various brain disorders [96]. The latest developments in stem cell technologies have provided new platforms for studying various human neurological disorders using human brain organoids [96]. The human brain has several types of cells derived from the neuroectodermal lineage, such as neurons, glial cells, oligodendrocytes, microglia, and vascular cells, which are organized into complex but distinct anatomical structures during the development of the human brain. Recent work has shown that stem-cell-derived 3D brain organoids can recapitulate the same organized structure of the brain *in vitro* [97].

The knowledge to comprehend the mechanism of cerebellar differentiation has promoted the development of *in vitro* generation of cerebellar neurons from ESCs and pluripotent stem cells. Muguruma *et al.* 2010 first reported the development of cerebellar Purkinje cells from mouse ESCs by recapitulating the self-inductive signaling microenvironment in 3D culture using fibroblast growth factor 2 (FGF2] and insulin [98]. Lancaster *et al.* 2013, in their Nature publication, demonstrated a 3-dimensional cerebral organoid model from a human pluripotent stem cell that developed various brain regions including a cerebral cortex, which produces mature cortical neuron subtypes. They have also shown that cerebral organoids can recapitulate an exact structure of human cortical development. They have further developed a 3-dimensional model for microcephaly by using RNA interference and patient-specific induced pluripotent stem cells and showed premature neuronal differentiation [99]. Nicholas *et al.* 2013 first time showed the use of differentiated hPSCs to get medial ganglionic eminence (MGE)-like progenitor cells, which are mature forebrain-type interneurons. They have further shown by both *in vitro* and *in vivo* transplantation that MGE-like cells develop into interneuron subtypes with mature physiological properties, indicating the importance of this model system for the study of human neural development and neurological disorders [100].

After that, Muguruma *et al.* 2015 used human ESC (hESCs) culture to generate human cerebellar tissue organoids *in vitro*, which gave rise to the generation of cerebellar cell type. They have used FGF19 and SDF1 promoting factors which developed cerebellar-type tissue organoids *in vitro* [101]. Pasca *et al.* 2015 have demonstrated a simple and reproducible 3D culture approach for generating human cortical spheroids (hCSs) from pluripotent stem cells, which contain neurons surrounded by nonreactive astrocytes. They have further suggested that these 3D cultures should allow us to study human cortical development, and function, and model disease conditions [102]. Sakaguchi *et al.* 2015 were the first to develop 3D functional hippocampal granule- and pyramidal-like neuron organoids using hESCs [103], which help study hippocampal-related disorders. Quin *et al.* 2016 generated a miniature spinning bioreactor to produce forebrain-specific organoids from human iPSCs. They have also developed protocols for midbrain and hypothalamic organoids. Thus, they have suggested brain-regio--specific organoids, which provide a versatile platform for modeling human brain development, disease mechanisms, and ZIKV antiviral drug testing [104]. Not until 2016 was a 3D organoid model of the midbrain with functioning midbrain dopaminergic (mDA) neurons reported.

Xiao *et al.* have developed a midbrain dopaminergic neuron model by differentiating human pluripotent stem cells. These 3D models have a large multicellular organoid-like structure and contain distinct layers of neuronal cells expressing characteristic markers of the human midbrain. Thus, their human midbrain *in vitro* model system may provide an excellent opportunity to study the human midbrain and its related diseases [105]. Human astrocytes play an important role in health and disease. Sloan *et al.* 2017 first time developed 3D astrocyte lineage cells by using hCS derived from pluripotent stem cells [106]. They found that hCS-derived glia cells look precisely like primary human fetal astrocytes, which formed mature astrocytes. Thus, this astrocyte 3D organoid model can be used for studying normal and developmental neurological disorders [106]. Quadrato *et al.* have studied gene expression in more than 80,000 individual cell types isolated from 31 human brain organoids and found that 3D brain organoids can generate a broad diversity of cells, which include cells from the cerebral cortex to the retinal cell types [107, 108]. Lancaster's group first described the perfect 3D brain organoid culture model, which is efficient and reliable in producing cortical and basal ganglia structures. The neurons within these organoids are functional and display network-like activities. They have further demonstrated the usefulness of this organoid system for *in vitro* modeling of the teratogenic effects of the Zika virus on the developing brain [108]. Kanton *et al.* 2019, in their Nature publication, for the first time analyzed stem cell-derived cerebral organoids from the pluripotency stage to neuroectoderm and neuroepithelial stages, using single-nucleus RNA sequencing analysis. They identified developmental differences in the adult brain, and thus this data provide a temporal cell atlas of significant ape forebrain development and show dynamic gene-regulatory features of the human brain [109].

Further, in 2020, Kanton *et al.* described the assay protocol for the single-cell genomic analysis of human cerebral organoids to correct the batch effects and check the variability in organoids and their gene expression analysis [110]. Fair *et al.* 2020 have profiled the neuroelectrical activities of cerebral organoids by using a multi-electrode array platform and assessed the maturation of several electrophysiologic properties of these organoids. They have also characterized the complex mature neuroelectrical properties with immunohistochemical and RNA single-cell sequencing and suggested that the cerebral organoid model system can be used to evaluate human brain development and neurological disorders [111]. Flacks *et al.* 2021 have developed a collection of computational tools (VoxHunt) to evaluate brain organoid patterning, developmental status, and cell identity by comparing with spatial and single-cell transcriptome reference datasets. They have used VoxHunt to analyze and visualize cell compositions utilizing single-cell and bulk genomic data generated from multiple organoid protocols modeling different brain structures [112].

ORGANOIDS AND CANCER

Despite excellent progress in cancer therapies such as personalized therapy, targeted therapy, and immunotherapy, cancer still has significant high risks of morbidity and mortality worldwide [1]. As cancer is a heterogeneous disease that involves various cell types and subtypes, the treatment program for curing cancer failed significantly. Therefore, there is a need for improved, pre-clinical *in vitro* model systems that can study inter- and intra-tumor heterogeneity. As we know, using human cancer 2D culture models for cancer cell lines and patient-derived xenografts (PDTXs) has several drawbacks. Thus, a 3D organoid culture model system that can almost overcome these challenges is the need of the hour. The last few years have seen incredible innovation in organoid technology and have made significant contributions to overall biomedical research, contributing to several drug discovery programs in cancer research and therapy [113]. Several investigators have developed various protocols to generate organoids from cancer tissues, thus opening up new opportunities in this area. Besides this, many investigators have further incorporated tumor microenvironment cell types into organoid cultures to study the exact role of immune cells in immunotherapy of metastatic cancers. In this book chapter, I am focusing on the development of various tumor organoids from breast cancer, lung cancer, colon cancer, *etc.*, and how they are developed and used to study various targeted drug combinations in therapies for metastatic cancers.

Cancer organoids can be generated from human ASCs obtained from surgical tissue specimens or needle biopsies of various primary or metastatic cancers, including bladder cancer [114, 115], brain cancer [116, 117], breast cancer [118], colon cancer [119 - 122], endometrium [123, 124], head and neck cancer [125, 126], kidney [127, 128], liver cancer [129], lung cancer [130], esophagus cancer [131], ovaries [131, 132], pancreas cancer [133, 134], prostate cancer [135], rectum cancer [136, 137], and stomach cancer [138, 139] (Fig. **2**). However, the major problem in developing these organoids is the overgrowth or contamination by normal epithelial cells [140]. Therefore, cancer organoid cultures can be established ideally from metastatic tissue and mainly avoided from lymph node biopsies, bone biopsies, or ascites fluid. It has been observed that cancer organoids often grow slower than normal organoids. As most of this research has been reported on carcinomas, *i.e.*, tumors raised from epithelial cells, future research should be aimed at developing organoid cultures from non-epithelial cell cancers such as glioblastoma and rhabdoid tumors of the kidney. Organoid cultures can be used to understand the molecular genetics of cancer through Whole Genome Sequencing (WGS), targeted sequencing for cancer gene mutations, and CRISPR-Cas9 gene editing technology [141]. Recently it has been reported that cancer organoid technology can be useful in personalized therapy of Colorectal Cancer (CRC) and Gastro-oesophageal cancers [86]. Ooft *et al.* 2019 have developed patient-derived tumor organoids (PDOs) from metastatic colorectal cancer to test the sensitivity of chemotherapy drugs for treating this cancer. They found that more than 80% of patients responded to the irinotecan-based therapies tested by these organoids [142]. Thereafter, Betge *et al.* 2019 also generated PDOs from colorectal cancer patients and treated them with more than 500 small molecules. They have used a confocal microscope as an imaging system to capture more than 3 million images and found diverse, but recurring phenotypes. Therefore, they have suggested a heterogeneous susceptibility of PDOs to the anticancer drugs used in this study [143].

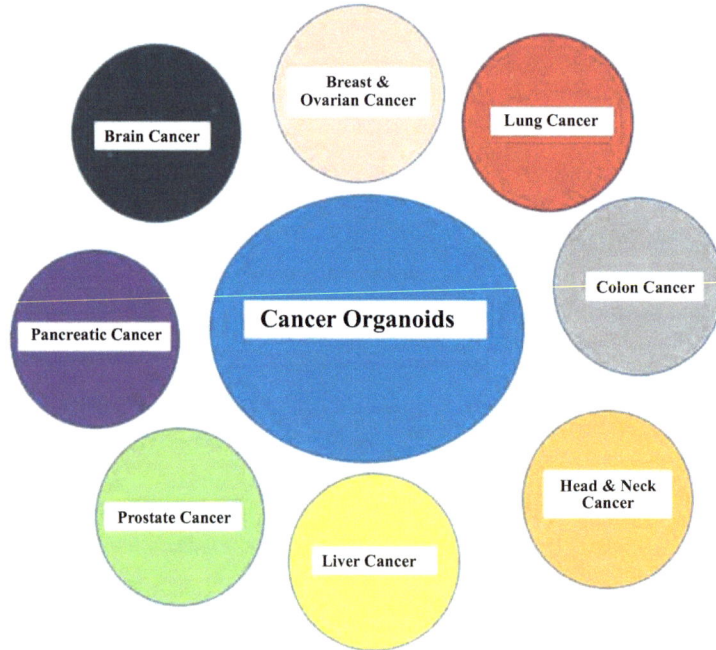

Fig. (2). Three-dimensional (3D) cancer organoids developed from various target organs of the human body.

Another critical factor that needs further attention is the tumor microenvironment (TEM). Hanahan *et al.* have described the hallmarks of cancer in their review published in Cell Journal. They highlighted the six major biological stages acquired during the multistep process of human tumor development and emphasized the tumor microenvironment (TEM) [144]. Junttila *et al.* 2013 in their Nature publication, stated that there is a tremendous therapeutic response depending on the heterogeneity of the tumor microenvironment [145]. So, there is a need for a better understanding of TEM, associated cell types, and mechanistic processes to guide anti-cancer therapy to minimize drug resistance in the management of cancer therapy. Dijkstra *et al.* 2018 have co-cultured autologous tumor organoids and enriched tumor-reactive T cells from peripheral blood of colorectal cancer and non-small-cell lung cancer and found that the co-cultured T cells can kill the matched tumor organoids. Thus, this experimentation can prove that the tumor-reactive T cells can attack tumor cells in tumor organoids and kill them [146]. Cattaneo CM *et al.* 2019 also described the same procedure for co-culturing tumor organoids and autologous tumor-reactive T cells from patients with non-small-cell lung cancer (NSCLC) and colorectal cancer (CRC). Their study established the *Ex vivo* test systems for T-cell-based immunotherapy for individual patients [147].

Neal *et al.* 2018 used an air-liquid interface (ALI) culture system to propagate patient-derived organoids (PDOs) from more than 100 human biopsies and co-cultured them with endogenous tumor-infiltrating lymphocytes (TILs) to form tumor microenvironment (TME) condition. Thus, they have suggested that the organoid-based primary tumor epithelium with endogenous immune stromal cells model system can be used to study drug discovery programs for immunotherapy of metastatic cancer [148]. The use of cancer organoids for cancer therapy still has several problems to overcome, which can be resolved by applying bioengineering approaches to develop hydrogels using artificial matrices. Gjorevsk *et al.* have suggested specific designer matrices for intestinal stem cell organoids for their growth and expansion. In their study, they used synthetic fibronectin-based adhesion as the key ECM component for the expansion and growth of intestinal stem cells (ISC) to form intestinal organoids. However, for differentiation of ISC and organoid formation, laminin-based soft matrix adhesion is required. Thus, they have concluded that at separate stages of growth and differentiation of ISC, there is a need for different mechanical environments and ECM components [149]. Lo *et al.* 2020 have highlighted some of the important methods and applications of organoids for cancer biology and precision medicine [150] Bredennoord *et al.* 2017 have very well described the current state of research and development in the field of cancer organoids and various ethical and regulatory implications for the drug discovery program in cancer therapeutics [151]. Thus, cancer organoid technology will significantly impact drug discovery programs in cancer therapeutics.

Colon Cancer Organoids

CRC is a common and dreadful cancer, and its major causative factors are genetic and environmental. However, it has been observed that the clinical outcome of this disease varies from patient to patient, and yet there is no satisfactory progress in the treatment of this disease [152, 153]. Therefore there is a dire need for developing *in vivo* or *in vitro* model systems that can accurately evaluate genetic diversity and specific tumor types to have more sensitive and effective therapy. Recently, 3D organoid cultures for colon cancer have been developed where Lgr5-expressing stem cells were supplemented with an R-spondin-based culture medium, which maintains a 3D ECM to form a 3D colon organoid culture [154]. Wetering *et al.* 2015 developed a living organoid biobank from 20 untreated CRC patients enrolled in this study. Most of these organoids are developed from normal uninvolved tissue of these patients. These organoids recapitulate the original tumor morphology and specific genetic changes of CRC. These CRC organoids can be used for drug screening for CRC treatment and allow us to prescribe precise, personalized therapy for colorectal cancer [155]. Fujii *et al.* 2016 have established different types of precancerous colon neoplasia in culture without

Wnt/R-spondin and suggested that there is no need for this activation as Wnt signaling can be activated in the niche growth factors secreted during the process of adenoma to carcinoma transformation [156]. Matano *et al.* 2015 used CRISPR-Cas9 genome-editing technology to introduce multiple such mutations in tumor suppressor genes and oncogenes. These mutations include APC, SMAD4, TP53 in tumor suppressor genes and KRAS, PIK3CA in an oncogene. They have modulated culture conditions suitable for the intestinal niche [157]. Drost *et al.* 2016 proposed a similar strategy carried out by Matano *et al.* 2015, where they utilized CRISPR/Cas9 technology to modify 4 targeted genes, mainly APC, P53, KRAS, and SMAD4, in human intestinal stem cells culture. These mutant organoids were selected by removing individual growth factors from the culture medium. The engineered CRC organoids showed substantial chromosomal instability (CIN) and aneuploidy [157]. Xie *et al.* 2016 successfully constructed a CRC organoid model that grew over 25 days in culture [158]. Fumagalli *et al.* 2017 suggested that in the adenoma-carcinoma sequence, the intestinal polyps go through various sets of defined mutations toward metastatic CRC. They have demonstrated this adenoma-carcinoma sequence *in vivo* using an orthotopic CRC organoid model of the human colon. They have further demonstrated these oncogenic mutations in the Wnt, EGFR, P53, and TGF-β signaling pathways, which promote tumor growth, migration, and metastasis [159]. Drost *et al.* 2017 used CRISPR-Cas9 gene editing technology for mutation in the mismatch repair gene MLH1, and thus this became an accurate model for mutation profiles observed in mismatch repair-deficient colorectal cancers. They have further suggested that this same strategy can be applied to cancer predisposition gene NTHL1, which encodes a base excision repair protein in CRC [160]. Verissimo *et al.* 2016 have developed patient-derived CRC organoid libraries for evaluating drug combinations for therapy. They have used normal, and RAS mutated tumor organoids by introducing oncogenic KRAS mutation by CRISPR-Cas9 technology to evaluate RAS pathway inhibitors. They have shown that the mutant RAS tumor organoids were highly resistant to drug combinations used in this study [161]. PDOs have been recently introduced as robust preclinical models for drug testing; however, their potential to predict clinical outcomes in colon cancer patients has remained unclear. Vlachogiannis *et al.* 2018 have developed PDOs from pre-treated metastatic colorectal and gastroesophageal cancer patients recruited in phase I and II clinical trials. The phenotypic and genotypic profiling of PDOs have exhibited similarities to original patient tumors. The molecular profiling of tumor organoids was also matched to drug-screening results indicating that the PDO model can replicate patient responses in the clinic and might be implemented in personalized medicine programs for the treatment of colon cancer [162]. Ooft *et al.* 2019 developed PDOs, which can predict response to chemotherapy in metastatic CRC patients. Their prospective clinical study has shown the feasibility of generating and testing PDOs for evaluating sensitivity to chemotherapy. They further suggested that the PDOs failed to predict the outcome for treatment with 5-fluorouracil plus oxaliplatin, but PDOs might be used to prevent cancer patients from inefficient irinotecan-based chemotherapy [163].

Rectal cancer (RC) is a challenging disease to treat through conventional chemotherapy, radiation therapy, and surgery. So far, no accurate RC model exists to answer fundamental research questions relevant to these cancer patients. Ganesh *et al.* from the Memorial Sloan Kettering Cancer Centre established a biorepository of 65 RC organoids (RCOs) from primary, metastatic, and recurrent cancer cases. The RC organoids retained molecular characteristics of the tumors from which they were derived, and their *Ex vivo* responses to clinically relevant chemo- and radiation therapy correlated well with the clinical responses noted in individual patients' tumors, suggesting that this model is crucial for monitoring therapeutic manipulation in RC [164]. Yao *et al.* 2020, in their recent co-clinical trial data from Fudan University, Shanghai Cancer Centre, have generated a living organoid biobank from patients with locally advanced RCs treated with neoadjuvant chemoradiation for a phase III clinical trial. They have indicated that their RCO model closely recapitulates the pathophysiology of corresponding tumors [165]. Therefore, PDOs have great potential in CRC research. These model systems can help make clinical decisions to improve patient survival with appropriate CRC treatment. However, these organoids in CRC research need further investigation to bring these model systems mainly in clinical trials.

Lung Cancer Organoids

According to GLOBOCAN 2020, lung cancer is a leading cause of cancer death globally, with less than 5% of patients surviving for 5 years. It is mainly due to the late diagnosis of this disease and the non-availability of a proper drug monitoring system to check drug resistance [166]. Therefore developing *in vitro* and *in vivo* model systems that will accurately recapitulate patients' tumors from their biological profiling will be vital to finding effective new therapies. 3D culture technologies with the addition of specific growth factors mimicking the stem cell niche led to the development of 3D organoids [20]. Several investigators have also established lung cancer organoids (LCOs) [167, 168]. These organoids retain their morphological and many other characteristics present in the original tumor tissue; therefore, this model system can be used for drug discovery programs in cancer therapeutics. Sach *et al.* 2019, developed human airway lung organoids from broncho-alveolar tissue and lavage material, which showed the pseudostratified epithelial with basal cells, ciliated cells, and mucus-producing secretory cells. These organoids were established from metastatic lung cancer biopsies and retained tumor histopathology and cancer gene mutations. Thus they concluded that the human airway organoids could be used as an *in vitro* model system for drug discovery programs for metastatic lung cancer therapy [167].

PDOs represent a promising preclinical cancer model to evaluate the targeted drugs for lung cancer treatment. Takahashi *et al.* 2019 produced PDOs tumor organoids from human lung tumors to recapitulate the tissue architecture and function *in vitro* for evaluating molecularly targeted drugs to be used for the treatment of lung cancer patients, which include small-molecule inhibitors, monoclonal antibodies, and an antibody-drug conjugate. They have evaluated the epidermal growth factor receptor and human epidermal growth factor receptor 2 (HER2) inhibitors for effective lung cancer treatment. Overall they have suggested that lung tumor PDO is a suitable *in vitro* system for evaluating molecular targeted drugs used for therapy of lung cancer [169]. Kim *et al.* 2019 have established lung cancer organoids and normal bronchial organoids from individual patient tissues consisting of five histological subtypes. They have shown that normal bronchial organoids maintain cellular components of the normal bronchial mucosa, and LCOs respond to drugs based on their genomic alterations to respective targeted therapies, *i.e.*, olaparib, erlotinib, and crizotinib. They have further suggested that this LCO-based model system may be useful for predicting patient-specific drug responses for personalized therapy for lung cancer patients [170].

Microfluidic devices provide a potential alternative to animal experiments due to their ability to mimic physiological parameters. Microfluidic devices accurately recapitulate the complex organ-level physiological and pharmacological responses. Jung *et al.* 2019 have developed a one-stop microfluidic device system for the first time, allowing us to perform both 3D LCO culturing and drug sensitivity tests directly on a microphysiological system (MPS). They have demonstrated the production of LCOs from patients with small-cell lung cancer, which can rapidly proliferate and exhibit disease-specific characteristics on MPS. They have suggested that this microfluidic lung cancer organoid system may provide important guidelines for therapeutic approaches to lung cancer at the preclinical level [171]. Shi *et al.* 2020 took thirty surgically resected NSCLC primary patient tissues and 35 previously established patient-derived xenograft (PDX) models and processed them for organoid culture establishment. Organoids were characterized by cellular and molecular pathology, and these organoids were subjected to drug testing using EGFR, FGFR, and MEK-targeted therapies. Their study showed that NSCLC organoids closely recapitulate the genomics and biology of patient tumors and are a potential platform for drug testing and biomarker validation [172].

Clinical applications of tumor organoids for personalized medicine require pure tumor organoid cultures. Dijkstra *et al.* 2020 have tried to develop pure organoid cultures from 70 NSCLC samples by evaluating several methods to identify the tumor purity of organoids established. They found that only 17% of pure NSCLC organoids are developed without overgrowth of the normal airway organoids. Therefore, current methods are inadequate to generate pure NSCLC organoids from intrapulmonary lesions, and therefore further essential steps should be taken to prevent overgrowth by normal airway organoids in further experimentation [173]. Li *et al.* 2020 have established 12 PDOs from lung adenocarcinoma (LADC) and confirmed that tumor organoids developed by them had retained the histological, genomic, and gene expression profiles of their parental tumors. These patient-derived LCOs can be used to monitor drug response in lung cancer patients in personalized medicine [174]. Chen *et al.* 2020 have established and characterized the consistency between primary tumors in NSCLC and PDOs. They have used 26 antineoplastic drugs and tested them in these PDOs, which have retained lung cancer tumors' morphological and genetic configuration. They have demonstrated that these PDOs are useful in predicting treatment response for personalized medicine in NSCLC [175]. Li *et al.* 2020 aimed to establish a living biobank of PDOs from NSCLC patients and study the responses to cancer drugs. They successfully developed a living NSCLC organoids biobank from ten NSCLC patients with similar pathological features to primary tumors and used it for testing various anticancer drugs. They have further suggested that the living biobank of PDOs from NSCLC patients might be helpful for high-throughput drug screening and personalized therapy [176].

Several investigators have developed lung cancer PDOs and the genomic profile of lung cancer. Kim *et al.* 2021 have evaluated the efficacy of PDOs to predict clinical responses to targeted therapies and to find effective anticancer therapies for novel molecular targets for individual lung cancer patients. They have generated eighty-four organoids from patients with advanced LADC and carried out whole-exome sequencing, and RNA sequencing for each organoid in response to mono or combination targeted therapies. They have demonstrated that PDOs in advanced LADC are an essential diagnostic program to help clinicians develop precise therapeutic strategies for lung cancer [177]. Yakota *et al.* 2021 used Airway Organoids (AOs) media developed by Cleavers laboratory to develop three lung tumoroid lines for long-term culture for almost 13 months from forty-one primary and metastatic lung cancer cases. They have used nutlin-3a for selecting lung tumoroids that have mutant p53 in order to eliminate normal lung epithelial organoids. They treated these tumoroids with trametinib/erlotinib for BRAFG469A, crizotinib/entrectinib for TPM3-ROS1, and ABT-263/YM-155 for EGFR L858R/RB1E737 targeted drugs and showed that there was a significant suppression of growth of each lung tumoroid line by these therapies. They have suggested that long-term lung tumoroid culture is feasible to identify NGS-based therapeutic targets and determine the response to targeted therapies [178].

Hu *et al.* 2021 produced long-term LCOs from tumor biopsy tissues and showed that they could recapitulate the histological and genetic features of parental tumors. They have used an integrated superhydrophobic microwell array chip (InSMAR-chip) and demonstrated hundreds of LCOs to produce clinically meaningful drug responses within a week. This LCO model, coupled with the microwell device, provides patient-specific drug responses in clinical settings [179]. Cisplatin is known to develop drug resistance in cancer patients. Li *et al.* 2021 have shown that Halofuginone (HF) plays a vital role as an antitumor agent and cisplatin sensitizer. They have combined cisplatin with HF and demonstrated that HF could sensitize the cisplatin-resistant patient-derived LCOs to cisplatin treatment, indicating that HF is a cisplatin sensitizer and a dual pathway inhibitor that may improve the prognosis of patients with cisplatin-resistant lung cancer [180]. Liu *et al.* 2021 have established a Superhydrophobic Microwell array chip (SMART-chip) for LOC based drug testing. This is a novel *in situ* cryopreservation technology that preserves the viability of these organoids for further testing. They have demonstrated that using this chip, the cryopreserved organoids can be recovered better, eliminate the harvesting and centrifugation steps, and make the whole freeze-thaw process easier for drug sensitivity testing [181].

Ma *et al.* 2021 have decided to identify crucial genes involved in LUAD and lung squamous cell carcinoma (LUSC) using lung cell lines and lung cancer organoids by differentially expressed gene analysis. They have shown that in LUAD carcinogenesis, CDK1, CCNB2, and CDC25A are the major differentiated genes and thus can be used as potential biomarkers to target LUAD [182]. So, we have seen here that LCOs have the most promising applications in personalized medicine, which can resolve the drug-resistant problem in lung cancer therapies. However, there are still several challenges and problems that need to be addressed to improve the LCOs models, including the optimal formula of culture medium, and the second is to improve the methods to derive LCOs and prevent normal growth of airway organoids in culture. Thus, this model will be useful in understanding research and development for discovering new agents for precise lung cancer treatment.

Breast Cancer Organoids

Breast cancer (BC) is the most common cancer worldwide. In 2020, more than 2.3 million women were diagnosed with BC worldwide, and 685,000 died from this cancer. To further the understanding of the mechanism of BC tumorigenesis and progression, it is emphasized that in the human BC organoid models, there should be an understanding of the breast tumor tissue microenvironment, which could help elucidate the underlying mechanisms of BC risk factors and also suggest the precisely targeted therapy to cure this cancer. Recent developments in 3D BC organoids suggest that they can accurately recapitulate the *in vivo* breast microenvironment and allow us to examine factors that affect signal transduction, gene expression, and tissue remodeling during BC development and progression. Walsh *et al.* 2014 used Optical Metabolic Imaging (OMI) to detect metabolic changes that occur with cellular transformation after the treatment with anti-cancer drugs in primary human tumor-derived BC organoids. It has predicted the therapeutic response of xenografts and measured anti-tumor drug responses within 24 hours after the treatment with clinically relevant anti-cancer drugs. They have suggested that optical metabolic imaging can potentiate a high-throughput screening to test the efficacy of a panel of anti-cancer drugs to select optimal drug combinations in BC therapy [183].

Sokol *et al.* 2016 developed primary human BC organoids using 3D hydrogels in a serum-free growth medium and the growth of these organoids was observed under bright-field and fluorescence microscopy. They have demonstrated that the 3D hydrogel scaffolds support the growth of complex mammary tissues and will be a useful model system for studying human mammary gland development [184]. Sowder *et al.* 2017, in their review article, highlighted the importance of the 3D BC organoid model system in BC research and therapy. They have stated that this model system can be utilized for BC drug discovery programs to find appropriate and precise therapy for BC. BC organoids could open new avenues of reliable drug discovery and may be useful in resolving the therapy for triple-negative BC [185]. It is known that BC consists of multiple distinct subtypes that differ genetically, pathologically, and clinically. Sach *et al.* 2019 developed more than a hundred primary and metastatic BC organoid lines and showed that all BC organoids have major gene-expression-based classification and allow *in vitro* drug screening. This study further describes a collection of well-characterized BC organoids that are available for drug development and assess drug response *in vitro* for personalized medicine [186]. Studies have shown that in breast cancer, there is an interaction between BRCA1 and PR (Progesterone). However, there is no information about how BRCA1 influences PR signaling in breast tissue.

Davaadelger *et al.* 2019 have developed a BC organoid model system that can reveal the role of PR in the BRCA1 gene in BC. Thus, this model system will help understand the molecular mechanism of BRCA1 and PR crosstalk in BC patients [187]. The development of PDOs from normal and cancerous breast tissue is urgently required today for drug discovery programs for BC therapy. Mazzucchelli *et al.* 2019 developed a new protocol to establish PDO from normal and tumor biopsy samples of BC. This study recruited 33 BC patients and processed them using the specific protocol developed to get PDO from cancerous and healthy mammary tissues. They have reported that the success rate from healthy samples was about 20.83%, while it was 87.5% for cancer tissue. They have confirmed that this is a novel, simple, and quick method to get PDO from normal and tumor biopsies of BC [188]. Djomehri *et al.* 2019, described the production of large scaffold-free, spherical organoids using an MCF10A cell line. These organoids recapitulate the normal human breast with a hollow lumen and secondary acini. When subjected to TGF-β, CoCl2 treatment, or co-culture with MSCs, these organoids produced collagen I. This scaffold-free, 3D breast organoid model can be used as a standardized 3D cellular model and has several advantages in studying tumor microenvironmental factors influencing breast tumorigenesis and therapy [189]. Goldhammer *et al.* 2019 have developed an assay to distinguish between malignant and non-malignant structures in primary BC organoid cultures. They found that there were residual non-malignant tissues in primary human BCs in five biopsies out of ten biopsies. They have used various molecular markers to differentiate malignant and non-malignant organoids derived from primary human BC and concluded that organoid cultures of human BCs are most representative of the origin of tissue in primary culture [190].

The ECM and TME in tumor tissues are important cell growth and differentiation mediators. So far, no *in vitro* model systems have been developed incorporating tissue-specific microenvironments. Mollica *et al.* 2019, described a novel mammary-specific culture protocol using hydrogel and decellularized rat or human breast tissues. These ECM hydrogels retained unique structural and signaling profiles of mammary-specific organoids. Using their bioprinter, they have established large organoids from all mammary-derived hydrogel by using bioprinter. These organoids in a tissue-specific matrix provide a suitable platform to study ECM and epithelial cancer cell interaction [191]. It is now established that TME cell types significantly influence cancer cell progression and metastasis. However, no clear-cut evidence has been postulated in this regard so far.

Truong *et al.* 2019 have used previously developed microfluidic devices for generating 3D *in vitro* organotypic models to study tumor interactions by replicating the spatial organization of the TME on a chip. They have cocultured BC cells and patient-derived fibroblast cells and studied cancer cell migration. They have shown that cancer-associated fibroblasts (CAF) enhance invasion by the expression of the glycoprotein nonmetastatic B (GPNMB) gene in BC organoids, which is responsible for increasing the speed of CAF cells. However, by knocking down the GPNMB gene, the migration speed of CAF was reduced significantly, indicating the ability of this model to recapitulate patient-specific tumor microenvironments to investigate the cellular and molecular mechanism in tumor-stromal interactions in BC [192]. A similar study was carried out by Reid *et al.* 2019, to study TME cell-type interaction in BC. They have described a low-cost bioprinting platform. Bioprinters can significantly increase organoid formation in 3D collagen gels. Thus they have first time demonstrated their 3D bioprinting platform to study tumorigenesis and tumor microenvironment cell types in BC organoids [193]. Nayak *et al.* 2019 established a 3D culture platform for primary BC cells by decellularizing CAFs cultured on 3D microporous polymer scaffolds to recapitulate tumor behavior and drug response. They have shown that the presence of the CAF-derived ECM deposited on the polycaprolactone scaffold promotes cell attachment and viability. They have further studied the response of two chemotherapeutic drugs, doxorubicin, and mitoxantrone, which varies significantly across patient samples. Thus this model system could be used as an *Ex vivo* platform to culture primary BC cells toward developing effective and personalized therapy for BC patients [194].

Due to the heterogeneity of BC, it is always challenging to treat these patients. Dekkers *et al.* 2020 have demonstrated that breast epithelial organoids can be generated from human reduction mammoplasties. They have used regularly clustered interspaced short palindromic repeats CRISPR-Cas9 for targeted knockout of four BC-associated tumor suppressor genes, *i.e.*, P53, PTEN, RB1, and NF1 in mammary progenitor cells, which recapitulate *de novo* oncogenesis. These mutant organoids gained long-term culturing capacity, formed estrogen-receptor-positive luminal tumors, and responded to endocrine or chemotherapy. Thus they have suggested the potential utility of this model system to enhance our understanding of the molecular mechanism involved in specific subtypes of BC [195]. Recently Yu *et al.* 2020 have very well reviewed the development of BC organoids culture system, their sources and clinical applications of these organoids in drug discovery programs for BC. They have stated that there is a significant gap between traditional cell lines and BC solid tumor organoids.

However, patient-derived tumor cells, combined with 3D culture technology by adding cytokines, promote the BC stem cell proliferation and inhibition of their apoptosis and may resemble tumors in the body. This model can be more prominent in clinical application and screening of drugs by high-throughput analysis for personalized medicine and immunotherapy [196]. Thus 3D BC organoids are now a promising tool for precision cancer medicine. Campaner *et al.* 2020 have established PDOs from 4 different BC subtypes, which recapitulate similar morphology and genetic configuration with parental tumors. They have further shown that the activation of the Yes-associated protein 1 (YAP1) gene was responsible for restoring the chemosensitivity of drug-resistant breast cancer organoids. Thus, they have suggested that these breast cancer organoids are an important model for identifying drug-resistant populations within breast cancer tumors [197].

Rosenbluth *et al.* 2020 have provided an extensive evaluation of the capability of organoid culture technology to retain complex stem/progenitor and differentiated cell types of normal human mammary tissues. They have shown that basal/stem and luminal progenitor cells can be differentiated in culture to generate basal and luminal cell types, including ER+ cells. Single-cell analyses by mass cytometry for 38 markers have provided multiple mammary epithelial cell types in the organoids and demonstrated that protein expression patterns of the tissue of origin could be preserved in culture. Thus, organoid cultures can provide a valuable platform for studies of mammary differentiation, transformation, and BC risk [198]. Recently, Dekkers *et al.* 2021 have provided an optimized, highly versatile protocol for long-term culture of forty-five bio-banked organoid samples derived from either normal human breast tissues or triple-negative, estrogen receptor-positive, progesterone receptor-positive, and HER2-positive BC tissues. These organoids were genetically modified and transplanted orthotopically in mice. They observed that the organoid derived from tissue fragments until the first split takes 7-21 days, genetically manipulated clonal organoid cultures take 14-21 days and organoid expansion for xenotransplantation takes more than 4 weeks [199]. Thus, 3D BC organoids will have tremendous applications in drug discovery programs for BC therapeutics and the model systems can be modified to develop in such a way that we can study the role of TME cell types involved in BC development and resistance to cancer therapies.

CONCLUSION

GLOBOCAN 2020 reported that 20 million people are currently suffering from cancer, and almost 10 million people are dying from cancer even though many innovative technologies and therapeutics protocols have been established to cure cancer. However, there are still many issues that must be addressed in order to improve cancer therapy. Therefore, the exact understanding of these complex phenomena is the most crucial fact in order to design precise and efficient therapies. The heterogeneity of tumor cells is the major key factor responsible for the development of resistant phenotypes in the management of cancer therapies. Therefore, most cancer research programs focus on finding new and efficient therapies that can target cancer stem cells and related drug-resistant cells to reduce drug resistance in these cancer patients. 3D cell culture technology has been introduced as a revolutionary platform in recent years, enabling cells to grow and interact with each other in all 3-dimensions. Organoids are 3D structures that are grown from stem cells *in vitro*. They recapitulate many structural, functional, and genetic aspects of their *in vivo* counterpart organs. It is now feasible to produce indefinitely expanding organoids from tumor tissue of patients suffering from a range of carcinomas. TME is one of the critical areas to be examined while evaluating cancer therapeutics.

Cancer organoids become a perfect model for studying the TME by co-culturing immune cells and fibroblasts within cancer organoid technology, enabling immune-oncology applications. Several recent studies have shown excellent documentation about the role of cancer organoids in accurately predicting drug responses in personalized medicine. However, several limitations to these cancer organoid models need to be addressed and resolved to get a perfect system for cancer drug evaluation. Several scientists are working on it by developing standardized protocols and reagents to generate individual tissue organoids. Besides, the major developments in technologies, such as organoids-on-chips, 3D bioprinting, and imaging techniques, improved the handling of these organoids more precisely. Some recent studies have shown that the CRISPR-Cas9-based gene editing technology could be useful to engineer normal organoid models by introducing any combination of cancer gene alterations to derive cancer organoids to evaluate the role of that particular gene in developing drug resistance in respective cancer cells. This book chapter has mainly covered specifically targeted cancers, such as lung cancer, colon cancer, BC, and the role of organoids in drug discovery programs for cancer therapeutics. Overall this book chapter has highlighted several such studies to improve the cancer organoid technology as a more powerful tool for elucidating its role in evaluating drug combinations for precise treatment of cancer without any side effects developed due to drug resistance in these cancer patients.

ACKNOWLEDGEMENTS

I am thankful to Dr. Manash Paul, Ph.D. senior scientist and the principal investigator from UCLA, USA, for giving me this opportunity to write a book chapter on such an important topic of "Organoids and Cancer".

REFERENCES

[1]   GLOBOCAN 2020: New Global Cancer Data https://gcoiarcfr/today/data/factsheets/populations/900-world-fact-sheetspdf 2020.

[2]   Li H, Zhao SL, Cohen P, Yang DH. Editorial: New Technologies in Cancer Diagnostics and Therapeutics. Front Pharmacol 2021; 12760833
[http://dx.doi.org/10.3389/fphar.2021.760833] [PMID: 34552497]

[3]   Ryan J. Introduction to Animal Cell Culture. Technical Bulletin Corning 2003; 2003: 1-8.

[4]   Sato T, Vries RG, Snippert HJ, *et al.* Single Lgr5 stem cells build crypt-villus structures *in vitro* without a mesenchymal niche. Nature 2009; 459(7244): 262-5.
[http://dx.doi.org/10.1038/nature07935] [PMID: 19329995]

[5]   Rheinwatd JG, Green H. Seria cultivation of strains of human epidemal keratinocytes: the formation

keratinizin colonies from single cell is. Cell 1975; 6(3): 331-43.
[http://dx.doi.org/10.1016/S0092-8674(75)80001-8] [PMID: 1052771]

[6]   Kapałczyńska M, Kolenda T, Przybyła W, *et al.* 2D and 3D cell cultures – a comparison of different types of cancer cell cultures. Arch Med Sci 2016; 14(4): 910-9.
[http://dx.doi.org/10.5114/aoms.2016.63743] [PMID: 30002710]

[7]   Azar J, Bahmad HF, Daher D, *et al.* The Use of Stem Cell-Derived Organoids in Disease Modeling: An Update. Int J Mol Sci 2021; 22(14): 7667.
[http://dx.doi.org/10.3390/ijms22147667] [PMID: 34299287]

[8]   Takahashi K, Yamanaka S. Induction of pluripotent stem cells from mouse embryonic and adult fibroblast cultures by defined factors. Cell 2006; 126(4): 663-76.
[http://dx.doi.org/10.1016/j.cell.2006.07.024] [PMID: 16904174]

[9]   John B. 2012. https://www.nobelprize.org/prizes/medicine/2012/press-release/ Gurdon and Shinya Yamanaka(2012). The Nobel Prize in Physiology or Medicine 2012 for the discovery that mature cells can be reprogrammed to become pluripotent

[10]   Doss MX, Sachinidis A. Current Challenges of iPSC-Based Disease Modeling and Therapeutic Implications. Cells 2019; 8(5): 403.
[http://dx.doi.org/10.3390/cells8050403] [PMID: 31052294]

[11]   Ben Jehuda R, Shemer Y, Binah O. Genome Editing in Induced Pluripotent Stem Cells using CRISPR/Cas9. Stem Cell Rev 2018; 14(3): 323-36.
[http://dx.doi.org/10.1007/s12015-018-9811-3] [PMID: 29623532]

[12]   McCauley HA, Wells JM. Pluripotent stem cell-derived organoids: using principles of developmental biology to grow human tissues in a dish. Development 2017; 144(6): 958-62.
[http://dx.doi.org/10.1242/dev.140731] [PMID: 28292841]

[13]   Clevers H. Modeling Development and Disease with Organoids. Cell 2016; 165(7): 1586-97.
[http://dx.doi.org/10.1016/j.cell.2016.05.082] [PMID: 27315476]

[14]   Schutgens F, Clevers H. Human Organoids: Tools for Understanding Biology and Treating Diseases. Annu Rev Pathol 2020; 15(1): 211-34.
[http://dx.doi.org/10.1146/annurev-pathmechdis-012419-032611] [PMID: 31550983]

[15]   Clevers HC. Organoids: Avatars for Personalized Medicine. Keio J Med 2019; 68(4): 95.
[http://dx.doi.org/10.2302/kjm.68-006-ABST] [PMID: 31875622]

[16]   Li Y, Tang P, Cai S, Peng J, Hua G. Organoid based personalized medicine: from bench to bedside. Cell Regen (Lond) 2020; 9(1): 21.
[http://dx.doi.org/10.1186/s13619-020-00059-z] [PMID: 33135109]

[17]   Xinaris C, Brizi V, Remuzzi G. Organoid Models and Applications in Biomedical Research. Nephron J 2015; 130(3): 191-9.
[http://dx.doi.org/10.1159/000433566] [PMID: 26112599]

[18]   Sasai Y. Next-generation regenerative medicine: organogenesis from stem cells in 3D culture. Cell Stem Cell 2013; 12(5): 520-30.
[http://dx.doi.org/10.1016/j.stem.2013.04.009] [PMID: 23642363]

[19]   Lancaster MA, Renner M, Martin CA, *et al.* Cerebral organoids model human brain development and microcephaly. Nature 2013; 501(7467): 373-9.
[http://dx.doi.org/10.1038/nature12517] [PMID: 23995685]

[20]   Sato T, *et al.* Paneth cells constitute the niche for Lgr5 stem cells in intestinal crypts. Nature 2011; 469(): 415-8.
[http://dx.doi.org/10.1038/nature09637]

[21]   Sasaki N, Sachs N, Wiebrands K, *et al.* Reg4 [+] deep crypt secretory cells function as epithelial niche for Lgr5 [+] stem cells in colon. Proc Natl Acad Sci USA 2016; 113(37): E5399-407.

[http://dx.doi.org/10.1073/pnas.1607327113] [PMID: 27573849]

[22] Clevers H. Wnt/b-catenin signaling indevelopment and disease. Cell 2006; 127(): 469-80.
[http://dx.doi.org/10.1016/j.cell.2006.10.018]

[23] Chia LA, Kuo CJ. The intestinal stem cell. Prog Mol Biol Transl Sci 2010; 96: 157-73.
[http://dx.doi.org/10.1016/B978-0-12-381280-3.00007-5] [PMID: 21075344]

[24] Farin HF. Visualization of a short-rangeWnt gradient in the intestinal stem-cell niche. Nature 2016; 530: 340-3.
[http://dx.doi.org/10.1038/nature16937]

[25] Fair KL, Colquhoun J, Hannan NRF. Intestinal organoids for modelling intestinal development and disease. Philos Trans R Soc Lond B Biol Sci 2018; 373(1750)20170217
[http://dx.doi.org/10.1098/rstb.2017.0217] [PMID: 29786552]

[26] Sato T, van Es JH, Snippert HJ, *et al.* Paneth cells constitute the niche for Lgr5 stem cells in intestinal crypts. Nature 2011; 469(7330): 415-8.
[http://dx.doi.org/10.1038/nature09637] [PMID: 21113151]

[27] Dekkers JF, Wiegerinck CL, de Jonge HR, *et al.* A functional CFTR assay using primary cystic fibrosis intestinal organoids. Nat Med 2013; 19(7): 939-45.
[http://dx.doi.org/10.1038/nm.3201] [PMID: 23727931]

[28] Sato T, Stange DE, Ferrante M, *et al.* Long-term expansion of epithelial organoids from human colon, adenoma, adenocarcinoma, and Barrett's epithelium. Gastroenterology 2011; 141(5): 1762-72.
[http://dx.doi.org/10.1053/j.gastro.2011.07.050] [PMID: 21889923]

[29] Yui S, Nakamura T, Sato T, *et al.* Functional engraftment of colon epithelium expanded in vitro from a single adult Lgr5+ stem cell. Nat Med 2012; 18(4): 618-23.
[http://dx.doi.org/10.1038/nm.2695] [PMID: 22406745]

[30] Spence JR, Mayhew CN, Rankin SA, *et al.* Directed differentiation of human pluripotent stem cells into intestinal tissue in vitro. Nature 2011; 470(7332): 105-9.
[http://dx.doi.org/10.1038/nature09691] [PMID: 21151107]

[31] Hannan NR, Fordham RP, Syed YA, *et al.* Generation of multipotent foregut stem cells from human pluripotent stem cells. Stem Cell Rep 2013; 1(): 293-306.
[http://dx.doi.org/10.1016/j.stemcr.2013.09.003]

[32] McCracken KW, Catá EM, Crawford CM, *et al.* Modelling human development and disease in pluripotent stem-cell-derived gastric organoids. Nature 2014; 516(7531): 400-4.
[http://dx.doi.org/10.1038/nature13863] [PMID: 25363776]

[33] Forbester JL, Goulding D, Vallier L, Hannan N. Theinteraction ofSalmonella entericaSerovartyphimurium with intestinal organoids derivedfrom human induced pluripotent stem cells. Infect Immun 2015; 83(): 2926-34.
[http://dx.doi.org/10.1128/IAI.00161-15]

[34] Lees EA, Forbester JL, Forrest S, Kane L, Goulding D, Dougan GJ. Using Human Induced Pluripotent Stem Cell-derived Intestinal Organoids to Study and Modify Epithelial Cell Protection Against Salmonella and Other Pathogens. Vis Exp 2019; 147
[http://dx.doi.org/10.3791/59478] [PMID: 31132035]

[35] Hill DR, Huang S, Tsai Y-H, Spence JR, Young VB. Real-time measurement of epithelial barrierpermeability in human intestinal organoids. J VisExp 2017.
[http://dx.doi.org/10.3791/56960]

[36] Wallach TE, Bayrer JR. Intestinal Organoids. J Pediatr Gastroenterol Nutr 2017; 64(2): 180-5.
[http://dx.doi.org/10.1097/MPG.0000000000001411] [PMID: 27632431]

[37] Watson CL, Mahe MM, Múnera J, *et al.* An *in vivo* model of human small intestine using pluripotent stem cells. Nat Med 2014; 20(11): 1310-4.

[http://dx.doi.org/10.1038/nm.3737] [PMID: 25326803]

[38] Schwank G. Functional repair of CFTR byCRISPR/Cas9 in intestinal stem cell organoids ofcystic fibrosis patients. Cell Stem Cell 2013; 13: 653-8.
[http://dx.doi.org/10.1016/j.stem.2013.11.002]

[39] Almeqdadi M, Mana MD, Roper J, Yilmaz ÖH. Gut organoids: mini-tissues in culture to study intestinal physiology and disease. Am J Physiol Cell Physiol 2019; 317(3): C405-19.
[http://dx.doi.org/10.1152/ajpcell.00300.2017] [PMID: 31216420]

[40] Rahmani S, Breyner NM, Su H-M, Verdu EF, Didar TF. Intestinal organoids: A new paradigm for engineering intestinal epithelium in vitro, Biomaterials 2019; 194(): 195-214.
[http://dx.doi.org/10.1016/j.biomaterials.2018.12.006]

[41] Asrani SK, Devarbhavi H, Eaton J, Kamath PS. Burden of liver diseases in the world. J Hepatol 2019; 70(1): 151-71.
[http://dx.doi.org/10.1016/j.jhep.2018.09.014] [PMID: 30266282]

[42] Zhou WC, Zhang QB, Qiao L. Pathogenesis of liver cirrhosis. World J Gastroenterol 2014; 20(23): 7312-24.
[http://dx.doi.org/10.3748/wjg.v20.i23.7312] [PMID: 24966602]

[43] Akbari S, Arslan N, Senturk S, Erdal E. Next-Generation Liver Medicine Using Organoid Models. Front Cell Dev Biol 2019; 7: 345.https://www.frontiersin.org/article/10.3389/fcell.2019.00345 DOI=10.3389/fcell.2019.00345
[http://dx.doi.org/10.3389/fcell.2019.00345] [PMID: 31921856]

[44] Boost KA, Auth MKH, Woitaschek D, *et al.* Long-term production of major coagulation factors and inhibitors by primary human hepatocytes *in vitro* : perspectives for clinical application. Liver Int 2007; 27(6): 832-44.
[http://dx.doi.org/10.1111/j.1478-3231.2007.01472.x] [PMID: 17617127]

[45] Hay DC, Pernagallo S, Diaz-Mochon JJ, *et al.* Unbiased screening of polymer libraries to define novel substrates for functional hepatocytes with inducible drug metabolism. Stem Cell Res (Amst) 2011; 6(2): 92-102.
[http://dx.doi.org/10.1016/j.scr.2010.12.002] [PMID: 21277274]

[46] Baxter M, Withey S, Harrison S, *et al.* Phenotypic and functional analyses show stem cell-derived hepatocyte-like cells better mimic fetal rather than adult hepatocytes. J Hepatol 2015; 62(3): 581-9.
[http://dx.doi.org/10.1016/j.jhep.2014.10.016] [PMID: 25457200]

[47] Si-Tayeb K, Lemaigre FP, Duncan SA. Organogenesis and Development of the Liver. Dev Cell 2010; 18(2): 175-89.
[http://dx.doi.org/10.1016/j.devcel.2010.01.011]

[48] Takebe T, Sekine K, Enomura M, *et al.* Vascularized and functional human liver from an iPSC-derived organ bud transplant. Nature 2013; 499(7459): 481-4.
[http://dx.doi.org/10.1038/nature12271] [PMID: 23823721]

[49] Camp JG, Sekine K, Gerber T, *et al.* Multilineage communication regulates human liver bud development from pluripotency. Nature 2017; 546(7659): 533-8.
[http://dx.doi.org/10.1038/nature22796] [PMID: 28614297]

[50] Guan Y, Xu D, Garfin PM, *et al.* Human hepatic organoids for the analysis of human genetic diseases. JCI Insight 2017; 2(17)e94954
[http://dx.doi.org/10.1172/jci.insight.94954] [PMID: 28878125]

[51] Wu F, Wu D, Ren Y, *et al.* Generation of hepatobiliary organoids from human induced pluripotent stem cells. J Hepatol 2019; 70(6): 1145-58.
[http://dx.doi.org/10.1016/j.jhep.2018.12.028] [PMID: 30630011]

[52] Akbari S, Sevinç GG, Ersoy N, *et al.* Robust, long-term culture of endoderm-derived hepatic organoids for disease modelling. Stem Cell Reports 2019; 13(4): 627-41.

[http://dx.doi.org/10.1016/j.stemcr.2019.08.007] [PMID: 31522975]

[53]   Collin de l'Hortet A, Takeishi K, Guzman-Lepe J, *et al.* Generation of Human Fatty Livers Using Custom-Engineered Induced Pluripotent Stem Cells with Modifiable SIRT1 Metabolism. Cell Metab 2019; 30(2): 385-401.e9.
[http://dx.doi.org/10.1016/j.cmet.2019.06.017] [PMID: 31390551]

[54]   Ouchi R, Togo S, Kimura M, *et al.* (2019) Modeling steatohepatitis in humans with pluripotent stem cell-derived organoids. Cell Metab 2019; 30(2): 374-384.e6.
[http://dx.doi.org/10.1016/j.cmet.2019.05.007] [PMID: 31155493]

[55]   Ramli MNB, Lim YS, Koe CT, *et al.* (2020) Human pluripotent stem cell-derived organoids as models of liver disease. Gastroenterology 2020; 159(4): 1471-1486.e12.
[http://dx.doi.org/10.1053/j.gastro.2020.06.010] [PMID: 32553762]

[56]   Lam DTUH, Dan YY, Chan YS, Ng HH. Emerging liver organoid platforms and technologies. Cell Regen (Lond) 2021; 10(1): 27.
[http://dx.doi.org/10.1186/s13619-021-00089-1] [PMID: 34341842]

[57]   Furuyama K, Kawaguchi Y, Akiyama H, *et al.* Continuous cell supply from a Sox9-expressing progenitor zone in adult liver, exocrine pancreas and intestine. Nat Genet 2011; 43(1): 34-41.
[http://dx.doi.org/10.1038/ng.722] [PMID: 21113154]

[58]   Dorrell C, Erker L, Schug J, *et al.* Prospective isolation of a bipotential clonogenic liver progenitor cell in adult mice. Genes Dev 2011; 25(11): 1193-203.
[http://dx.doi.org/10.1101/gad.2029411] [PMID: 21632826]

[59]   Español-Suñer R, Carpentier R, Van Hul N, *et al.* Liver progenitor cells yield functional hepatocytes in response to chronic liver injury in mice. Gastroenterology 2012; 143(6): 1564-1575.e7.
[http://dx.doi.org/10.1053/j.gastro.2012.08.024] [PMID: 22922013]

[60]   Kim KA, Kakitani M, Zhao J, *et al.* Mitogenic influence of human R-spondin1 on the intestinal epithelium. Science 2005; 309(5738): 1256-9.
[http://dx.doi.org/10.1126/science.1112521] [PMID: 16109882]

[61]   Carmon KS, Gong X, Lin Q, Thomas A, Liu Q. R-spondins function as ligands of the orphan receptors LGR4 and LGR5 to regulate Wnt/β-catenin signaling. Proc Natl Acad Sci USA 2011; 108(28): 11452-7.
[http://dx.doi.org/10.1073/pnas.1106083108] [PMID: 21693646]

[62]   Huch M, Dorrell C, Boj SF, *et al. in vitro* expansion of single Lgr5+ liver stem cells induced by Wnt-driven regeneration. Nature 2013; 494(7436): 247-50.
[http://dx.doi.org/10.1038/nature11826] [PMID: 23354049]

[63]   Huch M, Gehart H, van Boxtel R, *et al.* Long-term culture of genome-stable bipotent stem cells from adult human liver. Cell 2015; 160(1-2): 299-312.
[http://dx.doi.org/10.1016/j.cell.2014.11.050] [PMID: 25533785]

[64]   Raven A, Lu WY, Man TY, *et al.* Cholangiocytes act as facultative liver stem cells during impaired hepatocyte regeneration. Nature 2017; 547(7663): 350-4.
[http://dx.doi.org/10.1038/nature23015] [PMID: 28700576]

[65]   Grompe M. Liver stem cells, where art thou? Cell Stem Cell 2014; 15(3): 257-8.
[http://dx.doi.org/10.1016/j.stem.2014.08.004] [PMID: 25192457]

[66]   Schaub JR, Malato Y, Gormond C, Willenbring H. Evidence against a stem cell origin of new hepatocytes in a common mouse model of chronic liver injury. Cell Rep 2014; 8(4): 933-9.
[http://dx.doi.org/10.1016/j.celrep.2014.07.003] [PMID: 25131204]

[67]   Wang B, Zhao L, Fish M, Logan CY, Nusse R. Self-renewing diploid Axin2+ cells fuel homeostatic renewal of the liver. Nature 2015; 524(7564): 180-5.
[http://dx.doi.org/10.1038/nature14863] [PMID: 26245375]

[68]   Hu H, Gehart H, Artegiani B, *et al.* Long-term expansion of functional mouse and human hepatocytes as 3D organoids. Cell 2018; 175(6): 1591-1606.e19.
[http://dx.doi.org/10.1016/j.cell.2018.11.013] [PMID: 30500538]

[69]   Peng WC, Logan CY, Fish M, *et al.* Inflammatory cytokine TNFalpha promotes the long-term expansion of primary hepatocytes in 3D culture. Cell 2018; 175(6): 1607-1619.e15.
[http://dx.doi.org/10.1016/j.cell.2018.11.012] [PMID: 30500539]

[70]   Vyas D, Baptista PM, Brovold M, *et al.* Self-assembled liver organoids recapitulate hepatobiliary organogenesis in vitro. Hepatology 2018; 67(2): 750-61.
[http://dx.doi.org/10.1002/hep.29483] [PMID: 28834615]

[71]   Dianat N, Dubois-Pot-Schneider H, Steichen C, *et al.* Generation of functional cholangiocyte-like cells from human pluripotent stem cells and HepaRG cells. Hepatology 2014; 60(2): 700-14.
[http://dx.doi.org/10.1002/hep.27165] [PMID: 24715669]

[72]   Ogawa M, Ogawa S, Bear CE, *et al.* Directed differentiation of cholangiocytes from human pluripotent stem cells. Nat Biotechnol 2015; 33(8): 853-61.
[http://dx.doi.org/10.1038/nbt.3294] [PMID: 26167630]

[73]   Sampaziotis F, Cardoso de Brito M, Madrigal P, *et al.* Cholangiocytes derived from human induced pluripotent stem cells for disease modeling and drug validation. Nat Biotechnol 2015; 33(8): 845-52.
[http://dx.doi.org/10.1038/nbt.3275] [PMID: 26167629]

[74]   McCauley KB, Hawkins F, Serra M, Thomas DC, Jacob A, Kotton DN. Efficient derivation of functional human airway epithelium from pluripotent stem cells *via* temporal regulation of Wnt signaling. Cell Stem Cell 2017; 20(6): 844-57.

[75]   Jacob A, Morley M, Hawkins F, McCauley KB, Jean JC, Heins H, *et al.* Differentiation of human pluripotent stem cells into functional lung alveolar epithelial cells. Cell Stem Cell 2017; 21(4): 472-88.
[http://dx.doi.org/10.1016/j.stem.2017.08.014]

[76]   Chen YW, Huang SX, de Carvalho ALRT, *et al.* A three-dimensional model of human lung development and disease from pluripotent stem cells. Nat Cell Biol 2017; 19(5): 542-9.
[http://dx.doi.org/10.1038/ncb3510] [PMID: 28436965]

[77]   Kong J, Wen S, Cao W, *et al.* Lung organoids, useful tools for investigating epithelial repair after lung injury. Stem Cell Res Ther 2021; 12(1): 95.
[http://dx.doi.org/10.1186/s13287-021-02172-5] [PMID: 33516265]

[78]   Miller AJ, Dye BR, Ferrer-Torres D, *et al.* Generation of lung organoids from human pluripotent stem cells in vitro. Nat Protoc 2019; 14: 518-40.
[http://dx.doi.org/10.1038/s41596-018-0104-8]

[79]   Shannon JM, Mason RJ, Jennings SD. Functional differentiation of alveolar type II epithelial cells in vitro: Effects of cell shape, cell-matrix interactions and cell-cell interactions. Biochim Biophys Acta Mol Cell Res 1987; 931(2): 143-56.
[http://dx.doi.org/10.1016/0167-4889(87)90200-X] [PMID: 3663713]

[80]   Köpf-Maier P, Zimmermann B. Organoid reorganization of human tumors under *in vitro* conditions. Cell Tissue Res 1991; 264(3): 563-76.
[http://dx.doi.org/10.1007/BF00319046] [PMID: 1868523]

[81]   Kaartinen L, Nettesheim P, Adler KB, Randell SH. Rat tracheal epithelial cell differentiation in vitro. in Vitro Cell Dev Biol 1993; 29(6): 481-92.
[http://dx.doi.org/10.1007/BF02639383] [PMID: 7687243]

[82]   Potdar PD, Andrews KL, Nettesheim P, Ostrowski LE. Expression and regulation of γ-glutamyl transpeptidase-related enzyme in tracheal cells. Am J Physiol Lung Cell Mol Physiol 1997; 273(5): L1082-9.
[http://dx.doi.org/10.1152/ajplung.1997.273.5.L1082] [PMID: 9374738]

[83]   Ostrowski LE, Andrews K, Potdar P, Matsuura H, Jetten A, Nettesheim P. Cloning and characterization of KPL2, a novel gene induced during ciliogenesis of tracheal epithelial cells. Am J Respir Cell Mol Biol 1990; 20(4): 675-83.
[http://dx.doi.org/10.1165/ajrcmb.20.4.3496]

[84]   Ostrowski LE, Andrews KL, Potdar PD, Nettesheim P. Ciliated-cell differentiation and gene expression. Protoplasma 1999; 206(4): 245-8.
[http://dx.doi.org/10.1007/BF01288212]

[85]   Kaya L. which encodes a novel ph domain-containing protein, is induced during ciliated cell differentiation of rat tracheal epithelial cells. Experimental Lung Research, 2000; 26(4)

[86]   Rock JR, Onaitis MW, Rawlins EL, *et al.* Basal cells as stem cells of the mouse trachea and human airway epithelium. Proc Natl Acad Sci USA 2009; 106(31): 12771-5.
[http://dx.doi.org/10.1073/pnas.0906850106] [PMID: 19625615]

[87]   Barkauskas CE, Cronce MJ, Rackley CR, *et al.* Type 2 alveolar cells are stem cells in adult lung. J Clin Invest 2013; 123(7): 3025-36.
[http://dx.doi.org/10.1172/JCI68782] [PMID: 23921127]

[88]   Gotoh S, Ito I, Nagasaki T, *et al.* Generation of alveolar epithelial spheroids *via* isolated progenitor cells from human pluripotent stem cells. Stem Cell Reports 2014; 3(3): 394-403.
[http://dx.doi.org/10.1016/j.stemcr.2014.07.005] [PMID: 25241738]

[89]   Dye BR, Hill DR, Ferguson MAH, *et al.* in vitro generation of human pluripotent stem cell derived lung organoids. eLife 2015; 4e05098
[http://dx.doi.org/10.7554/eLife.05098] [PMID: 25803487]

[90]   de Carvalho ALRT, Strikoudis A, Liu HY, *et al.* Glycogen synthase kinase 3 induces multilineage maturation of human pluripotent stem cell-derived lung progenitors in 3D culture. Development 2019; 146(2)dev171652
[PMID: 30578291]

[91]   Wilkinson DC, Alva-Ornelas JA, Sucre JMS, *et al.* Development of a Three-Dimensional Bioengineering Technology to Generate Lung Tissue for Personalized Disease Modeling. Stem Cells Transl Med 2017; 6(2): 622-33.
[http://dx.doi.org/10.5966/sctm.2016-0192] [PMID: 28191779]

[92]   Wilkinson DC, Mellody M, Meneses LK, Hope AC, Dunn B, Gomperts BN. Development of a Three-Dimensional Bioengineering Technology to Generate Lung Tissue for Personalized Disease Modeling. Curr Protoc Stem Cell Biol 2018; 46(1)e56
[http://dx.doi.org/10.1002/cpsc.56] [PMID: 29927098]

[93]   Fulcher ML, Randell SH. Human nasal and tracheo-bronchial respiratory epithelial cell culture. Methods Mol Biol 2012; 945: 109-21.
[http://dx.doi.org/10.1007/978-1-62703-125-7_8] [PMID: 23097104]

[94]   Konishi S, Gotoh S, Tateishi K, *et al.* (2016) Directed induction of functional multi-ciliated cells in proximal airway epithelial spheroids from human pluripotent stem cells. Stem Cell Reports 2016; 6(1): 18-25.
[http://dx.doi.org/10.1016/j.stemcr.2015.11.010] [PMID: 26724905]

[95]   Huang SXL, Green MD, de Carvalho AT, *et al.* The in vitro generation of lung and airway progenitor cells from human pluripotent stem cells. Nat Protoc 2015; 10(3): 413-25.
[http://dx.doi.org/10.1038/nprot.2015.023] [PMID: 25654758]

[96]   Agboola OS, Hu X, Shan Z, Wu Y, Lei L. Brain organoid: a 3D technology for investigating cellular composition and interactions in human neurological development and disease models in vitro. Stem Cell Res Ther 2021; 12(1): 430.
[http://dx.doi.org/10.1186/s13287-021-02369-8] [PMID: 34332630]

[97]   Xu J, Wen Z. 2021.Brain Organoids: Studying Human Brain Development and Diseases in a Dish.

[http://dx.doi.org/10.1155/2021/5902824]

[98]   Muguruma K, Nishiyama A, Ono Y, *et al.* Ontogeny-recapitulating generation and tissue integration of ES cell–derived Purkinje cells. Nat Neurosci 2010; 13(10): 1171-80.
[http://dx.doi.org/10.1038/nn.2638] [PMID: 20835252]

[99]   Lancaster MA, Renner MC. Cerebral organoids model human brain development and microceph- aly," Nature. 2013; 501: pp. (7467)373-9.

[100] Nicholas CR, Chen J, Tang Y, *et al.* Functional maturation of hPSC-derived forebrain interneurons requires an extended timeline and mimics human neural development. Cell Stem Cell 2013; 12(5): 573-86.
[http://dx.doi.org/10.1016/j.stem.2013.04.005] [PMID: 23642366]

[101] Muguruma K, Nishiyama A, Kawakami H, Hashimoto K, Sasai Y. Self-organization of polarized cerebellar tissue in 3D culture of human pluripotent stem cells. Cell Rep 2015; 10(4): 537-50.
[http://dx.doi.org/10.1016/j.celrep.2014.12.051] [PMID: 25640179]

[102] Paşca MS. Functional corti- cal neurons and astrocytes from human pluripotent stem cells in 3D culture, Nature Methods. 2015; 12: pp. (7)671-8.

[103] Sakaguchi H, Kadoshima T, Soen M, *et al.* Generation of functional hippocampal neurons from self-organizing human embryonic stem cell-derived dorsomedial telencephalic tissue. Nat Commun 2015; 6(1): 8896.
[http://dx.doi.org/10.1038/ncomms9896] [PMID: 26573335]

[104] Qian X, Nguyen HN, Song MM, *et al.* (2015) "Brain-region-specific organoids using mini-bioreactors for modelling ZIKV exposure. Cell 2016; 165(5): 1238-54.
[http://dx.doi.org/10.1016/j.cell.2016.04.032] [PMID: 27118425]

[105] Jo J, Xiao Y, Sun AX, *et al.* (2016). "Midbrain-like organoids from human pluripotent stem cells contain functional dopaminer- gic and neuromelanin-producing neurons. Cell Stem Cell 2016; 19(2): 248-57.
[http://dx.doi.org/10.1016/j.stem.2016.07.005] [PMID: 27476966]

[106] Sloan SA, Darmanis S, Huber N, *et al.* Human astrocyte maturation captured in 3D cerebral cortical spheroids derived from pluripotent stem cells Neuron,. 2017; 95: pp. (4)779-90.
[http://dx.doi.org/10.1016/j.neuron.2017.07.035]

[107] Quadrato G, Nguyen T. Cell diversity and network dynamics in photosensitive human brain orga- noids," Nature. 2017; 545: pp. (7652)48-53.

[108] Renner M, Lancaster MA, Bian S, *et al.* Self-organized developmental patterning and differentiation in cerebral organoids. EMBO J 2017; 36(10): 1316-29.
[http://dx.doi.org/10.15252/embj.201694700] [PMID: 28283582]

[109] Kanton S, Boyle MJ, He Z, *et al.* Organoid single-cell genomic atlas uncovers human-specific features of brain development. Nature 2019; 574(7778): 418-22.
[http://dx.doi.org/10.1038/s41586-019-1654-9] [PMID: 31619793]

[110] Kanton S, Treutlein B, Camp JG. Single-cell genomic analysis of human cerebral organoids. Methods Cell Biol 2020; 159: 229-56.
[http://dx.doi.org/10.1016/bs.mcb.2020.03.013] [PMID: 32586444]

[111] Tanaka Y, Cakir B, Xiang Y, Sullivan GJ, Park IH. Synthetic analyses of single-cell transcriptomes from multiple brain organoids and fetal brain," Cell Reports. 2020; 30: pp. (6)1682-9.
[http://dx.doi.org/10.1016/j.celrep.2020.01.038]

[112] Fleck JS, Sanchís-Calleja F, He Z, *et al.* Resolving organoid brain region identities by mapping single-cell genomic data to reference atlases. Cell Stem Cell 2021; 28(6): 1148-1159.e8.
[http://dx.doi.org/10.1016/j.stem.2021.02.015] [PMID: 33711282]

[113] Kretzschmar K. Cancer research using organoid technology. J Mol Med (Berl) 2021; 99(4): 501-15.

[http://dx.doi.org/10.1007/s00109-020-01990-z] [PMID: 33057820]

[114] Mullenders J, de Jongh E, Brousali A, *et al.* Mouse and human urothelial cancer organoids: A tool for bladder cancer research. Proc Natl Acad Sci USA 2019; 116(10): 4567-74.
[http://dx.doi.org/10.1073/pnas.1803595116] [PMID: 30787188]

[115] Lee SH, Hu W, Matulay JT, *et al.* Tumor evolution and drug response in patient-derived organoid models of bladder cancer. Cell 2018; 173(2): 515-528.e17.
[http://dx.doi.org/10.1016/j.cell.2018.03.017] [PMID: 29625057]

[116] Jacob F, Salinas RD, Zhang DY, *et al.* A patient-derived glioblastoma organoid model and biobank recapitulates inter- and intra-tumoral heterogeneity. Cell 2020; 180(1): 188-204.e22.
[http://dx.doi.org/10.1016/j.cell.2019.11.036] [PMID: 31883794]

[117] Hubert CG, Rivera M, Spangler LC, *et al.* A three-dimensional organoid culture system derived from human glioblastomas recapitulates the hypoxic gradients and cancer stem cell heterogeneity of tumors found *in vivo.* Cancer Res 2016; 76(8): 2465-77.
[http://dx.doi.org/10.1158/0008-5472.CAN-15-2402] [PMID: 26896279]

[118] Sachs N, de Ligt J, Kopper O, *et al.* A living biobank of breast cancer organoids captures disease heterogeneity. Cell 2018; 172(1-2): 373-386.e10.
[http://dx.doi.org/10.1016/j.cell.2017.11.010] [PMID: 29224780]

[119] Sato T, Stange DE, Ferrante M, *et al.* Long-term expansion of epithelial organoids from human colon, adenoma, adenocarcinoma, and Barrett's epithelium. Gastroenterology 2011; 141(5): 1762-72.
[http://dx.doi.org/10.1053/j.gastro.2011.07.050] [PMID: 21889923]

[120] van de Wetering M, Francies HE, Francis JM, *et al.* Prospective derivation of a living organoid biobank of colorectal cancer patients. Cell 2015; 161(4): 933-45.
[http://dx.doi.org/10.1016/j.cell.2015.03.053] [PMID: 25957691]

[121] Fujii M, Shimokawa M, Date S, *et al.* A colorectal tumor organoid library demonstrates progressive loss of niche factor requirements during tumorigenesis. Cell Stem Cell 2016; 18(6): 827-38.
[http://dx.doi.org/10.1016/j.stem.2016.04.003] [PMID: 27212702]

[122] Weeber F, van de Wetering M, Hoogstraat M, *et al.* Preserved genetic diversity in organoids cultured from biopsies of human colorectal cancer metastases. Proc Natl Acad Sci USA 2015; 112(43): 13308-11.
[http://dx.doi.org/10.1073/pnas.1516689112] [PMID: 26460009]

[123] Turco MY, Gardner L, Hughes J, *et al.* Long-term, hormone-responsive organoid cultures of human endometrium in a chemically defined medium. Nat Cell Biol 2017; 19(5): 568-77.
[http://dx.doi.org/10.1038/ncb3516] [PMID: 28394884]

[124] Boretto M, Maenhoudt N, Luo X, *et al.* Patient-derived organoids from endometrial disease capture clinical heterogeneity and are amenable to drug screening. Nat Cell Biol 2019; 21(8): 1041-51.
[http://dx.doi.org/10.1038/s41556-019-0360-z] [PMID: 31371824]

[125] Driehuis E, Kolders S, Spelier S, *et al.* Oral mucosal organoids as a potential platform for personalized cancer therapy. Cancer Discov 2019; 9(7): 852-71.
[http://dx.doi.org/10.1158/2159-8290.CD-18-1522] [PMID: 31053628]

[126] Driehuis E, Kretzschmar K, Clevers H. Establishment of patient-derived cancer organoids for drug-screening applications. Nat Protoc 2020; 15(10): 3380-409.
[http://dx.doi.org/10.1038/s41596-020-0379-4] [PMID: 32929210]

[127] Calandrini C, Schutgens F, Oka R, *et al.* An organoid biobank for childhood kidney cancers that captures disease and tissue heterogeneity. Nat Commun 2020; 11(1): 1310.
[http://dx.doi.org/10.1038/s41467-020-15155-6] [PMID: 32161258]

[128] Grassi L, Alfonsi R, Francescangeli F, *et al.* Organoids as a new model for improving regenerative medicine and cancer personalized therapy in renal diseases. Cell Death Dis 2019; 10(3): 201.
[http://dx.doi.org/10.1038/s41419-019-1453-0] [PMID: 30814510]

[129] Broutier L, Mastrogiovanni G, Verstegen MMA, *et al.* Human primary liver cancer–derived organoid cultures for disease modeling and drug screening. Nat Med 2017; 23(12): 1424-35.
[http://dx.doi.org/10.1038/nm.4438] [PMID: 29131160]

[130] Sachs N, Papaspyropoulos A, Zomer-van Ommen DD, *et al.* Long-term expanding human airway organoids for disease modeling. EMBO J 2019; 38(4)e100300
[http://dx.doi.org/10.15252/embj.2018100300] [PMID: 30643021]

[131] Kopper O, de Witte CJ, Lõhmussaar K, *et al.* An organoid platform for ovarian cancer captures intra- and interpatient heterogeneity. Nat Med 2019; 25(5): 838-49.
[http://dx.doi.org/10.1038/s41591-019-0422-6] [PMID: 31011202]

[132] Hill SJ, Decker B, Roberts EA, *et al.* Prediction of DNA repair inhibitor response in short-term patient- derived ovarian cancer organoids. Cancer Discov 2018; 8(11): 1404-21.
[http://dx.doi.org/10.1158/2159-8290.CD-18-0474] [PMID: 30213835]

[133] Boj SF, Hwang CI, Baker LA, *et al.* Organoid models of human and mouse ductal pancreatic cancer. Cell 2015; 160(1-2): 324-38.
[http://dx.doi.org/10.1016/j.cell.2014.12.021] [PMID: 25557080]

[134] Seino T, Kawasaki S, Shimokawa M, *et al.* Human Pancreatic Tumor Organoids Reveal Loss of Stem Cell Niche Factor Dependence during Disease Progression. Cell Stem Cell 2018; 22(3): 454-467.e6.
[http://dx.doi.org/10.1016/j.stem.2017.12.009] [PMID: 29337182]

[135] Tiriac H, Belleau P, Engle DD, *et al.* Organoid profiling identifies common responders to chemotherapy in pancreatic cancer. Cancer Discov 2018; 8(9): 1112-29.
[http://dx.doi.org/10.1158/2159-8290.CD-18-0349] [PMID: 29853643]

[136] Gao D, Vela I, Sboner A, *et al.* Organoid cultures derived from patients with advanced prostate cancer. Cell 2014; 159(1): 176-87.
[http://dx.doi.org/10.1016/j.cell.2014.08.016] [PMID: 25201530]

[137] Yao Y, Xu X, Yang L, *et al.* Patient-derived organoids predict chemoradiation responses of locally advanced rectal cancer. Cell Stem Cell 2020; 26(1): 17-26.e6.
[http://dx.doi.org/10.1016/j.stem.2019.10.010] [PMID: 31761724]

[138] Yan HHN, Siu HC, Law S, *et al.* comprehensive human gastric cancer organoid biobank captures tumor subtype heterogeneity and enables therapeutic screening. Cell Stem Cell 2018; 23(6): 882-897.e11.
[http://dx.doi.org/10.1016/j.stem.2018.09.016] [PMID: 30344100]

[139] Nanki K, Toshimitsu K, Takano A, *et al.* Divergent routes toward Wnt and R-spondin niche independency during human gastric carcinogenesis. Cell 2018; 174(4): 856-869.e17.
[http://dx.doi.org/10.1016/j.cell.2018.07.027] [PMID: 30096312]

[140] Vlachogiannis G, Hedayat S, Vatsiou A, *et al.* Patient-derived organoids model treatment response of metastatic gastrointestinal cancers. Science 2018; 359(6378): 920-6.
[http://dx.doi.org/10.1126/science.aao2774] [PMID: 29472484]

[141] Gopal S, Rodrigues AL, Dordick JS. Exploiting CRISPR Cas9 in Three-Dimensional Stem Cell Cultures to Model Disease. Front Bioeng Biotechnol 2020; 8: 692.
[http://dx.doi.org/10.3389/fbioe.2020.00692]

[142] Ooft SN, Weeber F, Dijkstra KK, *et al.* Patient-derived organoids can predict response to chemotherapy in metastatic colorectal cancer patients. Sci Transl Med 2019; 11(513)eaay2574
[http://dx.doi.org/10.1126/scitranslmed.aay2574] [PMID: 31597751]

[143] Betge J, *et al.* 2019.Multiparametric phenotyping of compound effects on patient derived organoids. bioRxiv 660993

[144] Hanahan D, Weinberg RA. Hallmarks of cancer: the next generation. Cell 2011; 144(5): 646-74.
[http://dx.doi.org/10.1016/j.cell.2011.02.013] [PMID: 21376230]

[145] Junttila MR, de Sauvage FJ. Influence of tumour micro-environment heterogeneity on therapeutic response. Nature 2013; 501(7467): 346-54.
[http://dx.doi.org/10.1038/nature12626] [PMID: 24048067]

[146] Dijkstra KK, Cattaneo CM, Weeber F, *et al.* Schumacher T.N., Voest E.E(2018) Generation of tumor-reactive T cells by co-culture of peripheral blood lymphocytes and tumor organoids. Cell 2018; 174(6): 1586-1598.e12.
[http://dx.doi.org/10.1016/j.cell.2018.07.009] [PMID: 30100188]

[147] Cattaneo CM, Dijkstra KK, Fanchi LF, *et al.* Tumor organoid–T-cell coculture systems. Nat Protoc 2020; 15(1): 15-39.
[http://dx.doi.org/10.1038/s41596-019-0232-9] [PMID: 31853056]

[148] Neal JT, Li X, Zhu J, *et al.* Organoid modeling of the tumor immune microenvironment. Cell 2018; 175(7): 1972-1988.e16.
[http://dx.doi.org/10.1016/j.cell.2018.11.021] [PMID: 30550791]

[149] Gjorevski N, Sachs N, Manfrin A, *et al.* Designer matrices for intestinal stem cell and organoid culture. Nature 2016; 539(7630): 560-4.
[http://dx.doi.org/10.1038/nature20168] [PMID: 27851739]

[150] Lo YH, Karlsson K, Kuo CJ. Applications of organoids for cancer biology and precision medicine. Nat Can 2020; 1(8): 761-73.
[http://dx.doi.org/10.1038/s43018-020-0102-y] [PMID: 34142093]

[151] Bredenoord AL, Clevers H, Knoblich JA. Human tissues in a dish: The research and ethical implications of organoid technology. Science 2017; 355(6322)eaaf9414
[http://dx.doi.org/10.1126/science.aaf9414] [PMID: 28104841]

[152] Linnekamp JF, Wang X, Medema JP, Vermeulen L. Colorectal cancer heterogeneity and targeted therapy: a case formolecular disease subtypes. Cancer Res 2015; 75: 245-9.
[http://dx.doi.org/10.1158/0008-5472.CAN-14-2240]

[153] Ji DB, Wu AW. Organoid in colorectal cancer: progress and challenges. Chin Med J (Engl) 2020; 133(16): 1971-7.
[http://dx.doi.org/10.1097/CM9.0000000000000882] [PMID: 32826461]

[154] Sato T, Stange DE, Ferrante M, *et al.* Long-term expansion of epithelial organoids from human colon, adenoma, adenocarcinoma, and Barrett's epithelium. Gastroenterology 2011; 141(5): 1762-72.
[http://dx.doi.org/10.1053/j.gastro.2011.07.050] [PMID: 21889923]

[155] van de Wetering M, Francies HE, Francis JM, Bounova G, Iorio F, Pronk A, *et al.* Prospective derivation of a living organoid biobank of colorectal cancer patients. Cell 2015; 161: 933-45.
[http://dx.doi.org/10.1016/j.cell.2015.03.053]

[156] Fujii M, Shimokawa M, Date S, *et al.* A colorectal tumor organoid library demonstrates progressive loss of niche factor requirements during tumorigenesis. Cell Stem Cell 2016; 18(6): 827-38.
[http://dx.doi.org/10.1016/j.stem.2016.04.003] [PMID: 27212702]

[157] Matano M, Date S, Shimokawa M, *et al.* Modeling colorectal cancer using CRISPR-Cas9–mediated engineering of human intestinal organoids. Nat Med 2015; 21(3): 256-62.
[http://dx.doi.org/10.1038/nm.3802] [PMID: 25706875]

[158] Xie BY, Wu AW. Organoid culture of isolated cells from patient- derived tissues with colorectal cancer. Chin Med J (Engl) 2016; 129(20): 2469-75.
[http://dx.doi.org/10.4103/0366-6999.191782] [PMID: 27748340]

[159] Fumagalli A, Drost J, Suijkerbuijk SJE, *et al.* Genetic dissection of colorectal cancer progression by orthotopic transplantation of engineered cancer organoids. Proc Natl Acad Sci USA 2017; 114(12): E2357-64.
[http://dx.doi.org/10.1073/pnas.1701219114] [PMID: 28270604]

[160] Drost J, van Boxtel R, Blokzijl F, *et al.* Use of CRISPR-modified human stem cell organoids to study the origin of mutational signatures in cancer. Science 2017; 358(6360): 234-8. [http://dx.doi.org/10.1126/science.aao3130] [PMID: 28912133]

[161] Verissimo CS, Overmeer RM, Ponsioen B, *et al.* Targeting mutant RAS in patient-derived colorectal cancer organoids by combinatorial drug screening. eLife 2016; 5e18489 [http://dx.doi.org/10.7554/eLife.18489] [PMID: 27845624]

[162] Vlachogiannis G, Hedayat S, Vatsiou A, *et al.* Patient-derived organoids model treatment response of metastatic gastrointestinal cancers. Science 2018; 359(6378): 920-6. [http://dx.doi.org/10.1126/science.aao2774] [PMID: 29472484]

[163] Ooft SN, Weeber F, Dijkstra KK, *et al.* Patient-derived organoids can predict response to chemotherapy in metastatic colorectal cancer patients. Sci Transl Med 2019; 11(513)eaay2574 [http://dx.doi.org/10.1126/scitranslmed.aay2574] [PMID: 31597751]

[164] Ganesh K, Wu C, O'Rourke KP, *et al.* A rectal cancer organoid platform to study individual responses to chemoradiation. Nat Med 2019; 25(10): 1607-14. [http://dx.doi.org/10.1038/s41591-019-0584-2] [PMID: 31591597]

[165] Yao Y, Xu X, Yang L, Zhu J, Wan J, Shen L, *et al.* Patient-derived organoids predict chemoradiation responses of locally advanced rectal cancer. Cell Stem Cell 2019; 26: 17-26.

[166] Sung H, Ferlay J, Siegel RL, Laversanne M, *et al.* Global cancer statistics 2020: GLOBOCAN estimates of incidence and mortality worldwide for 36 cancers in 185 countries CA: Cancer J Clin 2021; 394-424.

[167] Sachs N, Papaspyropoulos A, Zomer-van Ommen DD, *et al.* Long-term expanding human airway organoids for disease modeling. EMBO J 2019; 38(4)e100300 [http://dx.doi.org/10.15252/embj.2018100300] [PMID: 30643021]

[168] Ma H, Zhu Y, Zhou R, Yu Y, Xiao Z, Zhang H. Lung cancer organoids, a promising model still with long way to go, Critical Reviews in Oncology/Haematology 2022; 171: 1040-8428. [http://dx.doi.org/10.1016/j.critrevonc.2022.103610]

[169] Takahashi N, Hoshi H, Higa A, *et al.* An *in vitro* system for evaluating molecular targeted drugs using lung patient-derived tumor organoids. Cells 2019; 8(5): 481. [http://dx.doi.org/10.3390/cells8050481] [PMID: 31137590]

[170] Kim M, Mun H, Sung CO, *et al.* Patient-derived lung cancer organoids as *in vitro* cancer models for therapeutic screening. Nat Commun 2019; 10(1): 3991. [http://dx.doi.org/10.1038/s41467-019-11867-6] [PMID: 31488816]

[171] Jung DJ, Shin TH, Kim M, Sung CO, Jang SJ, Jeong GS. A one-stop microfluidic-based lung cancer organoid culture platform for testing drug sensitivity. Lab Chip 2019; 19(17): 2854-65. [http://dx.doi.org/10.1039/C9LC00496C] [PMID: 31367720]

[172] Shi R, Radulovich N, Ng C, *et al.* Organoid cultures as preclinical models of non-small cell lung cancer. Clin Cancer Res 2020; 26(5): 1162-74. [http://dx.doi.org/10.1158/1078-0432.CCR-19-1376] [PMID: 31694835]

[173] Dijkstra KK, Monkhorst K, Schipper LJ, *et al.* Challenges in establishing pure lung cancer organoids limit their utility for personalized medicine. Cell Rep 2020; 31(5)107588 [http://dx.doi.org/10.1016/j.celrep.2020.107588] [PMID: 32375033]

[174] Li Z, Qian Y, Li W, *et al.* Human lung adenocarcinoma-derived organoid models for drug screening. iScience 2020; 23(8)101411 [http://dx.doi.org/10.1016/j.isci.2020.101411] [PMID: 32771979]

[175] Chen JH, Chu XP, Zhang JT, *et al.* Genomic characteristics and drug screening among organoids derived from non-small cell lung cancer patients. Thorac Cancer 2020; 11(8): 2279-90. [http://dx.doi.org/10.1111/1759-7714.13542] [PMID: 32633046]

[176] Li YF, Gao Y, Liang BW, *et al.* Patient-derived organoids of non-small cells lung cancer and their application for drug screening. Neoplasma 2020; 67(2): 430-7.
[http://dx.doi.org/10.4149/neo_2020_190417N346] [PMID: 31973535]

[177] Kim SY, Kim SM, Lim S, *et al.* Modeling clinical responses to targeted therapies by patient-derived organoids of advanced lung adenocarcinoma. Clin Cancer Res 2021; 27(15): 4397-409.
[http://dx.doi.org/10.1158/1078-0432.CCR-20-5026] [PMID: 34083237]

[178] Yokota E, Iwai M, Yukawa T, *et al.* Clinical application of a lung cancer organoid (tumoroid) culture system. NPJ Precis Oncol 2021; 5(1): 29.
[http://dx.doi.org/10.1038/s41698-021-00166-3] [PMID: 33846488]

[179] Hu Y, Sui X, Song F, *et al.* Lung cancer organoids analyzed on microwell arrays predict drug responses of patients within a week. Nat Commun 2021; 12(1): 2581.
[http://dx.doi.org/10.1038/s41467-021-22676-1] [PMID: 33972544]

[180] Li H, Zhang Y, Lan X, *et al.* Halofuginone sensitizes lung cancer organoids to cisplatin *via* suppressing PI3K/AKT and MAPK signaling pathways. Front Cell Dev Biol 2021; 9773048
[http://dx.doi.org/10.3389/fcell.2021.773048] [PMID: 34901018]

[181] Liu Q, Zhao T, Wang X, Chen Z, Hu Y, Chen X. In situ vitrification of lung cancer organoids on a microwell array. Micromachines (Basel) 2021; 12(6): 624.
[http://dx.doi.org/10.3390/mi12060624] [PMID: 34071266]

[182] Ma X, Yang S, Jiang H, Wang Y, Xiang Z. Transcriptomic analysis of tumor tissues and organoids reveals the crucial genes regulating the proliferation of lung adenocarcinoma. J Transl Med 2021; 19(1): 368.
[http://dx.doi.org/10.1186/s12967-021-03043-6] [PMID: 34446056]

[183] Walsh AJ, Cook RS, Sanders ME, *et al.* Quantitative optical imaging of primary tumor organoid metabolism predicts drug response in breast cancer. Cancer Res 2014; 74(18): 5184-94.
[http://dx.doi.org/10.1158/0008-5472.CAN-14-0663] [PMID: 25100563]

[184] Sokol ES, Miller DH, Breggia A, Spencer KC, Arendt LM, Gupta PB. Growth of human breast tissues from patient cells in 3D hydrogel scaffolds. Breast Cancer Res 2016; 18(1): 19.
[http://dx.doi.org/10.1186/s13058-016-0677-5] [PMID: 26926363]

[185] Sowder ME, Ludwik KA, Pasic L, *et al.* Abstract P1-06-05: Breast cancer organoid cultures preserve intra-tumor heterogeneity and reveal intrinsically resistant phenotypes to standard chemotherapies. Cancer Res 2017; 77(4_Supplement)P1-06-05
[http://dx.doi.org/10.1158/1538-7445.SABCS16-P1-06-05]

[186] Sachs N, de Ligt J, Kopper O, *et al.* A Living Biobank of Breast Cancer Organoids Captures Disease Heterogeneity. Cell 2018; 172(1-2): 373-386.e10.
[http://dx.doi.org/10.1016/j.cell.2017.11.010] [PMID: 29224780]

[187] Davaadelger B, Choi MR, Singhal H, Clare SE, Khan SA, Kim JJ. BRCA1 mutation influences progesterone response in human benign mammary organoids. Breast Cancer Res 2019; 21(1): 124.
[http://dx.doi.org/10.1186/s13058-019-1214-0] [PMID: 31771627]

[188] Mazzucchelli S, Piccotti F, Allevi R, *et al.* Establishment and Morphological Characterization of Patient-Derived Organoids from Breast Cancer. Biol Proced Online 2019; 21(1): 12.
[http://dx.doi.org/10.1186/s12575-019-0099-8] [PMID: 31223292]

[189] Djomehri SI, Burman B, Gonzalez ME, Takayama S, Kleer CG. A reproducible scaffold-free 3D organoid model to study neoplastic progression in breast cancer. J Cell Commun Signal 2019; 13(1): 129-43.
[http://dx.doi.org/10.1007/s12079-018-0498-7] [PMID: 30515709]

[190] Goldhammer N, Kim J, Timmermans-Wielenga V, Petersen OW. Characterization of organoid cultured human breast cancer. Breast Cancer Res 2019; 21(1): 141.
[http://dx.doi.org/10.1186/s13058-019-1233-x] [PMID: 31829259]

[191] Mollica PA, Booth-Creech EN, Reid JA, *et al.* 3D bioprinted mammary organoids and tumoroids in human mammary derived ECM hydrogels. Acta Biomater 2019; 95: 201-13.
[http://dx.doi.org/10.1016/j.actbio.2019.06.017] [PMID: 31233891]

[192] Truong DD, Kratz A, Park JG, *et al.* A Human Organotypic Microfluidic Tumor Model Permits Investigation of the Interplay between Patient-Derived Fibroblasts and Breast Cancer Cells. Cancer Res 2019; 79(12): 3139-51.
[http://dx.doi.org/10.1158/0008-5472.CAN-18-2293] [PMID: 30992322]

[193] Reid JA, Palmer XL, Mollica PA, Northam N, Sachs PC, Bruno RD. A 3D bioprinter platform for mechanistic analysis of tumoroids and chimeric mammary organoids. Sci Rep 2019; 9(1): 7466.
[http://dx.doi.org/10.1038/s41598-019-43922-z] [PMID: 31097753]

[194] Nayak B, Balachander GM, Manjunath S, Rangarajan A, Chatterjee K. Tissue mimetic 3D scaffold for breast tumor-derived organoid culture toward personalized chemotherapy. Colloids Surf B Biointerfaces 2019; 180: 334-43.
[http://dx.doi.org/10.1016/j.colsurfb.2019.04.056] [PMID: 31075687]

[195] Dekkers JF, Whittle JR, Vaillant F, *et al.* Modeling Breast Cancer Using CRISPR-Cas9–Mediated Engineering of Human Breast Organoids. J Natl Cancer Inst 2020; 112(5): 540-4.
[http://dx.doi.org/10.1093/jnci/djz196] [PMID: 31589320]

[196] Yu J, Huang W. The Progress and Clinical Application of Breast Cancer Organoids. Int J Stem Cells 2020; 13(3): 295-304.
[http://dx.doi.org/10.15283/ijsc20082] [PMID: 32840232]

[197] Campaner E, Zannini A, Santorsola M, *et al.* Breast Cancer Organoids Model Patient-Specific Response to Drug Treatment. Cancers (Basel) 2020; 12(12): 3869.
[http://dx.doi.org/10.3390/cancers12123869] [PMID: 33371412]

[198] Rosenbluth JM, Schackmann RCJ, Gray GK, *et al.* Organoid cultures from normal and cancer-prone human breast tissues preserve complex epithelial lineages. Nat Commun 2020; 11(1): 1711.
[http://dx.doi.org/10.1038/s41467-020-15548-7] [PMID: 32249764]

[199] Dekkers JF, van Vliet EJ, Sachs N, *et al.* Long-term culture, genetic manipulation and xenotransplantation of human normal and breast cancer organoids. Nat Protoc 2021; 16(4): 1936-65.
[http://dx.doi.org/10.1038/s41596-020-00474-1] [PMID: 33692550]

CHAPTER 8

Current Advances in the use of Tumor Organoids in Lung Cancer Modeling and Precision Oncology

Bharti Bisht[1,*], Arkaprabha Basu[2], Keshav S. Moharir[3], Swati Tripathi[4], Rohit Gundamaraju[5], Jyotirmoi Aich[6], Soumya Basu[7] and Manash K. Paul[8,9,*]

[1] *Department of Microbiology, Kasturba Medical College, Manipal Academy of Higher Education, Manipal, Karnataka, 576104, India*

[2] *Harvard John A. Paulson School of Engineering and Applied Sciences, Harvard University, Cambridge, Boston, MA 02134, United States*

[3] *Department of Pharmaceutics, Gurunanak College of Pharmacy, Nagpur, Maharashtra, India*

[4] *Section of Electron Microscopy, National Institute of Physiological Sciences, Okazaki, Japan*

[5] *Division of Gastroenterology, Department of Medicine, Washington University School of Medicine, St Louis, MO, USA*

[6] *School of Biotechnology and Bioinformatics, Dr. D. Y. Patil Deemed to be University, CBD Belapur, Navi Mumbai, Maharashtra, 400 614, India*

[7] *Cancer and Translational Research Centre, Dr. D. Y. Patil Biotechnology and Bioinformatics Institute, Dr. D. Y. Patil Vidyapeeth, Pune, Maharashtra, 411 033, India*

[8] *Department of Radiation Biology and Toxicology, Manipal School of Life Sciences, Manipal Academy of Higher Education, Manipal, Karnataka, 576104, India*

[9] *Division of Pulmonary and Critical Care Medicine, Department of Medicine, David Geffen School of Medicine, UCLA, Los Angeles, CA, 90095, USA*

Abstract: Lung epithelium involves adult stem or progenitor cells that possess self-renewal, differentiation, and self-organizing potential and form the concoction of tissue-specific organoids. Researchers have used genetically modified lung organoids to study different aspects of lung tumorigenesis. Another approach is the patient-derived lung organoid to create a more representative lung cancer model with the tumor microenvironment, extracellular matrix, and immune component. The *In vitro* patient-derived organoids histologically and functionally mimic the related parent tumors. Lung cancer organoids and organoid-co-cultures can be used to dissect difficult-t--answer questions, especially regarding human lung cancer. Lung cancer organoids are

* **Corresponding authors Bharti Bisht and Manash K. Paul:** Department of Radiation Biology and Toxicology, Manipal School of Life Sciences, Manipal Academy of Higher Education, Manipal, Karnataka, 576104, India & Division of Pulmonary and Critical Care Medicine, David Geffen School of Medicine, University of California Los Angeles, Los Angeles, CA, 90095, USA & Department of Radiation Biology and Toxicology, Manipal School of Life Sciences, Manipal Academy of Higher Education, Manipal, Karnataka, 576104, India; E-mails: Bbisht@mednet.ucla.edu, manashp@ucla.edu

used not only for understanding tumor biology but also to undertake biomarker studies, and drug screening, evaluate immunotherapeutics, and target tumor microenvironment, and personalized medicine. Lung organoids can also be used to create organoid biobanks for future gene-specific pre-clinical trials and evaluation. This chapter will present an overview of the therapeutic areas in which lung cancer organoids are transforming therapeutic discovery and development, followed by a discussion of future prospects.

Keywords: Cancer modeling, Drug discovery, Patient-derived xenograft, Tumor microenvironment, Tumor organoid, Therapy.

INTRODUCTION

Lung cancer with an expected 1.8 million fatalities, is still the most significant cause of cancer-related mortality worldwide, affecting both smokers and nonsmokers, despite early identification and treatment advances. Men have the highest incidence and fatality rates for lung cancer, while women rank third in incidence and second in mortality [1]. Lung cancer is heterogeneous with numerous genetic and epigenetic alterations and has the lowest five-year survival rate (10 to 20%) among all malignancies [2]. Eighty to eighty-five percent of lung malignancies can be histologically categorized as non-small cell lung cancer (NSCLC), and approximately fifteen percent are small cell lung cancer (SCLC). Lung adenocarcinoma (LUAD) and lung squamous cell carcinoma (LUSC) are the most prevalent subtypes of NSCLC [3, 4]. Lung cancer patients have a poor prognosis and survival rate due to a lack of early detection and ineffective treatments. The treatment options for lung cancer are surgery, chemotherapy, radiation, or targeted therapy, but survival is dismal. Treatment exhibits poor therapeutic indices, varied health implications, and chemotherapeutic toxicity [2]. The accuracy, efficacy, and success are limited by tumor cell resistance, immune suppression, and inter-subject variation. For lung cancer therapeutics to be successful, tailored cancer models that accurately depict the heterogeneity of tumors mimicking that of the patient's tumor are required immediately [4, 5]. Recent advances in 3D culture techniques, single-cell transcriptomics, pharmacogenetics, imaging techniques, and immunotherapy have inspired better therapeutic strategies [4, 6, 7].

The existing lung cancer models, *i.e.*, 2D tumor cell culture and animal models suffer from multiple challenges, including the absence of verified driver genetic mutations, stepwise progression, ambiguous cell of origin, lack of human immune components, difficulties replicating the complex lung tumor microenvironment (TME) and predicting therapeutic response [8]. In recent years, the price of new anti-cancer medications has risen partly due to the increased complexity and limited success of clinical trials, especially in lung cancer. Therefore, the discovery of lung cancer treatments has faced a severe setback. Recent developments in 3D culture systems constitute an invaluable set of tools for studying cell biology, especially cancer. A 3D cell culture is an artificially created smart environment *In vitro* wherein biological cells may grow or interact in 3D [6]. Organoids are a type of 3D cell culture system and are self-renewing, self-organizing 3D cell aggregates produced from primary tissue-derived adult stem cells (ASCs), embryonic stem cells (ESCs), induced pluripotent stem cells (iPSCs), and tumor cells. It has been demonstrated that tissue-resident ASCs, ESCs, and iPSCs may self-organize into 3D architectures that resemble *In vivo* organs such as the breast, lung, colorectum, stomach, liver, pancreas, ovary, prostate, and brain. Organoids have a limited population of self-renewing stem cells that can develop into all major cell lineages, can be co-cultured with niche components, cryopreserved, and grown indefinitely. Organoids mimic the corresponding tissue's histological, genetic, and physiological state and are amenable to genetic modifications, making them highly relevant for basic and translational research applications. Though efficient, regular organoids lack the entire spectrum of cells and components, especially the immune repertoire seen in a patient's tumor, which is a disadvantage.

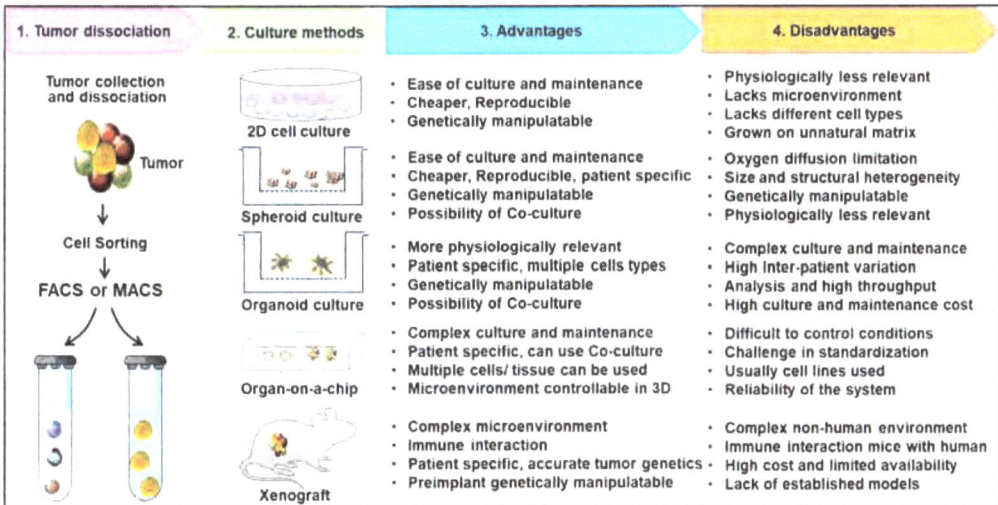

Fig. (1). Models used in cancer research. Patient-derived xenograft or cancer cells can be cultured using multiple model systems like 2D cell culture, spheroid culture, organoid culture, organ-on-a-chip, and rodent xenograft models. This figure also discusses the advantages and disadvantages of the different models. The figure is inspired by [10].

Tumoroids, on the other hand, are generated directly using freshly isolated patient tumor tissue and preserve the cellular complexity of the TME and native extracellular matrix (ECM), in contrast to stem cell-derived organoids. Patient-derived tumor organoids (PDTOs) are tumoroids and may keep tumor epithelial cells in a near-native state, retaining the disease heterogeneity and human tumor immune microenvironment (TIME), opening the door for targeted cancer immunotherapy [9]. The recent advances in 3D cell culture systems and gene modification open the possibility of using gene-modified organoids and tumoroids to evaluate and study cancer immunotherapies, such as antibody-based immunotherapy, oncolytic virus treatment, and adoptive cell transfer therapy. Thereby combining the strengths of PDTOs with cutting-edge multi-omics techniques, robotics, and machine learning (ML) can help develop cancer immunobiology and immunotherapy faster. Organoids and tumoroids can serve as unique platforms for preclinical cancer research. A detailed description of different *In vitro* models used in cancer research is schematically presented in Fig. (**1**) [10]. Fig. (**1**) summarizes these models, the culture methods, advantages, and disadvantages. This chapter highlights the current advances in the use of tumor organoids in lung cancer modeling and precision oncology.

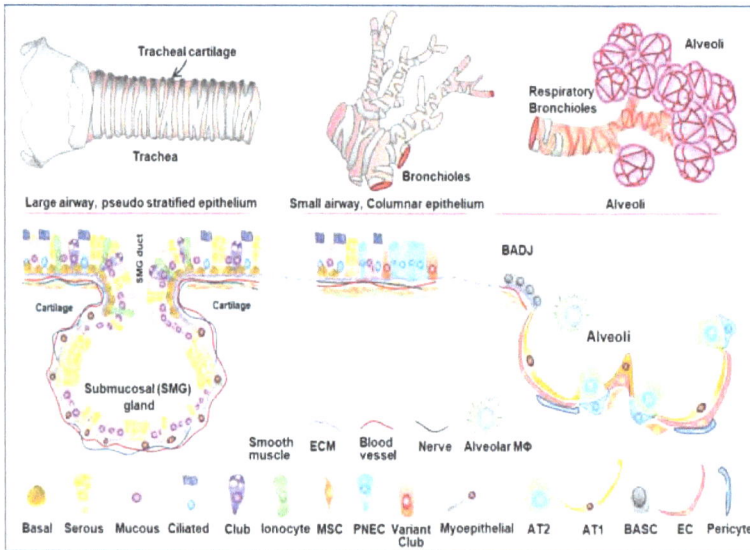

Fig. (2). Adult lung stem/progenitor cells and their location in the lung. The diagram illustrates the broad airway (pseudostratified epithelium), the minor airway (Columnar epithelium), the respiratory bronchioles, and the alveoli. Future lung epithelial stem/progenitor cells and differentiated pulmonary cells are spatially distributed from proximal to distal along the axis. Along the Proximal-to-Distal axis of the airway, there are many possible region-specific habitats for stem cells. The potential progenitor/stem cells are located in their local niches, where they preserve their stem/progenitor properties and the capacity to differentiate into various lung cell types. PNEC: Pulmonary neuroendocrine cell; MSC: Mesenchymal stem cell; AT1: Alveolar type I cell; AT2: Alveolar type II cell; EC: Endothelial cell; BASC: Bronchoalveolar stem cell; M: Macrophage; ECM: Extracellular matrix. The figure is used with permission [12].

LUNG ORGANOID AND MODELING TUMOR

The lung can be divided into the proximal conducting and the distal respiratory zone and is coated with a continuous epithelial layer and contains epithelial, immune, endothelial, and stromal cells. From the nasal passage, the trachea, principal bronchi, intra-pulmonary bronchi, and the bronchioles are considered the conducting zone. The respiratory zone comprises respiratory bronchioles, alveolar ducts, and alveolar sacs, where the lung's air and blood exchange gases. Lung epithelium has region-specific stem or progenitor cells and is subsequently used for growing lung organoids [11, 12]. Fig. (**2**) shows a schematic representation of the area-specific distribution of lung stem /progenitor cells used for generating lung organoids and especially model human lung-specific diseases. Lung organoids *In vitro* replicate the 3D structure and function of the lung tissue. Human lung organoids (HLO) may be used to research epithelial-mesenchymal cross-talk during lung development, infectious and non-infectious lung diseases, lung cancer, regenerative medicine, tissue engineering, and pharmacological safety and testing of the effectiveness of lung therapeutics [7].

Development of Lung Organoids

Lung organoid development started when Jennings and colleagues developed the lung organoid in 1987 by cultivating alveolar epithelial type 2 (AT2) cells *In vitro* [13]. In 2009, Rock *et al.* showed that mouse and human airway basal cells could make self-renewing, ciliated tracheospheres [14]. Hegab *et al.* 2012, isolated human basal and submucosal gland duct stem cells and showed differences in organoid formation ability [15, 16]. Barkauskas *et al.* in 2013 co-cultured pure AT2 cells with primary (PDGFR+) lung fibroblasts to form self-renewing alveolospheres and reasoned that AT2 cells are adult lung stem cells [17]. Gotoh and colleagues identified induced alveolar epithelial progenitor cells (AEPCs) from human ESCs and generated alveolar epithelial spheroids in 2014 [18]. Dye *et al.* generated HLOs using step-wise differentiation of human pluripotent stem cells (PSCs) in 2015 [19]. Using iPSC-derived mesenchymal cells, airway epithelial cells, umbilical vein endothelial cells, and collagen-coated alginate beads, Wilkinson and colleagues created 3D lung organoids in a bioreactor [20]. The bead-based technique helped

to generate HLOs in a scalable manner conducive to quick therapeutic screening and personalized medicine [21, 22]. Miller *et al.* in 2019 described a new protocol to differentiate human PSCs to generate organoids with alveolar cells surrounded by lung mesenchyme [23]. Recently Sachs *et al.* developed a methodology for the long-term expansion of HLOs and disease modeling [24]. Lung organoids may be readily genetically altered using standard gene modifying tools like CRISPR-

Cas9, allowing researchers to explore the effects of a single gene alteration or combinations in lung injury repair, mechanism of tumorigenesis, and screening for lung cancer therapeutics.

Lung Organoids for Studying Lung Tumorigenesis and Therapeutic Screening

Lung organoids are promisingly being used to model lung cancer. Cigarette smoke is considered a critical reason for lung cancer initiation, and organoid-based studies can help understand the mechanism of such tumorigenesis. Cigarette smoke includes several poisonous, carcinogenic, and mutagenic compounds and induces the production of reactive oxygen species (ROS) in the particulate and gas phases, all of which have the ability to cause oxidative damage to living organisms. Paul and colleagues used airway basal stem cell tracheospheres to demonstrate that ROS levels can activate Nrf2, which triggers the Notch pathway to drive ABSC self-renewal and may result in lung cancer premalignancy in mice [25]. To examine the impact of cigarette smoke on mesenchymal-tumor cell interaction, fibroblasts were seeded with mouse pulmonary EpCam+ cells in Matrigel, and spheroids were produced [26]. Data suggest cigarette smoke affects cellular crosstalk between stromal mesenchymal cells and epithelial counterparts, thereby restricting lung repair [27].

Tamela and colleagues employed lung cancer Kras^{G12D} overexpression and p53 ablated mice LUAD to produce tumors and subsequently generated tumor-derived spheroids to demonstrate that a Wnt-producing niche promotes the expansion and development of LUAD [28]. Tata *et al.* isolated EpCAM+ cells from Kras overexpression, Sox2 overexpression, and NkX2-1 inactivationed murine tumor model and grew organoids to mimic LUSC [29]. Hai and colleagues deleted tumor suppressors (Pten, p53, and p16) in SOX2 Cre-dependent lung organoids, evaluated the therapeutic efficacy and immunologic consequences of PD-1 blocking and WEE1 inhibition, and hypothesized that Anti–PD-1 and DNA damage-inducing treatments might aid in the treatment of LSCC [30]. Dost *et al.* employed alveolar epithelial organoids to recreate early-stage LUAD in a KRAS^{G12D} background and used single-cell RNA sequencing to evaluate gene expression, revealing a loss of AT2 cell differentiation markers and concomitant activation of tumor markers [31]. Tien *et al.* used adenoviral Cre to induce lung-specific expression of a KRAS^{G12D} activation and inactivation of p53 and Ago2 created nodule-derived organoids and proved that AGO2 increased tumor growth in KRAS-driven mice models [32]. Using murine alveolar type 2 cells (AT2), Naranjo *et al.* generated organoid-based models of KRAS and ALK-mutant LUAD and demonstrated that Kras or Alk in combination with p53 deletion permitted organoid expansion in the lack of growth factors [33].

Lung organoids can be an attractive model for studying epithelial-immune cell interaction and its role in lung tumorigenesis. Recently, Vazquez-Armendariz and colleagues used mouse multi-lineage bronchioalveolar stem cells (EpCAMhighCD24lowSca-1+) and cultured them with resident mesenchymal cells, producing bronchioalveolar lung organoids (BALO). BALOs were microinjected with tissue-resident macrophages, leading to enhanced differentiation of alveolar epithelial type I cells (AECI) and ciliated cells [34]. To address existing challenges surrounding lung cancer, it is necessary to do further research using human organoids as the platform.

LUNG TUMOR ORGANOIDS AND TUMOR MODELING

Cancer cells are usually grown in 2D, on plastic plates using a culture medium, or as xenografts, which do not accurately replicate the complexity of human cancers due to the lack of tumor-initiating cells, a human-specific TME, reciprocating ECM, while introducing long-term *In vitro* culture-associated genetic variance [35]. Patient-derived xenograft (PDX) models are hindered by sample availability, time taking, logistical and budgetary challenges, absence of human-specific TME, and ethical considerations. As an alternative, tumor organoids have addressed multiple challenges and achieved high success in lung cancer modeling and developing cancer therapeutics [36]. Tumoroids imitate primary tissues in architecture, function, histopathology, genetic profile, mutational landscape, and therapeutic response. NSCLC-3D cultures create solid spheres without apico-basal polarity compared to lung stem cell-derived organoids, which exhibit a hollow lumen, suggesting that they retain the physiological properties of tumor cells [37]. Tumor organoids help study carcinogenesis and cancer development *In vitro* and show great translational promise. To facilitate fundamental immuno-oncology research and clinical translation, an increasing number of studies are focusing on the essential features of organoid formation and innovative organoid culture techniques.

Patient tumor collection and optimal isolation of single cells or cell clusters play a crucial role in an efficient organoid generation; hence, multiple groups have studied this aspect. Tumor specimens as small as 1-4 cm^3 [38, 39], and even smaller biopsy samples (0.2 to 0.3 cm^3) have also been used to generate lung tumor organoids [40, 41]. LCOs are also generated from pleural effusion-derived tumor cells [42], circulating tumor cells, and also from PDX. Multiple techniques may be used in the creation of tumor organoids. Single-cell types from a tumor cell line or primary cells can be grown into spheroids. Again, organotypic cocultures can be established using gene-modified ESC/ iPSC/ ASC-derived epithelial organoids and co-cultured with stromal and other cells embedded in an ECM (*e.g.*, Matrigel). Another approach can be to dissociate tumors, isolate primary cells from the fresh tumor without cell enrichment, and grow them together in 3D cultures as tumor organoids. A rapid and optimized protocol for the generation of patient-derived lung tumor organoids can help acquire the characteristics and distinctiveness of their corresponding parental tumors, allowing for the validation of personalized therapies closely linked to the organoid characteristics and will help in the creation of a lung cancer organoid (LCO) biobank for future research.

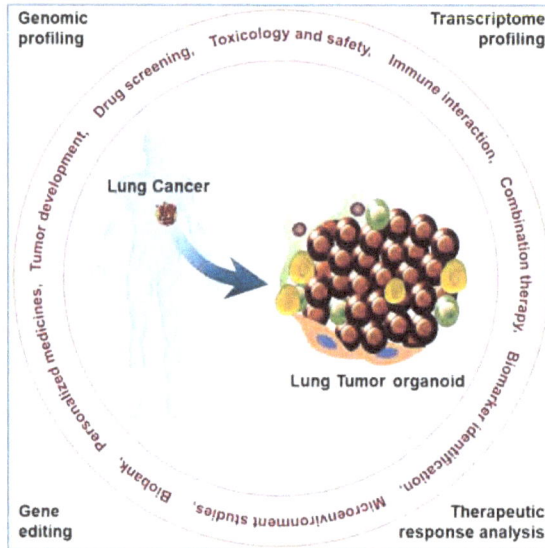

Fig. (3). Tumor organoid model in lung cancer research. Several key areas can be addressed using surgically resected/biopsied lung tumor tissue, circulating tumor cells, and cancer cell line-derived organoids, as shown in the circle. Tumor development, drug screening, toxicology and safety evaluation, immune interaction studies, combination therapy, biomarker identification, tumor microenvironment associated studies, lung tumor organoid biobank, and personalized lung cancer therapeutics. Several techniques can help comprehend the lung tumor and lung tumor-derived organoids, including genomic and transcriptomic profiling, gene editing, therapeutic response analysis, and in-depth data analysis.

A pertinent challenge is the tumor heterogeneity and clonal evolution during tumor organoid development [43]. To restrict the growth of normal epithelium, and amplify and investigate a particular mutational phenotype, protocols for generating tumor organoids with particular mutations protocols have been established. One technique employed was the TP53 stabilizing agent Nutlin-3a to specifically block the development of wild-type TP53 cells, thereby enriching mutant TP53+ tumor cells [24]. Recently, Kim *et al.* generated lung cancer organoids and normal lung organoids from surgically removed patient's lung tumor tissue utilizing Matrigel as ECM and minimal basal medium containing epidermal growth factor (EGF), basic fibroblast growth factor, insulin, and transferrin. The minimal basal media did not support normal airway epithelial expansion, thereby allowing tumor cell-derived organoids to grow [38]. Another problem is the patient-matched control, and this challenge was overcome by adding WNT3A, TGF, and BMP signaling pathway inhibitors to the basal media while growing lung tumor organoids [24, 38]. Fig. (**3**) shows a schematic representation of the lung tumor organoid model and its critical applications and innovative techniques that can be linked with this organoid model to advance human lung cancer research and drug discovery.

Lung Cancer Organoid and Drug Screening

The lung cancer organoids may serve as an efficient platform for drug screening and biomarker identification, thereby allowing for more accurate predictions of patient-specific therapeutic responses. Zhang *et al.* created lung cancer patient (Stage I/II)-derived tumor spheroids (PDS), which retained the genetic and cytological features and phenocopied the parent tumor, these PDSs were subsequently used for drug screening [44]. Shi *et al.* used patient tissue and 35 previously established PDXs for generating organoids which were subcutaneously injected in NOD/Scid mice. Using these organoids, they found FGFR1 overexpression and screened for specific inhibitors and their combinations [45]. Li *et al.*, set up an LUAD organoid biobank. Organoids retained the mutational spectrum of patients and used the same for biomarker discovery and drug screening [39]. Kim *et al.* established LCO and showed that organoids mimic the original lung tumor. Fig. (**4**) shows the histological comparison between LCOs and their parent lung tumors with evidence of LUAD, LUSC and a mixed phenotype, and were developed as a single cell clone [38]. Hu *et al.* adopted a mechanical sample processing approach to manufacture LCOs using surgically excised and biopsy samples that mimicked the histological and genetic characteristics of original tumors and possessed long-term growth capability. Using an integrated superhydrophobic microwell array chip (InSMAR-chip), they demonstrated that hundreds of LCOs may be created in one week to test for clinically significant therapeutic responses [46].

Artificially produced LUAD-derived lung organoid simulations, or aiLUNG-LUAD, were employed by Esmail and Danter to mimic LUAD, enabling biomarker identification and performing high throughput drug screening. They imitated lung adenocarcinoma by using DeepNEU machine learning algorithms for their stem cell and organoid simulation platform, including broad gene editing [47]. There is a dearth of studies on lung tumor organoids, which provides a broad scope of improvement in this domain [35]. PDOs are employed in drug development studies and may help discover innovative cancer therapies. Lung cancer has been studied using long-term tumor patient-derived organoid cultures. In addition to this, colorectal, pancreatic, breast, ovarian, prostate, bladder, and liver cancers have been used to create living organoid biobanks.

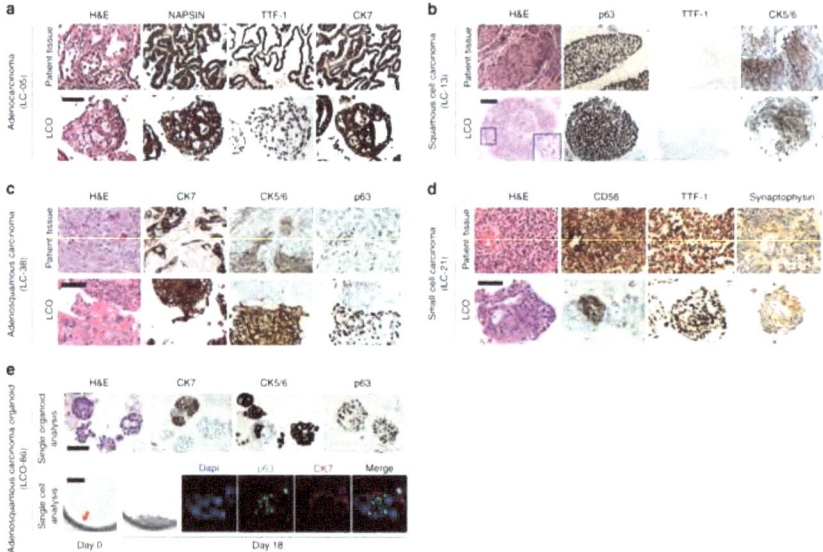

Fig. (4). LCOs replicate the features of the original tissues. a–d Images representing H&E and IHC stains of LCOs and their original lung cancer tissues. Individual squamous carcinoma cells show cytoplasmic keratinization, as shown in the magnified blue boxes of image section b. Scale bars are 100 μm. e H&E and IHC-stained images of an adenosquamous organoid (upper panel). Individual organoids displayed markers for either adenocarcinoma (CK7) or squamous cell carcinoma (CK5/6 or p63). Some organoids exhibited markers representing mixed cell markers, consisting of CK7+ and CK7- cells, CK5/6+ and CK5/6- cells, and p63+ and p63- cells. The scale bar represents 100 μm. Bright-field and immunofluorescence (IF) pictures show an adenosquamous carcinoma organoid derived from a single cell (lower panel). IF staining was carried out on Day 18 after seeding. A single cell organoid comprised p63+/CK7- cells and p63-/CK7+ cells. The red arrow denoted a single planted cell in a microwell. 100 μm scale bar is used in bright-field microscopy images. 20 μm scale bar in IF images. The image used is published by [38].

Lung Cancer Organoid and Immune-oncology

The lung TIME is complex and depends on the lung cancer type and genetic feature. Patient-derived xenografts have been extensively embraced for highly predictive *In vivo* therapy testing in translational research over the past two decades. PDX are created from tumor samples and thus retain the morphological, genetic, and pathophysiological identity of their *In vivo* counterparts, as well as mimicking their pharmacological profiles and treatment responsiveness. Recent data suggest that engraftment and serially passaged NSCLC PDXs remain histologically representative of the original patient tumor, with a conserved mutational profile in NOD/scid mice [48]. Genetically engineered mouse models (GEMMs) and syngeneic mouse models have also been used to understand tumor-immune interaction, but the absence of human immune repertoire limits these studies significantly in the context of immune interaction. Moreover, GEMMs may not accurately replicate human tumors phenotypically. Cancer-immune cell cocultures that simulate the reciprocal interaction between tumor cells and the immune component may assist in developing preclinical models for understanding immunotherapy response. Patient-derived epithelial-only PDTOs are advantageous for epithelial biology but lack the immune and other nonimmune components of the TIME and are not suitable for understanding tumor-matrix and tumor-immune interactions. Currently, two methodologies are used to address the specific and personalized tumor-TME interactions. In the reconstitution strategy, tumor organoids and immune cells are grown independently and are subsequently cocultured to study organoid-immune cell interactions. In comparison, the holistic approach employs tumor organoids produced directly from tumors while retaining the endogenous immune cells.

The tumor organoid culture system must be optimized for organoid-guided customized cancer immunotherapy. Dijkstra *et al.*, using autologous co-cultures of cancer organoids with peripheral blood mononuclear cells (PBMCs) from mismatch repair-deficient NSCLC and one LUSC patient, exhibited the capacity of T lymphocytes to recognize and destroy cancer cells. Interestingly, the tumor-sensitive T cells did not target healthy organoids [40]. Utilizing PDTOs and respective PBMCs, Takahashi *et al.* evaluated immune-checkpoint inhibitors (ICIs). The combination of nivolumab and pembrolizumab significantly increased PBMC-mediated PDTO cell death [49]. This work expands the potential use of PDTOs for antibody-dependent cellular toxicity (ADCC), immune-synapse formation, and personalized ICI treatment. Recent improvements in complicated tumor organoids demonstrate that the PDTO system may be utilized to evaluate cancer immunotherapy and identify innovative combination treatment strategies. The immune-tumor organoids system may help discover new antibody-based cancer therapies. TAK1/IKKe inhibitors and CDK4/6 inhibitors have been shown to synergize with PD-1 inhibition to improve tumor killing. The complicated cancer organoids were employed in the discovery of an innovative cibisatamab treatment strategy.

Lung tumor organoids are also used to study the role and mechanism of the oncolytic virus to target lung cancer. The immune checkpoint, PD-L1, is present on healthy and malignant cells, hence effector functions are diminished. To reduce ICI side effects, Hamdan used the oncolytic virus (OV) approach. OVs may selectively locate, infect, and lyse tumor cells, halting and reducing tumor growth. They either naturally target cancer cells or can be genetically altered to that effect and are increasingly becoming an exciting immunotherapy option for lung cancer patients. Hadman and colleagues created an Fc-fusion peptide against PD-L1 that contains constant sections of IgG1 and IgA1 to activate diverse immune effector populations, including neutrophils and natural killer cells. Adenoviruses cloned with this Fc-fusion peptide showed improved oncolytic effectiveness in complicated immune-organoids [50]. Yang *et al.* used a tumor necrosis factor-related apoptosis-inducing ligand (TRAIL) gene-armed OV (ZD55-TRAIL) and demonstrated significant tumor killing using A549 lung cancer cell spheroids. Further studies using PDTOs can be more informative and clinically applicable. OV has had a roller-coaster ride, with early excitement followed by a downturn and a recent return to center stage owing to improvements in molecular biology and advances in human LCO modelings.

Studying Lung Tumor Cell-microenvironment Interaction using Tumor Organoids

In an effort to simulate and better comprehend the interactions between tumor cells and stromal and immune cells, sophisticated tumor organoid cultures are being established. Cancer-associated fibroblasts (CAFs) are abundant in tumor stroma and play essential roles in TIME; therefore, CAFs are integrated into tumor organoid co-culture. Recently, PDTOs and CAFs were used in a 3D coculture system to study tumor cell-CAF interactions [51]. In a 3D coculture system of LUSC organoids with CAFs and ECM, Chen *et al.* demonstrated that stromal cell-derived stimuli might overcome tumor cell inherent oncogenic alterations, thereby modulating the disease phenotype. Thus these data obtained from organoid cocultures demonstrate the existence of dynamic feedback signaling loops amongst the LUSC tumor and stromal cells [51].

The modern organoid system also lacks vascular circulation; therefore, organoids reach a maximum size beyond which necrosis occurs. New advancements in organoid vascularization and perfusion are necessary to overcome this obstacle. A recent study by Seitlinger *et al.* attempted to improve the tumor organoid microvasculature. 3D NSCLC tumor spheroids were combined with human umbilical vein endothelial cells (HUVEC) and normal human lung fibroblasts in a fibrin gel to form a perfusable vasculature, leading to infiltration of endothelial cells and the formation of a vascularized network [52]. It has been shown that pre-vascularization using endothelial cells may stimulate the creation of a vascular network when organoids are grown on microfluidic chips. Another approach is bioprinting of endothelial cells along with lung tumor cells to form the organoids. The growth of lung tumors is associated with ECM remodeling, a potential area that needs to be investigated using the organoid model. However, large batch-t--batch variability, xenogenic contamination, poorly characterized ECM components, and inadequate mechanical property control hamper the knowledge of organoid-ECM interactions. Some disadvantages of engineered matrices include poor cultural efficiency and insufficient spatiotemporal control to represent TIME dynamics. More studies are needed to optimize ECM for tumor organoid growth and improve the understanding of cancer-ECM interactions *in vitro.*

LUNG ORGANOIDS AND FUTURE DIRECTIONS

Lung organoids and LCOs have created new paths for transferring fundamental cancer research into clinical treatment, but several recent scientific advances can help achieve future organoid technological breakthroughs for quick and efficient lung cancer therapeutic discovery. Achieving tumor organoid-based clinical decision-making requires further research and development. The CRISPR–Cas technology has enhanced genetic modification and screening in *In vitro* and *In vivo* human cancer models, uncovering previously undiscovered cancer drivers. Hans Clevers' group was the first to use CRISPR–Cas in an organoid model in 2013 [53]. They effectively edited the CFTR gene in intestinal organoids. Recent data using tumor organoids provides strong evidence that CRISPR-Cas9 genome editing can be used for efficient gene modification of lung cancer patient-derived organoids for large-scale genetic screening, drug discovery, and cancer immunotherapy [54].

Another exciting area is to identify actionable tumor antigens for patient-specific immunotherapy. MS-based immunopeptidomics identifies MHC-bound peptides *In vivo* with high throughput and has been used to analyze cell lines, tumors, and blood plasma. Poor resolution, the necessity for a large amount of sample, and the lack of high-throughput capacity are limiting constraints for MS-based proteomics. In 2017, Cristobal and colleagues completed the first deep proteome analysis of human colon organoids, identifying shared traits and individual diversity that may assist in customized cancer treatment [55]. Demmers and associates performed the first tumor organoid proteomics study in 2020 using single cell-derived colorectal organoids. Individual tumor organoids from the same patient revealed substantial HLA peptide presentation diversity [56]. Organoid-based proteomic investigations are already possible and may soon broaden the toolset for precision lung cancer treatment. The advent of adoptive cell therapy also offers a novel therapeutic strategy and fresh hope for lung cancer. Adoptive cell transfer treatment employs genetically modified T cells with chimeric antigen receptors (CARs) as an alternative to immune checkpoint inhibitors. While CAR-T cells targeting CD19 have shown success in hematological malignancies such as B cell lymphoma and AML, efficacy in solid tumors is yet unknown. Complex organoids have shown considerable promise as efficient platforms for CAR cell effectiveness testing. PDTOs might be used to increase tumor-reactive T cells and boost lung cancer specific immune responses.

Exosomes are nanosized (30-150 nm) extracellular vesicles (EVs) released by most eukaryotic cells and aid cell-cell communication [57]. Exosomes may deliver medications and diagnostic chemicals to tumor cells [58]. To facilitate intercellular communication, tumor cells release tumor-derived exosomes (TEX), harboring unique cargo suggestive of the parent tumor cells and have an immunomodulatory function. TEX induces immunosuppression and tumorigenesis by inducing apoptosis of T cells and natural killer cells, inhibiting dendritic cell differentiation, myeloid-derived suppressor cell expansion, promoting myofibroblast formation, angiogenesis, tumor growth, and metastasis [59]. Exosomes from tumor organoids may help us better understand immunosurveillance and cancer immunotherapy. Exosomal miR-25 and miR-210 may promote an oncogenic phenotype in gastroids when cocultured with esophageal adenocarcinoma-derived exosomes [60]. Recent research suggested that PDTO-derived exosomal miRNAs might be used to detect precancerous lesions in colorectal cancer. Cancer detection and treatment may be aided by PDTO-derived exosome manufacturing platforms that are standardized and scalable. More research on the use of organoid-derived exosomes for cancer immunotherapy is ongoing, with findings expected soon. Genome-wide CRISPR screens, immunoproteomics, and exosomes in lung tumor organoids hold enormous promise for future fundamental and translational cancer research.

CONCLUSION

A faithful preclinical model that permits improved translation from bench to bedside is urgently needed to improve the effectiveness of lung cancer therapeutics, especially lung cancer immunotherapy. Immunotherapy for cancer has shown promising results in treating blood malignancies and melanomas. Immune infiltration in lung tumors is linked to tumor mutational burden, tumor mutational profile and is associated with poor prognosis. Further research is needed to employ patient-relevant, extensible, and genetically stable models. Using co-cultures of autologous PBMC or tumor-infiltrating T cells with matching tumor and normal organoids may predict the cytotoxicity of T cells towards lung PDTOs and test the effectiveness of checkpoint inhibitors, thereby predicting patient response *in vivo*. Conjugating patient-produced organoids with other cells like fibroblasts or macrophages provides an exciting new *In vitro* platform for patient-relevant therapeutic development targeting the TME. Also, improvements are needed in protocols for the long-term preservation of patient-derived immune cells and CAFs. Tumor-derived organoids can simulate the effects of immunotherapy. While the therapeutic use of organoid technology is appealing, considerable obstacles remain. Tumor organoids are typically produced from biopsies representing only a tiny portion of the tumor. Intratumoral heterogeneity may hamper clinical translation if the intricacy of the underlying malignant lesion is underestimated. Many clinical studies are presently underway to evaluate PDTOs in precision cancer therapy. Complex tumor organoid culture methods may help researchers, and clinicians better understand the dynamic connections between cancer and the immune system. Tumor organoids are currently the most accurate *In vitro* technique for recreating human cancer tissues. Genome-wide CRISPR screens, proteomics, and exosomes in tumor organoids hold enormous promise for future fundamental and translational cancer research. Clinical trials in a dish using patient-derived organoids strongly predict patient response and permit agnostic or genetically customized evaluation of a novel therapy plan. These investigations can improve clinical therapeutic candidates by predicting patient response more precisely and help deliver quick and less expensive therapeutics for lung cancer patients.

ABBREVIATIONS

ASCs	Adult stem cells
AT2	Alveolar epithelial type 2 cells
AEPCs	Alveolar epithelial progenitor cells
AECI	Alveolar epithelial type I cells
ADCC	Antibody-dependent cellular toxicity
BALO	Bronchioalveolar lung organoids
CAF	Cancer-associated fibroblasts
CARs	Chimeric antigen receptors
ESCs	Embryonic stem cells
ECM	Extracellular matrix
EGF	Epidermal growth factor
EVs	Extracellular vesicles
GEMMs	Genetically engineered mouse models
HLO	Human lung organoids
iPSCs	Induced pluripotent stem cells
ICIs	Immune-checkpoint inhibitors
LUAD	Lung adenocarcinoma
LUSC	Lung squamous cell carcinoma
LCO	Lung cancer organoid
NSCLC	Non-small cell lung cancer
OV	Oncolytic virus
PDTOs	Patient-derived tumor organoids
PSCs	Pluripotent stem cells
PDX	Patient-derived xenograft
PDS	Patient-derived tumor spheroids

PBMCs   Peripheral blood mononuclear cells

SCLC   Small cell lung cancer

TRAIL   Tumor-necrosis factor related apoptosis-inducing ligand

TME   Tumor microenvironment

TIME   Tumor immune microenvironment

TEX   Tumor-derived exosomes

ACKNOWLEDGEMENTS

BB acknowledges The Department of Biotechnology, Ministry of Science and Technology, Government of India for the Ramalingaswami Re-entry Fellowship. SB acknowledges Intramural Grants, Dr. D. Y. Patil Vidyapeeth (DPU), Pimpri, Pune, India to S. Basu [DPU/644-43/2021].

REFERENCES

[1]   Sung H, Ferlay J, Siegel RL, *et al.* Global Cancer Statistics 2020: Globocan Estimates of Incidence and Mortality Worldwide for 36 Cancers in 185 Countries. CA Cancer J Clin 2021; 71(3): 209-49.
[http://dx.doi.org/10.3322/caac.21660] [PMID: 33538338]

[2]   Ruwali M, Moharir K, Singh S, Aggarwal P. K Paul M. Updates in Pharmacogenetics of Non-Small Cell Lung Cancer. Pharmacogenetics 2021.

[3]   Mukherjee A, Paul M, Mukherjee S. Recent Progress in the Theranostics Application of Nanomedicine in Lung Cancer. Cancers (Basel) 2019; 11(5): 597.
[http://dx.doi.org/10.3390/cancers11050597] [PMID: 31035440]

[4]   Salehi-Rad R, Li R, Paul MK, Dubinett SM, Liu B. The Biology of Lung Cancer. Clin Chest Med 2020; 41(1): 25-38.
[http://dx.doi.org/10.1016/j.ccm.2019.10.003] [PMID: 32008627]

[5]   Xu H, Jiao D, Liu A, Wu K. Tumor organoids: applications in cancer modeling and potentials in precision medicine. J Hematol Oncol 2022; 15(1): 58.
[http://dx.doi.org/10.1186/s13045-022-01278-4] [PMID: 35551634]

[6]   Mukherjee A, Sinha A, Maibam M, Bisht B. K Paul M. Organoids and Commercialization. Organoids 2022. [Working Title]

[7]   Mahapatra C, Lee R, Paul MK. Emerging role and promise of nanomaterials in organoid research. Drug Discov Today 2022; 27(3): 890-9.
[http://dx.doi.org/10.1016/j.drudis.2021.11.007] [PMID: 34774765]

[8]   Neal JT, Li X, Zhu J, *et al.* Organoid Modeling of the Tumor Immune Microenvironment. Cell 2018; 175(7): 1972-1988.e16.
[http://dx.doi.org/10.1016/j.cell.2018.11.021] [PMID: 30550791]

[9]   Finnberg NK, Gokare P, Lev A, *et al.* Application of 3D tumoroid systems to define immune and cytotoxic therapeutic responses based on tumoroid and tissue slice culture molecular signatures. Oncotarget 2017; 8(40): 66747-57.
[http://dx.doi.org/10.18632/oncotarget.19965] [PMID: 28977993]

[10]   Porter RJ, Murray GI, McLean MH. Current concepts in tumour-derived organoids. Br J Cancer 2020; 123(8): 1209-18.
[http://dx.doi.org/10.1038/s41416-020-0993-5] [PMID: 32728094]

[11]   Cunniff B, Druso JE, van der Velden JL. Lung organoids: advances in generation and 3D-visualization. Histochem Cell Biol 2021; 155(2): 301-8.
[http://dx.doi.org/10.1007/s00418-020-01955-w] [PMID: 33459870]

[12]   Swami D, Aich J, Bisht B, Paul MK. Reconstructing the lung stem cell niche in vitro. 2022.
[http://dx.doi.org/10.1016/bs.asn.2022.05.001]

[13]   Shannon JM, Mason RJ, Jennings SD. Functional differentiation of alveolar type II epithelial cells in vitro: Effects of cell shape, cell-matrix interactions and cell-cell interactions. Biochim Biophys Acta Mol Cell Res 1987; 931(2): 143-56.
[http://dx.doi.org/10.1016/0167-4889(87)90200-X] [PMID: 3663713]

[14]   Rock JR, Onaitis MW, Rawlins EL, *et al.* Basal cells as stem cells of the mouse trachea and human airway epithelium. Proc Natl Acad Sci USA 2009; 106(31): 12771-5.
[http://dx.doi.org/10.1073/pnas.0906850106] [PMID: 19625615]

[15]   Hegab AE, Ha VL, Darmawan DO, *et al.* Isolation and *in vitro* characterization of basal and submucosal gland duct stem/progenitor cells from human proximal airways. Stem Cells Transl Med 2012; 1(10): 719-24.
[http://dx.doi.org/10.5966/sctm.2012-0056] [PMID: 23197663]

[16]   Hegab AE, Ha VL, Bisht B, *et al.* Aldehyde dehydrogenase activity enriches for proximal airway basal stem cells and promotes their proliferation. Stem Cells Dev 2014; 23(6): 664-75.
[http://dx.doi.org/10.1089/scd.2013.0295] [PMID: 24171691]

[17]   Barkauskas CE, Cronce MJ, Rackley CR, *et al.* Type 2 alveolar cells are stem cells in adult lung. J Clin Invest 2013; 123(7): 3025-36.
[http://dx.doi.org/10.1172/JCI68782] [PMID: 23921127]

[18]   Gotoh S, Ito I, Nagasaki T, *et al.* Generation of alveolar epithelial spheroids *via* isolated progenitor cells from human pluripotent stem cells. Stem Cell Reports 2014; 3(3): 394-403.
[http://dx.doi.org/10.1016/j.stemcr.2014.07.005] [PMID: 25241738]

[19]   Dye BR, Hill DR, Ferguson MAH, *et al. In vitro* generation of human pluripotent stem cell derived lung organoids. eLife 2015; 4e05098
[http://dx.doi.org/10.7554/eLife.05098] [PMID: 25803487]

[20]   Wilkinson DC, Alva-Ornelas JA, Sucre JMS, *et al.* Development of a Three-Dimensional Bioengineering Technology to Generate Lung Tissue for Personalized Disease Modeling. Stem Cells Transl Med 2017; 6(2): 622-33.
[http://dx.doi.org/10.5966/sctm.2016-0192] [PMID: 28191779]

[21]   Wilkinson DC, Mellody M, Meneses LK, Hope AC, Dunn B, Gomperts BN. Development of a Three-Dimensional Bioengineering Technology to Generate Lung Tissue for Personalized Disease Modeling. Curr Protoc Stem Cell Biol 2018; 46(1)e56
[http://dx.doi.org/10.1002/cpsc.56] [PMID: 29927098]

[22]   Sucre JMS, Vijayaraj P, Aros CJ, *et al.* Posttranslational modification of β-catenin is associated with pathogenic fibroblastic changes in bronchopulmonary dysplasia. Am J Physiol Lung Cell Mol Physiol 2017; 312(2): L186-95.
[http://dx.doi.org/10.1152/ajplung.00477.2016] [PMID: 27941077]

[23]   Miller AJ, Dye BR, Ferrer-Torres D, *et al.* Generation of lung organoids from human pluripotent stem cells in vitro. Nat Protoc 2019; 14(2): 518-40.
[http://dx.doi.org/10.1038/s41596-018-0104-8] [PMID: 30664680]

[24]   Sachs N, Papaspyropoulos A, Zomer-van Ommen DD, *et al.* Long-term expanding human airway organoids for disease modeling. EMBO J 2019; 38(4)e100300
[http://dx.doi.org/10.15252/embj.2018100300] [PMID: 30643021]

[25]   Paul MK, Bisht B, Darmawan DO, *et al.* Dynamic changes in intracellular ROS levels regulate airway basal stem cell homeostasis through Nrf2-dependent Notch signaling. Cell Stem Cell 2014; 15(2):

199-214.
[http://dx.doi.org/10.1016/j.stem.2014.05.009] [PMID: 24953182]

[26] Barkauskas CE, Chung MI, Fioret B, Gao X, Katsura H, Hogan BLM. Lung organoids: current uses and future promise. Development 2017; 144(6): 986-97.
[http://dx.doi.org/10.1242/dev.140103] [PMID: 28292845]

[27] Khedoe PPSJ, Ng-Blichfeldt J-P, Van Schadewijk A, Marciniak SJ, Koenigshoff M, Gosens R, *et al.* Impairment of lung organoid formation by cigarette smoke treatment of mesenchymal cells. Airway Cell Biology and Immunopathology 2018.
[http://dx.doi.org/10.1183/13993003.congress-2018.PA4274]

[28] Tammela T, Sanchez-Rivera FJ, Cetinbas NM, *et al.* A Wnt-producing niche drives proliferative potential and progression in lung adenocarcinoma. Nature 2017; 545(7654): 355-9.
[http://dx.doi.org/10.1038/nature22334] [PMID: 28489818]

[29] Tata PR, Chow RD, Saladi SV, *et al.* Developmental History Provides a Roadmap for the Emergence of Tumor Plasticity. Dev Cell 2018; 44(6): 679-693.e5.
[http://dx.doi.org/10.1016/j.devcel.2018.02.024] [PMID: 29587142]

[30] Hai J, Zhang H, Zhou J, *et al.* Generation of Genetically Engineered Mouse Lung Organoid Models for Squamous Cell Lung Cancers Allows for the Study of Combinatorial Immunotherapy. Clin Cancer Res 2020; 26(13): 3431-42.
[http://dx.doi.org/10.1158/1078-0432.CCR-19-1627] [PMID: 32209571]

[31] Dost AFM, Moye AL, Vedaie M, *et al.* Organoids Model Transcriptional Hallmarks of Oncogenic KRAS Activation in Lung Epithelial Progenitor Cells. Cell Stem Cell 2020; 27(4): 663-678.e8.
[http://dx.doi.org/10.1016/j.stem.2020.07.022] [PMID: 32891189]

[32] Tien JCY, Chugh S, Goodrum AE, *et al.* AGO2 promotes tumor progression in KRAS-driven mouse models of non–small cell lung cancer. Proc Natl Acad Sci USA 2021; 118(20)e2026104118
[http://dx.doi.org/10.1073/pnas.2026104118] [PMID: 33972443]

[33] Naranjo S, Cabana CM, LaFave LM, Westcott PMK, Romero R, Ghosh A, *et al.* 2021.

[34] Vazquez-Armendariz AI, Heiner M, El Agha E, *et al.* Multilineage murine stem cells generate complex organoids to model distal lung development and disease. EMBO J 2020; 39(21)e103476
[http://dx.doi.org/10.15252/embj.2019103476] [PMID: 32985719]

[35] Ma H, Zhu Y, Zhou R, Yu Y, Xiao Z, Zhang H. Lung cancer organoids, a promising model still with long way to go. Crit Rev Oncol Hematol 2022; 171103610
[http://dx.doi.org/10.1016/j.critrevonc.2022.103610] [PMID: 35114386]

[36] Li Y, Chan JWY, Lau RWH, *et al.* Organoids in Lung Cancer Management. Front Surg 2021; 8753801
[http://dx.doi.org/10.3389/fsurg.2021.753801] [PMID: 34957199]

[37] Fessart D, Begueret H, Delom F. Three-dimensional culture model to distinguish normal from malignant human bronchial epithelial cells. Eur Respir J 2013; 42(5): 1345-56.
[http://dx.doi.org/10.1183/09031936.00118812] [PMID: 23349442]

[38] Kim M, Mun H, Sung CO, *et al.* Patient-derived lung cancer organoids as *in vitro* cancer models for therapeutic screening. Nat Commun 2019; 10(1): 3991.
[http://dx.doi.org/10.1038/s41467-019-11867-6] [PMID: 31488816]

[39] Li Z, Qian Y, Li W, *et al.* Human Lung Adenocarcinoma-Derived Organoid Models for Drug Screening. iScience 2020; 23(8)101411
[http://dx.doi.org/10.1016/j.isci.2020.101411] [PMID: 32771979]

[40] Dijkstra KK, Cattaneo CM, Weeber F, *et al.* Generation of Tumor-Reactive T Cells by Co-culture of Peripheral Blood Lymphocytes and Tumor Organoids. Cell 2018; 174(6): 1586-1598.e12.
[http://dx.doi.org/10.1016/j.cell.2018.07.009] [PMID: 30100188]

[41]   Gmeiner WH, Miller LD, Chou JW, *et al.* Dysregulated Pyrimidine Biosynthesis Contributes to 5-FU Resistance in SCLC Patient-Derived Organoids but Response to a Novel Polymeric Fluoropyrimidine, CF10. Cancers (Basel) 2020; 12(4): 788.
[http://dx.doi.org/10.3390/cancers12040788] [PMID: 32224870]

[42]   Mazzocchi A, Devarasetty M, Herberg S, *et al.* Pleural Effusion Aspirate for Use in 3D Lung Cancer Modeling and Chemotherapy Screening. ACS Biomater Sci Eng 2019; 5(4): 1937-43.
[http://dx.doi.org/10.1021/acsbiomaterials.8b01356] [PMID: 31723594]

[43]   Gómez-López S, Whiteman ZE, Janes SM. Mapping lung squamous cell carcinoma pathogenesis through *in vitro* and *in vivo* models. Commun Biol 2021; 4(1): 937.
[http://dx.doi.org/10.1038/s42003-021-02470-x] [PMID: 34354223]

[44]   Zhang Z, Wang H, Ding Q, *et al.* Establishment of patient-derived tumor spheroids for non-small cell lung cancer. PLoS One 2018; 13(3)e0194016
[http://dx.doi.org/10.1371/journal.pone.0194016] [PMID: 29543851]

[45]   Shi R, Radulovich N, Ng C, *et al.* Organoid Cultures as Preclinical Models of Non–Small Cell Lung Cancer. Clin Cancer Res 2020; 26(5): 1162-74.
[http://dx.doi.org/10.1158/1078-0432.CCR-19-1376] [PMID: 31694835]

[46]   Hu Y, Sui X, Song F, *et al.* Lung cancer organoids analyzed on microwell arrays predict drug responses of patients within a week. Nat Commun 2021; 12(1): 2581.
[http://dx.doi.org/10.1038/s41467-021-22676-1] [PMID: 33972544]

[47]   Esmail S, Danter WR. 2022.

[48]   Hao C, Wang L, Peng S, *et al.* Gene mutations in primary tumors and corresponding patient-derived xenografts derived from non-small cell lung cancer. Cancer Lett 2015; 357(1): 179-85.
[http://dx.doi.org/10.1016/j.canlet.2014.11.024] [PMID: 25444907]

[49]   Takahashi N, Hoshi H, Higa A, *et al.* An *In Vitro* System for Evaluating Molecular Targeted Drugs Using Lung Patient-Derived Tumor Organoids. Cells 2019; 8(5): 481.
[http://dx.doi.org/10.3390/cells8050481] [PMID: 31137590]

[50]   Hamdan F, Ylösmäki E, Chiaro J, *et al.* Novel oncolytic adenovirus expressing enhanced cross-hybrid IgGA Fc PD-L1 inhibitor activates multiple immune effector populations leading to enhanced tumor killing in vitro, *in vivo* and with patient-derived tumor organoids. J Immunother Cancer 2021; 9(8)e003000
[http://dx.doi.org/10.1136/jitc-2021-003000] [PMID: 34362830]

[51]   Chen S, Giannakou A, Wyman S, *et al.* Cancer-associated fibroblasts suppress SOX2-induced dysplasia in a lung squamous cancer coculture. Proc Natl Acad Sci USA 2018; 115(50): E11671-80.
[http://dx.doi.org/10.1073/pnas.1803718115] [PMID: 30487219]

[52]   Seitlinger J, Nounsi A, Idoux-Gillet Y, *et al.* Vascularization of Patient-Derived Tumoroid from Non-Small-Cell Lung Cancer and Its Microenvironment. Biomedicines 2022; 10(5): 1103.
[http://dx.doi.org/10.3390/biomedicines10051103] [PMID: 35625840]

[53]   Schwank G, Koo BK, Sasselli V, *et al.* Functional repair of CFTR by CRISPR/Cas9 in intestinal stem cell organoids of cystic fibrosis patients. Cell Stem Cell 2013; 13(6): 653-8.
[http://dx.doi.org/10.1016/j.stem.2013.11.002] [PMID: 24315439]

[54]   Artegiani B, Hendriks D, Beumer J, *et al.* Fast and efficient generation of knock-in human organoids using homology-independent CRISPR–Cas9 precision genome editing. Nat Cell Biol 2020; 22(3): 321-31.
[http://dx.doi.org/10.1038/s41556-020-0472-5] [PMID: 32123335]

[55]   Cristobal A, van den Toorn HWP, van de Wetering M, Clevers H, Heck AJR, Mohammed S. Personalized Proteome Profiles of Healthy and Tumor Human Colon Organoids Reveal Both Individual Diversity and Basic Features of Colorectal Cancer. Cell Rep 2017; 18(1): 263-74.
[http://dx.doi.org/10.1016/j.celrep.2016.12.016] [PMID: 28052255]

[56]   Demmers LC, Kretzschmar K, Van Hoeck A, *et al.* Single-cell derived tumor organoids display diversity in HLA class I peptide presentation. Nat Commun 2020; 11(1): 5338.
[http://dx.doi.org/10.1038/s41467-020-19142-9] [PMID: 33087703]

[57]   Paul MK. Introductory Chapter: Role of Extracellular Vesicles in Human Diseases and Therapy. Extracellular Vesicles - Role in Diseases, Pathogenesis and Therapy. Physiology 2022.

[58]   Mukherjee A, Bisht B, Dutta S, Paul MK. Current advances in the use of exosomes, liposomes, and bioengineered hybrid nanovesicles in cancer detection and therapy. Acta Pharmacol Sin 2022; 43(11): 2759-76.
[http://dx.doi.org/10.1038/s41401-022-00902-w] [PMID: 35379933]

[59]   S. Chauhan D, Mudaliar P, Basu S, Aich J, K, Paul M. Tumor-Derived Exosome and Immune Modulation. Extracellular Vesicles - Role in Diseases, Pathogenesis and Therapy. Physiology. 2022.

[60]   Ke X, Yan R, Sun Z, *et al.* Esophageal Adenocarcinoma–Derived Extracellular Vesicle MicroRNAs Induce a Neoplastic Phenotype in Gastric Organoids. Neoplasia 2017; 19(11): 941-9.
[http://dx.doi.org/10.1016/j.neo.2017.06.007] [PMID: 28968550]

CHAPTER 9

Additive Manufacturing and Organoids

Shivaji Kashte[1], **Shahabaj Mujawar**[1], **Tareeka Sonawane**[2], **Atul Kumar Singh**[3] and **Sachin Kadam**[4,*]

[1] *Department of Stem Cell and Regenerative Medicine, Centre for InterdisciplinaryResearch, D. Y. Patil Education Society (Institution Deemed to be University), Kolhapur416006, India*

[2] *Amity Institute of Biotechnology, Amity University, Pune Expressway, Bhatan, Mumbai 410221, India*

[3] *Central Research Facility, IIT-Delhi Sonipat Campus, Rajiv Gandhi Education City, Sonipat, Haryana 131029, India*

[4] *Manipal Center for Biotherapeutics Research, Manipal Academy of Higher Education(Institute of Eminence Deemed to be University), Manipal 576104, India*

Abstract: Additive manufacturing (AM) is a rapid and efficient process of creating complex geometries or structures using a digital three-dimensional (3D) printing process. AM has many diverse applications in aerospace, automotive, defense, manufacturing industries, education, and research, most notably in the healthcare and bio-medical industries. 3D bioprinting allows us to create tissue-specific architecture with precise geometries limited to conventional fabrication methods. In this chapter, we have discussed the generalized process of 3D printing of objects in various organoid cultures, focusing on the advantages and limitations of AM technology. Further, we have discussed the major challenges and future direction in the context of organoid bioprinting.

Keywords: Additive Manufacturing, Bioprinting, Organoid culture, Tissue mimetics, 3D printing.

INTRODUCTION

Additive manufacturing (AM) or additive layer manufacturing (ALM) is the industrial production name for three-dimensional (3D) printing [1]. AM is a computer-controlled technique that progressively deposits different layers of materials to create a complex structural prototype. The process starts with generating a 3D computer model of a prototype obtained from the Computer-Aided Design (CAD) system and is split and cut using computer software. This process first creates a two-dimensional (2D) outline that determines whether or

* **Corresponding Author Sachin Kadam**: Manipal Center for Biotherapeutics Research, Manipal Academy of Higher Education (Institute of Eminence Deemed to be University), Manipal 576104, India; Email: kadamsachin@gmail.com

not material is added to each layer. Then every material layer is processed sequentially and compiled or stalked from the bottom to the top portion of the final structure.

AM is a rapid prototyping technology that allows the creation of bespoke parts with complex geometries using the digital process [2]. The use of 3D CAD at the beginning of the product designing process and conversion of the data throughout AM is coherent, making design alterations quick, error-free, and efficient during the conversion or translation of the design intent. Another significant advantage of AM is that it reduces the number of steps involved, decreasing material waste. A reduction in material waste reduces the price of high-value components. Unlike conventional complex designing methods, AM machine production is done in a single step irrespective of the complexity of the product's design [3, 4], thereby reducing the lead times and providing improved strength and durability to the final product. Reduction in labor costs and shortening of the supply chain make AM a revolutionary market process.

These advantages not only make an excellent socio-economic and environmental (controlling resource consumption, use of hazardous material, pollution) impact [5]but also enhance product quality and society's wellbeing [6]. It is anticipated that by 2025, the worldwide market for AM products may bloom up to 230-550 billion dollars due to its great potential in medical, aerospace, and tools manufacturing [7].

AM has many diverse applications in aerospace, automotive, defense, manufacturing industries, education, and research, most notably in the healthcare and biomedical industries.AM technologies are used in the healthcare and medical sector to fabricate patient-specific medical constructs such as anatomical implants, surgery tools, or those, prostheses, dental casts, tissue engineering scaffolds, drug delivery devices, *etc* [8]. In addition to the biofabrication of these advanced medical devices, materials, structural grafts, and prosthetic devices, the use of AM also encompasses a versatile and expanding array of technologies like organs and organoids synthesis by selectively distributing cells, bioactive materials, and cytokines which are otherwise difficult to fabricate due to their complex geometries and heterogeneous material distribution [9]. In 3D bioprinting, cells, growth factors, and biomaterials are combined in a 3D printer to create a biological component to mimic the properties of tissues and organs. Layer-by-layer construction with 3D bioprinting facilitates the creation of highly biomimetic and robust *In vitro* models for scientific research, therapeutics, drug delivery systems, and even organoid and tissue printing. Herein, we discuss the principles and methodologies used for constructing different organoids concerning AM technologies.

HISTORY OF ADDITIVE MANUFACTURING

The origin of AM dates back to almost 150 years to the concepts of topography and photo sculpture. These technologies use the principle of layer-wise 'cutting and slacking' to build a free-form object. From creating a 3D replica of an object, including human forms, capturing simultaneous surround photography [10] to the wax plates layering method of producing molds for the topographical maps, artisans have been using 3D modeling since the eighteenth century. It was in 1950 when writer Raymond Jones first described the3D printing as a 'molecular jet' in his story "Tools and Trade" in The Astounding Science Fiction Anthology [11]; the same was brought into a concept by David Jones (1974), in his regular column named 'Ariadne' in the journal 'New Scientist [12].

The first patent detailing 3D printing along with rapid prototyping and regulated on-demand manufacturing of specific patterns was filed by Johannes Gottwald in 1971 for the continuous Inkjet metal material device known as Liquid Metal Recorder, which was used for metal fabrication of a reusable surface [13]. However, the development of AM equipment and materials began in the 1980s. Hideo Kodama, in April 1980, invented two additive methods for the production of 3D plastic molds by using photo-hardening thermoset polymer [14]. A patent for the fabrication method of articles by sequential deposition was granted to Raytheon Technologies Corporation in 1982, which includes hundreds of thousands of layers of powdered metal and a source of laser energy [15]. Bill Matsers (the year 1984) patent for Computer Automated Manufacturing Process and the system was the first on record 3D printing patent in history at the United States Patent Trademark Office (USPTO). This invention further laid the foundation for the 3D printing system today [16]. Robert Howard developed a color inkjet 2D printer in 1984, which was further commercialized as Pixelmaster in 1986. Chuck Hall, in 1984 filed a patent for the Stereolithography (SLA) apparatus [17], and in 1986, his company 3D Systems became the first 3D printing company in the world to release the first commercial 3D printer named 'SLA-1' in 1987 [18].

The practical origin of 3D printing can be traced back to Professor Emanuel Sachs of Massachusetts Institute of Technology (MIT), with 3D printing as a powder bed process using standard and custom inkjet print heads in 1993. This achievement was made possible by spreading powder and binder material that selectively joins the powder to create a layer with the help of inkjet printing [19]. In the same year, Solidscape Inc. introduced a high-precision polymer jet fabrication system that used soluble support structures classified as the "dot-o--dot" technique. Here, the printer uses a predetermined path on which it deposits printing material point-wise to construct the cross-sectional area of the object or part to be manufactured [20]. In 1995, the Fraunhofer Society developed the Selective Laser Melting (SLM) technology. Here, a focused high-power optic laser melts the metal powder and fuses it to form a solid part of the component [21].

The first bioprinter developed in the early 2000s by Thomas Boland used an inkjet printer to deposit cells instead of ink. This technology could print a bioink composed of cells, culture media, and serum onto a cell culture plate [22]. Not only cells but the bioprinting approach by Gabor Forgacs allowed positioning the individual cell spheroid in the desired pattern, further creating thicker tissues by fusion on culture [23]. A similar approach was adopted in 2011 by the Japanese company Cyfuse for assembling spheroid-based 3D structures through needle arrays known as Kenzen [24]. In 2007, Gabor Forgacs formed the first bioprinting company Organovo and commercialized tissue models for drug screening and disease modeling [25].

The Swiss company regenHU was established in 2007 with a focus on extrusion-based 3D bioprinting devices known as 'Biofactory'. EnvisionTEC, a manufacturer of conventional 3D printers, similarly created their 3D-Bioplotter device. EnvisionTEC and regenHU developed advanced commercial bioprinter systems that allowed researchers to build their desired tissue construct and created a revolution in 3D bioprinting [25, 26]. By 2010, it was clear to engineers that AM was making a significant impact with its added advantages of AM and had more to offer in the manufacturing practices. Further, the Fused Deposition Modeling (FDM) technology patent expired in 2009, decreasing the price of FDM printers [27]. By 2018, Fused Deposition Modeling was the most commonly used 3D printing process. This material extrusion technique has been used about 46% of the time for all applications. Although FDM was discovered later than the other two technologies, namely Stereolithography and Selective Laser Sintering (SLS), FDM seems more popular because of its low cost [28]. These 3D printing technologies have recently evolved, and 3D printers have reached the level of excellent quality and reasonable pricing, allowing most people to enter the world of 3D printing. Decent pricing of suitable quality printers has been seen in entry-level machines. Also, FDM printers are usually more affordable than the rest [29]. A detail of the historical milestones in the evolution of AM is shown in Table **1**.

Table 1. Historical milestones in the evolution of additive manufacturing.

Year	Type of Invention/Milestones in the Development of AM	Inventor	References
1981	An automated method for fabrication of 3D plastic models with photo hardening	Hideo Kodama	[14]

(Table 1) cont.....

Year	Type of Invention/Milestones in the Development of AM	Inventor	References
1986	Patent for apparatus for production of 3D objects by Stereolithography	Charles Hull	[30]
1987	SLS (Selective Laser Sintering) method	Carl Deckard	[31]
1989	FDM (Fused Deposition Modeling) Process	S. ScottCrump	[32]
1994	Model Maker Wax 3D Printer	SolidScape Company, USA	[32]
1997	LAM (Laser Additive Manufacturing)	Aeromet Company, UK	[32]
1999	1St Ink-jet 3D Printer	Object Geometries Company, Israel	[33]
2001	World's first Desktop version of 3D printer	Solid Dimension Company, USA	[33]
2003	Multiple Color 3D printer	Z Corp company, USA	[33]
2003	Patent on Ink-jet printing of viable cells	Thomas Boland	[34]
2008	Darwin- self-replicating printer	Adrian Bowyer's RepRap Project	[35]
2009	Novogen-MMX: First commercial 3D bioprinter	Organovo Company, USA	[36]
2015	First commercialized bioink in market	Cellink Company, Sweden	[37]
2019	First 3D bio-printed heart with human cells	Tal Dvir, University of Tel Aviv, Israel	[38]
2020	Engineered soft rubbery brain implants	MIT, USA	[39]
2021	A new bioink was discovered, allowing small human-sized airways to be 3D-bioprinted.	Lund University, USA	[39]

ADDITIVE MANUFACTURING: GENERAL PRINCIPLES AND WORKING PROCEDURE

A generalized process to 3D print an object in AM is discussed here. The development of any product starts with its conceptualization and design and then manufacturing it. The steps in printing an object can be modified or vary according to the type of AM process, technology, type of material, and application of the product. The generalized process of AM is divided into seven steps that are described as follows (Fig. **1**).

Conceptualization

It is the first step in product development. It is an idea of the type of product, its appearance, morphology, structure, and applications. It could be text, narrative descriptions, sketches, or representative models. While using AM, this conceptualized product description should be digital [40].

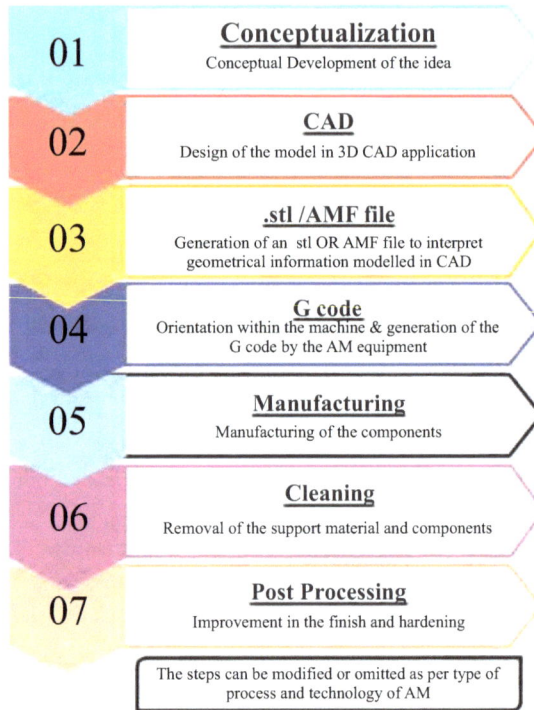

Fig. (1). Flowchart of generalized AM process. Adapted from [44].

Computer-Aided Design (CAD)

The next step is to convert the digital description of the product into a CAD file without which the AM functioning is not possible. Therefore, AM is also called a "direct or streamlined Computer-Aided Design to Computer-Aided Manufacturing (CAD/CAM) process." The CAD file can be created from source data with the help of design experts *via* a user interface, commercial software, and scanning of an existing physical component or their combination [40].

A .stl File/AMF File

The term 'stl' was derived from STereoLithography. It is a straightforward way to describe a CAD file in its geometry alone. The conversion of CAD files to the .stl file is automatic in most CAD systems. The .stl files contain an ordered collection of triangle vertices and surfaces normal vectors. The .stl files have limitations such as no units, color, material, or other feature information. Therefore, these limitations are overcome using the new Additive Manufacturing File (AMF) format. AFM file stores information using curved triangulations [41]. It is now the International Organization for Standardization (ISO) and American Society for Testing and Materials (ASTM) standard format as an extension of the .stl file. The

.stl or AMF file may have holes, directional vectors, self-intersections, noise shells, and manifold errors. These errors should be fixed before the next step [42].

G Code

The next step in the AM process is generating a 'G' code. After creating the .stl/AMF file, it is processed with specialized software like "Slicer" to transform the models into a series of thin layers. It generates a G-code file containing instructions modified for the specific type of 3D printer. This G-code file is printable using 3D printing client software, which can interpret the G-code to command the 3D printer during the 3D printing process [43].

Manufacturing

The 3D printer is correctly set up for suitable build parameters like material constraints, energy source, layer thickness, and duration of printing. After this, the automated process of building the product is initiated, and layer-by-layer addition is continued until the product is made. The process should be supervised to avoid any errors [40].

Cleaning

Once the process is finished and the building product is ready, it is removed and cleaned. The build parts are separated from supporting materials and the build platform. The parts are removed from the printer, taking all safety precautions like high temperatures or moving parts. The cleaning of the building part is entirely manual and significantly varies with the type of material, process, technology, and applications [40, 44].

Post-processing

It finishes building materials for application purposes. Finishing of the product is performed as per the type of its application. It may involve cutting, curing, abrasive finishing such as polishing, sandpapering or coating of the materials, chemical or thermal treatment, *etc.* After post-processing, the product is ready to use for its application [40, 44].

ADDITIVE MANUFACTURING PROCESSES

There are seven primary additive manufacturing processes according to the ISO and ASTM. Every function is distinct because of the type of material used, method of material deposition, and different machinery and techniques used for manufacturing. These additive techniques are utilized according to the material used, and the product manufactured [45, 46].

Powder Bed Fusion

A laser or electron beam is used to melt the powdered material to form a layer-b-
-layer deposition in this additive manufacturing technique. 3D dispensing, 3D
bioprinting, Direct metal laser sintering (DMLS), Selective laser sintering (SLS),
Selective laser melting (SLM), and Electron beam machining (EBM) techniques
are examples of powder bed fusion. These are less costly and used for various
materials like polymers, ceramics, metals, and composites. Nevertheless, they
have disadvantages like an absence of structural integrity, limited size, high power
requirement, and are relatively slow. The resolution, quality, and reusability of
unsintered powder depend on the laser diameter [47].

Directed Energy Deposition (DED)

A focused heat energy source such as an electron beam or laser is used to melt the
materials. It is mainly used for metals. Laser deposition (LD), laser-engineered
net shaping (LENS), electron beam, and plasma arc melting are

examples of DED. It provides a high degree of control over grain size. It can be
used with multi-material systems having different porosity. A balance between
surface quality and speed is required for optimum performance in DED. There is a
low efficiency of powder after surface processing [40].

Sheet Lamination/ laminated Additive Manufacturing

In this AM process, material sheets are bonded to form the final product part. The
material sheet is connected to the previous layer of other sheets by using an
adhesive material. The required shape is cut from these bonded layer sheets by a
laser beam or cutting knife to form the final product. Laminated object
manufacturing (LOM), Ultrasound consolidation (UC), or Ultrasound additive
manufacturing (UAM) are examples of sheet lamination. It is a high-speed, low-
cost process and offers easy material handling. However, the strength and
integrity of constructs highly depend on the type of adhesive used. The finished
products require post-processing [45, 46, 48].

Binder Jetting

In binder jetting, layers of a powder material are selectively dispersed, and a
liquid binder is used to bond selective areas of deposited layers of powder
together. Final structures are processed through sintering and debinding to get a
more dense and hard network where the final object shrinks about 20-25% of the
original structure [49]. 3D inkjet technology is an example of binder jetting. It is a
low-cost, high-speed process. It offers freedom of designs of constructs and

provides larger build volume, and does not require any support or substrate. However, it involves post-processing, and the constructs have poor mechanical stability [50, 51].

Material Extrusion

This additive manufacturing process involves an extrusion system in which the thermoplastic or filamentous metal or synthetic and natural polymers-like materials are heated till it gets melted. Then this molten material is extruded through a nozzle of the printing assembly to form parts in a layer-by-layer manner according to the CAD of the product part [52]. Fused deposition modeling (FDM), Fused filament fabrication (FFF), and Fused layer modeling (FLM) are examples of material extrusion. It is a low-cost and scalable process and does not require a platform. It has limitations like the vertical anisotropy and step-structured surface. It is used for a limited type of materials like polymers and composites due to the requirement of a molten phase of the material [50, 51].

Material Jetting

It is similar to 2D inkjet printing as it is used to create constructs by the jetting process. The droplets of materials are jetted forcefully onto a build platform using the Drop on Demand (DOD) or continuous approach. After deposition, the material is hardened or cured using ultraviolet light [53]. 3D inkjet technology and Direct ink writing are examples of material jetting. It has high accuracy for droplet deposition, and it forms low waste. However, it requires support material, and its use is limited to photopolymers and thermoset resins [50, 51].

Vat Photopolymerization

This type of AM uses a vat (a cylindrical container) containing liquid photopolymer resins for layer-by-layer product formation. A build platform from the upside is descended in the liquid resin vat, and layers of different thicknesses and shapes are built up. A beam of ultraviolet (UV) light is passed throughout the resin build structure, which cures and hardens it wherever necessary according to design; this process is called photopolymerization. Finally remaining liquid resin is drained, and the object is removed from the vat [54]. SLA, Digital light processing (DLP), and Multiphoton polymerization (MPP) are examples of vat polymerization. It provides the construction of large objects and has superior accuracy and surface finish. However, its use is limited to photopolymers only. The constructs have a low shelf life and mechanical stability [55, 56].

BIOFABRICATION

To overcome the shortage of autologous organs and tissue for transplantation, there is a need to develop biological substitutes. Considering this pressing need, a new tissue engineering strategy must be formulated [57]. Prior manufacturing techniques were unable to produce clinically effective tissue constructions with well-interconnected pores, complicated structures, patient-specific geometric morphologies, and heterogeneous material deposition [58]. 3D printing or AM was discovered in the 1980s and pioneered in creating objects from a computer-generated file. Similarly, it has been explored as a powerful tool in tissue engineering and regeneration therapy [59]. Over the years, 3D bioprinting has been useful in overcoming these limitations. This technology has enabled the fabrication of different tissue constructs with control over biomaterials, bioactive molecules cells, and their deposition, which leads to an increase in regenerative capacity after implantation [57, 60, 61]. In this tissue reconstruction or bio-printing technology, bioactive and bio-inert materials like metals, polymers, ceramics, *etc.*, are used to develop different tissue structures with precise deposition of cells and bioactive molecules [62]. The high density of cells is being patterned spatially with the layer-by-layer prescribed organization in cell-based bioprinting, which ultimately forms tissue-like constructs [63]. 3D bioprinting allows us to create tissue-specific architecture with precise geometries, which were limited in the conventional fabrication methods. Firstly, Landers *et al.* introduced 3D bioprinting, an extrusion-based approach. The continuous dispersion of cells within a hydrogel material, *i.e.*, the bioink from a dispenser head to a stage based on patterns, is designed *via* CAD/CAM tools [36].

PRINCIPLE OF THREE-DIMENSIONAL BIOPRINTING

Different 3D printing methods have been developed to produce bioengineered 3D human tissue and organ-like constructs for tissue engineering applications. Accuracy and successful tissue construct regeneration mainly depend on the chosen biomaterial and the targeted application [64]. 3D bioprinters consist of three main components: three (XYZ) axis stages, printing cartridges, and the dispenser head. A control system moves the printer head in X, Y, and Z directions. The cartridge used for printing is generally a syringe containing polymeric components for the scaffold or the cells containing hydrogel. These syringes have nozzles that determine the quantity of material to be dispensed at predetermined printing parameters. This system of dispensing material is the final component, which carries out materials deposition, and it is different with different printing techniques [65].

DIFFERENT TYPES OF 3D BIOPRINTING TECHNIQUES

3D bioprinting techniques are achieved by different means; some prominent techniques include extrusion printing, inkjet printing, laser-assisted bioprinting, and stereolithography bioprinting [55]. Among all these techniques, extrusion bioprinting has been widely studied due to its ability to print different biomaterials at substantially larger cell density [56] to generate artificial tissue constructs such as cartilages [66, 67], liver [68], and neural tissues [69].

Extrusion-based 3D bioprinting (Micro-extrusion)

Extrusion-based bioprinting is recently the most common bioprinting technique, and mechanically it is similar to a conventional 3D printing technique like Fused deposition modeling (FDM) [36]. In extrusion 3D bioprinting, different types of cells and biomaterials or bioinks are loaded into the cartridge and then extruded under pneumatic pressure or mechanical force from a nozzle to a predefined area on the collecting plate or fabrication platform [72] (Fig. **2**). Usually, the cartridges are fitted so they can move in the XYZ axis of the stage. Temperature and pressure (speed of piston and rotational speed in case screw or piston-driven) are controlled by computer software [73]. Extrusion-based bioprinting has its advantages as follows;

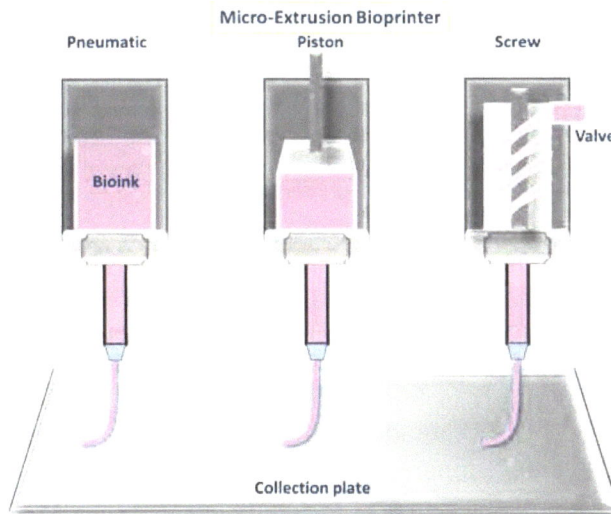

Fig. (2). Extrusion Bioprinter. Adapted from [70, 71].

Different types of extrusion-based bioinks have been developed in recent years as this technique can extrude higher viscosity bioinks or inks with high cell densities [74].

With the help of extrusion bioprinting, now operator can create heterogeneous models as the deposition of multiple types of cells at controlled locations with predesigned numbers and structure is possible. This property is significant when unique engineering is required for creating artificial tissue constructs or organs where the heterogeneity and organization of cells are to be controlled to perform their biological function [75].

This technology of extrusion-based bioprinting is comparatively affordable compared to other bioprinting techniques. It is possible to highly customize it with respect to different types of structure or bioinks. These commercially available extrusion-based bioprinters and their instrumentation can be modified according to specific applications that enable combined extrusion, core-shell printing, UV curing, electrospinning at the time of printing, and post-printing [76].

Extrusion-based bioprinting has been used to produce tumor models, biomedical implants, prostheses, and *In vitro* models of different diseases. One of the limitations of extrusion-based bioprinting models is that mammalian cell viability is low compared to the viability of cells after ink-jet-based bioprinted models (*e.g.*, 40% to 85%). This is mainly because of the shear and mechanical stress that occurs during high extrusion pressure [77]. The second notable limitation is the resolution of printing and printing speed. Currently, efforts are being made to reduce these limitations by controlling process parameters, optimizing instrumentation, and bioink formulations. This will eventually increase the resolution of printing, better cell deposition, enhanced cell viability, and model fidelity. This technique effectively produces soft tissue mimics and bone implants [78]. A comparison of the different 3D printing techniques is shown in Table **2**.

Droplet-Based 3D Bioprinting

Droplet-based bioprinting is another technique after extrusion-based bioprinting that uses droplets as a basic unit. Its basic principle is to stack discrete droplets into a 3D model (Fig. **3**). Based on this property of forming different droplets, droplet-based bioprinting can be divided majorly into inkjet bioprinting, laser-assisted bioprinting (LAB), and electrohydrodynamic jetting (EHDJ) [79]. Inkjet bioprinting uses physical properties like density, viscosity, and surface-tension of bioink to create a microsphere to be collected on a surface by deposition. It can be classified further into DOD inkjet bioprinting and continuous inkjet bioprinting (CIJ). DOD inkjet bioprinting, thermal inkjet bioprinting, and piezoelectric inkjet bioprinting techniques are used in tissue engineering. Inkjet printing has advantages like low production cost, high resolution, and precision printing due to its ability to control biomaterials precisely. It offers low damage to cells [80].

Table 2. Advantages and disadvantages of 3D bioprinting techniques [36, 57, 58, 61, 64, 65, 72 - 87].

Technique	Extrusion Based	Ink-Jet Based	Laser-assisted
Advantages	Short fabrication time	Non-contacting Nozzle	Nozzle less
	Higher mechanical properties	Higher cell viability	Even a single cell can be manipulated.
	Bioink with different viscosities can be printed.	Printing of different patterns of cells using another type of cells.	Bioink with higher viscosities can be used for printing.
	A wide range of biocompatible materials can be used.	Heterogeneous constructs with multi types of cells.	High resolution
		High throughput can be achieved.	Higher accuracy
		High-speed gelation	High-speed gelation
Disadvantages	Viability of cells affected due to shear and mechanical stress caused due to nozzle wall.	Low structural integrity and mechanical property.	Low mechanical strength
	Low accuracy	Long duration of fabrication.	Long duration of fabrication
	Death of cells due to change in dispensing pressure and bioink concentration	Only for low viscosity bioink.	The heat generated due to laser energy damages cells.
		Lower reproducibility	Aggregates formation in final tissue construct.
		Aggregation of cells	
		Frequent nozzle clogging	

Despite having such advantages, it has limitations like frequent blockage of nozzles while using cell-laden bioink, which makes the nozzle head prone to damage. Also, a jetting mechanism causes damage to living cells present in bioink. Hence, limited numbers of bioinks can be used in this bioprinting technique. Similarly, EHDJ jetting depends on applying a high voltage electric field to pull droplets out of the nozzle orifice. Hence, a change in droplet size can be observed if voltage fluctuation occurs. EHDJ can be suitable only for printing bioinks with higher weight/volume and a high number of cells. This technique is popularly used in skin and cartilage tissue applications like direct deposition of biomaterial onto skin and cartilage lesions, layered deposition of primary stem cells and other biomaterials [81].

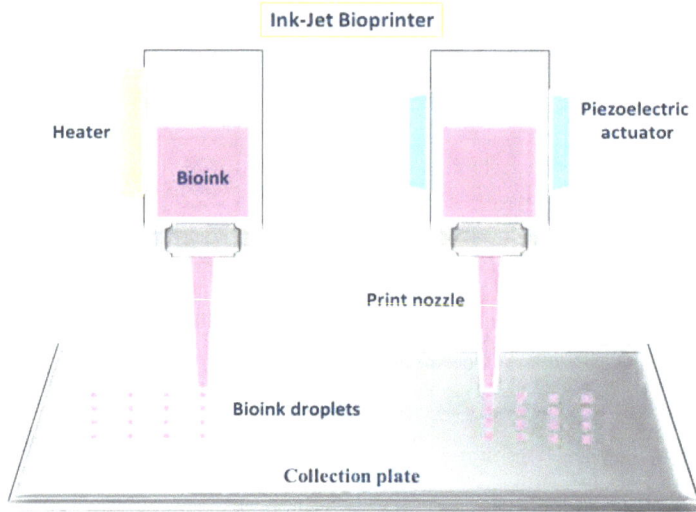

Fig. (3). Inkjet Bioprinter. Adapted from [70, 71].

Laser-based Bioprinting

The method of biomaterial deposition onto a substrate surface or a collecting plate using a laser as an energy source is laser-assisted bioprinting. In the beginning, this technique was used only for transferring metals. However, technological advancement is now used for biomaterials like proteins, hydrogel, ceramics, *etc.*, and different types of cells [82]. In this technique, the bioprinter includes; a pulsed laser beam, projector, and bioink containing receiver substrate mixed with biological material [83] (Fig. **4**).

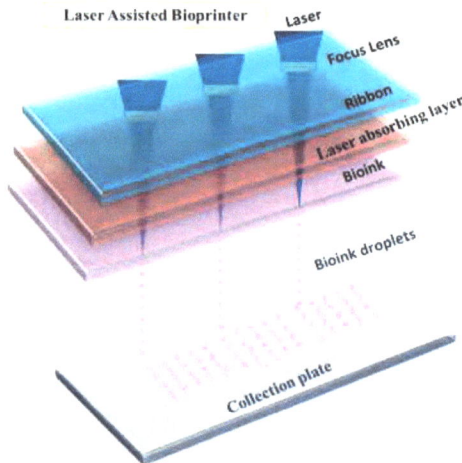

Fig. (4). Laser-Assisted Bioprinter. Adapted from [70, 71].

It is a contactless, nozzle-free printing approach that deposits bioink droplets with precision. Laser-guided direct writing (LGDW) guides the deposition of cells onto the collecting plate or receiver substrate. On this plate, laser-induced forward transfer (LIFT) uses a focused pulsed laser, which causes evaporation of the liquid present in the bioink. This bioink further reaches to receiver substrate in the form of droplets. Then on receiving substrate, there are biopolymers or cell-containing mediums, which additionally help in the attachment of cells and subsequent growth of the biomaterial [84]. The key feature of this method is that it retains higher cell viability, *i.e.*, up to 95%. It avoids clogging of nozzles, which increases its reproducibility. This technique has the medium speed of bioprinting but has high-resolution printing compared to inkjet bioprinting as it generates small-sized droplets. Human skin tissue constructs, dermal fibroblasts, breast cancer cells, *etc.*, can be bioprinted using LAB [85].

ADDITIVE MANUFACTURING FOR ORGANOIDS

Lung Organoid

Chronic obstructive pulmonary disease (COPD), pulmonary hypertension, cystic fibrosis, lung cancer, acute respiratory distress syndrome (ARDS), and viral infections like SARS-CoV-2 are the leading causes of lung disease-related mortality. Most lung diseases have no effective therapies [88, 89]. Researchers are attempting to develop 3D scaffolds using 3D printing to evaluate their suitability for lung cells and lung tissue engineering. Bioink prepared from chitosan and polycaprolactone (PCL) was 3D printed to create scaffolds for lung tissue engineering. MRC-5 cells were studied with these scaffolds and showed cell adhesion, low apoptosis, proliferation, and migration within these scaffolds. The microenvironment created by these 3D scaffolds could be promising for creating disease models for fibrosis, cancer, regeneration, and COVID-19 studies [90]. Extrusion-based 3D printing was used to develop lung tissue-mimicking hydrogel scaffolds from silk fibroin and 2,2,6,6-tetramethylpiperidine-1-oxyl-oxidized bacterial cellulose nanofibrils. Lung epithelial stem cells proliferated and maintained their phenotype up to one week of culture on these scaffolds. These scaffolds could be promising for further developing lung organoids [91].

A lung organoid is a 3D cell and hydrogel composite resembling the morphology and cellular composition of the human distal lung. Wilkinson *et al.* developed a four-step protocol for the generation of lung organoids. Firstly, alveolar scaffolds were generated from polydopamine and collagen-coated alginate beads; then, cells were incorporated onto these alginate beads. These cells encapsulated hydrogels incubated into a 3D printed silicone elastomer bioreactor, allowing it to aggregate and form organoids. The organoids containing only fibroblasts cells may take 24 hours, while a multicellular organoid may take seven days. The organoid formation can be evaluated based on its contraction and overall aggregation of organoid structure [92]. Polyethylene glycol (PEG) diacrylate hydrogels used in the 3 D print lung model have shown multivessel structure and breathing function. The experiment indicated that branching topology, hydrogel distension, and rerouting of fluid streams during breathing might enhance intravascular mixing and permit quicker volumetric oxygen absorption by RBCs that are well-mixed [93]. The complex structure of the lung makes it very difficult to print lung organoids. Most of the progress made in developing a pulmonary system is restricted to the trachea. The story of 3D-printed lung organoids is in the nascent stage. Further action is needed to overcome the challenges of multicellular layers, a hollow structure with an air-fluid interface, and a bioreactor with gas-fluid perfusion [94].

Kidney Organoid

Chronic kidney disease affects approximately 10% of the global population [95]. These patients have limited treatment options, and novel therapies are urgently needed. Bioprinting could potentially form kidney tissues and organoids used for drug screening, disease models, and repair or regeneration of kidney tissues [95]. Homan *et al.* reported a bioprinting method to develop perfusable 3D renal proximal tubules using an extracellular matrix. These structures maintained enhanced epithelial cell morphology and functional properties for more than two months compared with epithelial cells grown on 2D controls. This method provided new insights into the bioprinting of kidney organoids [96]. The drop-o--demand bioprinting fabricates renal spheroids and nephron-like tubules comprising lumen with predefined size and spatial localization using hydrogel matrix and renal epithelial cells. These structures showed higher cell viability and increased kidney-specific gene expression and tissue functionality [97]. Primary murine tubular cells and endothelial cells were embedded in the polysaccharide biomaterial ink. A core-shell hollow filament bioprinted construct as a 3D renal tubulointerstitium model was developed using a microfluidic 3D bioprinter. The cells in bioprinted constructs showed good cell viability and metabolic activity. This preliminary model could further study chronic kidney disease [95]. Higgins *et al.* used pluripotent stem cells to create kidney organoids using a bioprinter. The bioprinted kidney organoids showed higher reproducibility and fidelity between independent starting cell lines. They showed similar cellular complexity and gene expression to the manually developed kidney organoids using gold-standard protocols. The bioprinting method generated many kidney organoids from the same number of starting cells or even fewer cells, like 4000 cells. These kidney organoids showed the development of renal epithelium, glomeruli, stroma, and endothelium. These kidney organoids can be bioprinted in a 96-well and screened for drug toxicity as proof of concept [98]. Kidney organoids were generated in 6 or 96 well plates using iPSCs and STEMdiff APEL medium as bioink with extrusion-based bioprinting. These bioprinted kidney organoids were comparable to the morphology, component cell type, and gene expression level of those previously reported by manually generated kidney organoids. There were more nephrons and functional proximal tubular segments in the organoid by altering biophysical parameters like organoid size, cell number, and conformation. Further development of robust long-term mechanically stable structures [97] and entirely uniform differentiated kidney structures with functional albumin uptake into proximal tubules could lead to the creation of transplantable kidney tissues [99].

Liver Organoid

The liver is one of the most important organs, having more than five hundred known functions, including metabolism, detoxification, homeostasis, protein synthesis, and bile production. Approximately two million people every year die from liver disease [100 - 103]. Liver or hepatocyte transplantation is the last option to treat a diseased liver, but this choice is constrained by the lack of available organ donors or a low degree of engraftment. The emerging bioengineering strategies like bioprinting focus on liver microtissues or liver organoids or restore hepatic functions [101, 104].

Recently, embryonic stem cell (ESC)-derived hepatocyte-like cells (HLCs) were embedded in an alginate-based bioink, and 3D printed to form hepatic constructs. These structures showed hepatic functions like albumin secretion after 21 days of differentiation [105]. The rat hepatocytes, human umbilical vein endothelial cells (HUVECs), and human lung fibroblasts were embedded in collagen bio-ink and 3D printed on PCL to form a capillary-like network. The vascular formation and functional abilities of hepatocytes demonstrated that the heterotypic interaction among hepatocytes and non-parenchymal cells increased the survivability and functionality of hepatocytes within the collagen gel showing potential for functional liver tissue engineering [106]. The 3D liver tissues were bioprinted with human hepatic stellate cells, HUVECs, and primary hepatocytes embedded in NovoGel® 2.0 Hydrogel. These 3D liver tissues maintained these cells for up to four weeks in culture and showed drug metabolism [107]. The 3D hepatic constructs were bioprinted with HepG2/C3A cell spheroids, gelatine methacryloyl (GelMA), and 2-hydroxy-4'-(2-hydroxyethoxy)-2-methylpropiophenone hydrogel. These hepatic constructs were functional over 30 days *In vitro* and showed drug metabolism [108].

Goulart *et al.* used an extrusion-based bioprinter to fabricate liver spheroids by using autologous induced pluripotent stem (iPS) cells-derived parenchymal and iPS-derived hepatocyte-like cell spheroids, both in combination with non-parenchymal cells, mesenchymal and endothelial cells embedded in alginate-based bioink. The spheroid printed constructs showed better hepatic functions after 18 days of culture than iPS-derived parenchymal constructs [100]. 3D hepatic constructs fabricated by extrusion bioprinting using a thermoresponsive pluronic/alginate semisynthetic hydrogel and hepG2 cells. These hepatic constructs showed high viability of cells and liver-specific metabolic activity like urea synthesis, albumin, *etc.* Cells within the constructs showed increased sensitivity to acetaminophen, a hepatotoxic drug, compared to 2D constructs. These 3D hepatic constructs may be an *In vitro* alternative to animal models for drug-induced hepatotoxicity testing [109]. Wu *et al.* developed a 135ACG bioink containing alginate, cellulose nanocrystal, gelatin methacrylolyl (GelMA), and a separate GelMA bioink. The fibroblast NIH/3T3 cells with and without hepG2 cells embedded in 135ACG bioink and GelMA bioink, respectively, and liver lobules were fabricated using microextrusion bioprinting. The fibroblast cells were grown well in the matrix. The hepatic spheroids formed by NIH/3T3/hepG2 cells were smaller than spheroids formed by hepG2 cells in GelMA. The hepatic functions were improved in NIH/3T3/hepG2 cells [110].

Extrusion bioprinting was used to fabricate a hepatic lobule (~1mm), a small functional unit of the liver. The fabricated hepatic lobule includes hepatic cells, endothelial cells, and a lumen. The cells within these hepatic lobules showed cellular organization preserving structural integrity and improved cellular functions. Compared with hepatic and endothelial cell mixtures, these fabricated hepatic lobules showed higher hepatic functions like increased albumin secretion, urea production, albumin, *etc.*, compared with hepatic and endothelial cell mixtures [111]. Mao *et al.* developed a liver microtissue with a liver decellularised extracellular matrix (dECM) bioink using digital light processing bioprinting. The liver-specific bioinks were developed by combining photocurable GelMA with dECM and human-induced hepatocytes (hiHep cells) that were encapsulated to form cell-laden bioinks. The liver dECM improved both printability and cell viability of GelMA bioinks. The bioprinted liver microtissue showed better hepatic functions like albumin secretion and urea [104]. Hepatoorganoids were generated by extrusion bioprinting using HepaRG cells and alginate-based bioink. These hepato-organoids showed liver functions such as albumin secretion, drug metabolism, and glycogen storage after seven days of differentiation *in vitro*. When these hepato-organoids were transplanted into mice models with liver failure, mice survival and increased liver-specific protein synthesis were prolonged. The mice acquired human-specific drug metabolism activities. The functional vascular systems were also formed in these hepato-organoids after transplantation into mice. This bioprinted hepato-organoid model showed hepatic functions and alleviated liver failure *In vivo* [112]. Further development of such liver Organoids could generate human liver tissues as the alternative transplantation donor for the treatment of liver diseases.

Skin Organoid

The human skin is the body's largest organ. It plays a vital role in the primary immune system as it forms a physical barrier to external harmful physical, chemical and biological environmental constituents [113]. Chronic and non-healing skin wounds caused due to accidental trauma, skin infections, and disease conditions that cause damage to natural human skin [114] require cost-effective treatment [115]. Although split-thickness autograft is a gold standard of skin wound management, the availability of healthy skin donors limits its application [116]. Researchers are creating various engineered cellular and acellular skin substitutes to overcome these limitations to model skin for therapeutic purpose [117]. The *in situ* bioprinting of autologous skin cells speeds the healing of severe, full-thickness excisional lesions [118] and evaluates various pharmacological and cosmetic substances for their safety and efficacy [119]. In skin tissue engineering, cells like keratinocytes, stem cells, and extracellular matrix (ECM) components are isolated from part or full-thickness skin tissue with enzymatic digestion. These cellular and acellular components are seeded into a carrier material-support matrix called a scaffold. These scaffold materials are of biological, synthetic, and semisynthetic origin, providing support and nutrition to cells that further proliferate and differentiate to produce functional skin tissue [120]. Human skin was accurately fabricated using 3D bioprinting in a layer-b--layer manner. In this experiment, multiple layers of cells and matrix components were constructed. Human keratinocytes were seeded at the air to the liquid interface on human fibroblasts embedded collagen matrices. These cells represent the epidermis and dermis layer with collagen as a dermal matrix of human skin. Immuno-histological characterization of this 3D printed skin construct concluded that this tissue construct could be used as a morphological and biological representation of *In vivo* human skin [121]. Another group has fabricated full-thickness human skin substitutes by extrusion-based 3D bioprinting. Here two layers of dermal collagen and a resting acellular layer of polycaprolactone/collagen scaffold were synthesized. Further, bioprinting of keratinocytes was performed using sequential extrusion. The 3D printed human skin construct showed better cell viability, proliferation, and immuno-histochemical characteristics than the human full-thickness skin construct with manual cell seeding. This experiment showed the significance of 3D bioprinting for scalable and reproducible functional human full-thickness skin constructs [122].

An innovative approach was tried to treat severe skin burn treatment using 3D bioprinting of cells directly on to dermis of the affected wound area of a murine model of burnt skin tissue. Bioink was prepared from autologous cells of the murine model of burnt skin tissue through biopsy without enzymes for digestion. Immunohistochemical analysis of the bioprinted dermal skin showed reepidermalization and burn wound healing *In vivo* [123]. Human Endothelial Colony Forming Cells (HECFC)-derived from endothelial cells and placental pericytes used to construct skin grafts consisting of perfusable microvascular systems by 3D bioprinting. These grafts showed perfusion and host vasculature when implanted in immunosuppressed mice [124]. Further improvements in 3D bioprinting of skin organoids can potentially treat skin disease conditions and produce artificial skin models for studying drugs and cosmetics.

Pancreatic Organoid

The pancreas is an exocrine gland organ located in the abdomen and produces insulin, glucagon, and other essential enzymes like trypsin, chymotrypsin, amylase, and lipase required for the digestive system. The whole body may be impacted by pancreatic issues. There are a variety of disorders of the pancreas, including acute pancreatitis, chronic pancreatitis, hereditary pancreatitis, pancreatic cancer, and the most prevalent is diabetes mellitus (DM). For patients requiring total or subtotal pancreatectomy for benign disease of the pancreas, whole organ transplant or islet transplantation has paved the way, but the availability of cadaveric pancreas and diseases associated with xenotransplant limit its use [125, 126]. A substantial loss of islets and long-term impaired function have also been seen in many patients post islet transplantation [127, 128]. Bioartificial pancreas developed using various scaffolds packed with islets like cells derived from stem cells have gained more attention. 3D-bioprinting of an endocrine pancreas has shown encouraging results in the curative treatment for type 1 DM [129]. Using a scaffold for encapsulation of islets restricts the vascularisation and revascularisation post-transplantation, leading to hypoxic stress and responsiveness of islets to glucose [130].

Fabricating a bioartificial pancreas that mimics the natural pancreas with all the required complexity depends on the selection of biomaterials, cell delivery techniques, printing resolution, and printing method. The 3D printing extrusion method has been primarily deployed to generate bio-artificial pancreas. It has been observed that a requirement for low viscosity bioink eliminates several effective bioinks from being used 3D printing method. In the case of the 3D printed bioartificial pancreas, using various modified bioinks has always shown a supportive effect on printed islet cells. Marchioli *et al.*(2015) have successfully printed human islets and rat β-cell lines with the help of alginate-based bioink onto a predefined 3D scaffold without changing the cell viability and morphology [68]. Similarly, using an alginate/methylcellulose bioink, rat islets printed into macroporous 3D constructs showed intact cell viability and functions [131].

When co-cultured with endothelial cells (ECs), islets provide a natural cellular niche and enhance the islet's secretory function [132, 133]. The recent co-axial bioprinting showed the way of co-printing islets with supporting cells and bioactive factors for improved graft functionality. The use of co-axial printed pancreatic structure led to improved islet cell functionality. It helped in the overall encapsulation of islets further by precise control over the distribution of multiple cell types [134]. Careful consideration for the revascularization strategies and the transplantation site for the bioartificial pancreas contribute to the desired long-term clinical outcomes. Incorporating endothelial progenitor cells (EPCs) as supporting cells for islets in co-culture while 3D printing has positively affected vascularisation [134]. A well-summed pancreas will revolutionize regenerative medicine and the treatment of diabetes and create a new perimeter in organogenesis and disease modeling.

Intestinal Organoid

The Gastrointestinal (GI) tract or the digestive system consists of many key digestive organs, such as the mouth, esophagus, stomach, small and large intestine, rectum, and anus [135]. Several reports suggest that GI disorders such as gastric and colorectal cancers [136], gastric ulcers [137], and inflammatory bowel disease (IBD) are increasing worldwide [138]. To understand the causes of GI disorders, efforts have been made by scientists by way of study it on animal models [139 - 142] and develop primary cell cultures [143]. Still, both these approaches have their limitations [144]. Although the development of *In vitro* 2D models using cell monolayers, 3D intestinal organoids using adult stem cells (ASC), and intestinal epithelial cells (IEC) shows a better understanding of intestinal morphogenesis and homeostasis [145 - 150] but with limitations [151, 152].

To address the limitations of conventional *In vitro* systems, a bioprinting platform was used to construct a multicellular 3D primary human intestinal tissue model that more closely mimics the structure and function of genuine tissue [107, 147]. Sato *et al.* (2011) and Madden and his co-workers (2018) used epithelial cells from the human ileal tissue segment and thermo-responsive Novogel® bioink for bioprinting of the intestine. An intestinal model containing villi and capillary structure was 3D printed by Kim and Kim (2018). They have used a dual cell printing process using bioinks developed from Caco-2 cells and HUVECs [153]. This 3D printed structure functionally mimicked the actual intestinal structure with villi and showed cellular activities of functional intestine projecting as a promising start in constructing a fully functioning human intestine [153]. In 2019, Study by Justine reported the development of a 3D intestinal epithelium *In vitro* model using a new fabrication process combining photo polymerizable hydrogel that supports the growth of intestinal cell lines with high-resolution stereolithography 3D printing. This method allows for the creation of artificial 3D scaffolds that match the proportions and architecture of mice intestinal crypts and villi [154]. In the future, implanting an entire section of the 3D printed tubular structure of the intestine seems well within reach. However, reestablishing the endogenous lining from the proximal and distal sites of anastomosis in the case of long implants could emerge as a limitation in human implants [155]. Addressing this limit by determining the higher patch length limit will surely help successfully implant a 3D printed intestine.

Vascularized Organoid

The biggest hurdle faced in successful clinical outcomes of bioengineered tissue constructs is the inability to create required vascularisation before any human implantation application can happen [156]. The human body is highly vascularised. Each and every one of our body's tissues and organs is a vascularized structure. Vasculature allows the exchange of gaseous products and electrolytes, nutrients, and waste material [157]. It has been observed that tissue-engineered constructs of more than 3-6 millimeters have poor vascularisation, leading to inadequate cellular integration or cell death and necrosis due to mass transport and diffusional constraints [73]. They produce a blood vessel that mimics the endogenous system in terms of biocompatibility, physiological flow rates, and ability to withstand systemic pressure changes. It has been observed that vascularisation is a significant challenge in tissue engineering to fabricate 3D complex tissues with a suitable function. Several methods for enhancing vascularisation are currently under evaluation for bioprinting blood vessels. The strategies like direct printing of vascular scaffold [158] and controlled release of angiogenic factors, while 3D printing of tissue has shown the promising outcome of neovascularization [159, 160] even in the implantable material [158] have seen some success. In drug testing and metabolic assays, *In vitro* tissue samples with pre-printed vascular networks and the direct printing of blood vessels are frequently used [158].

Bioprinting of blood vessels was attempted to mimic a native blood vessel both in terms of cellular and structural way (adventitia) by including endothelial cells of the intima and smooth muscle cells of the media [161]. Weinberg and Bell in 1986 have created the first bioartificial blood vessel [162] using smooth muscle cells in collagen gel in an annular mold, covered by Dacron mesh with an outer fibroblast layer. However, insufficient burst pressure and limitation of autologous endothelial cells for adventitia were major setbacks for Weinberg's vessel. This study has paved the way for numerous additional future studies using a variety of synthetic materials like poly(L-lactic) acid (PLA) and poly (DL-lactic--o-glycolic) acid (PLGA) polyglycolic acid (PGA) *etc.* with cultured cells [155].

Based on previous works and current studies, bioprinting vasculature is achieved using extrusion, inkjet (droplet), and laser-based 3D printing. The most crucial part is bio-ink selection in 3D printing, as it is necessary to mimic the biological conduits and vascularisation to produce complete organ or extra-large-sized tissue. The bio-ink also differs depending on the tissue that needs to be printed. Once 3D printed, these vasculature structures require vascular connections for gaseous, waste, and nutrient exchange to prevent ischemic injury. Also, getting a proper neural connection to provide appropriate conductive signals to tissue is the biggest challenge in bioprinting of vasculature tissue [155].

LIMITATIONS AND FUTURE PROSPECTS

The shortage of organ donors is a critical challenge in treating end-stage organ failure. Even with bioengineering as the potential alternative to generate organs *in vitro*, it is still challenging to successfully bioprint a functional organ and use it for regenerative medicine applications. The key features like simplicity, flexibility, and agility for complex manufacturing have acclaimed 3D printing. The popularity of 3D printing has increased due to the superior printing of materials like metals, biopolymers, ceramics, composites, and hydrogels and their use in controls, sensing, and automation areas [163]. Though 3D printing is looking lucrative, it is not without any limitations. The 3D printing has challenges like void formation between subsequent layers, stair-stepping effect, or layering error affecting the quality of outer surfaces. These also affect the microstructure and mechanical stability of constructs [164].

One of the limitations of 3D printing is the low resolution. The higher resolutions of between 10 μm and 100 μm are achieved by the Jetting-based and extrusion-based bioprinters, respectively. Jetting-based bio-printing cannot use high-viscous bioink materials to generate droplets and printed structures. Therefore, it is challenging to fabricate durable 3D structures that can maintain their shape and withstand external stress after implantation. Jetting-based bio-printing cannot fabricate mechanically stable, clinically applicable tissue and organ-sized structures [104, 165]. The fabrication of 3D tissue and organ-sized structures is more frequent, using a tiny droplet size of 10 μm. Also, there is decreased cell viability due to the high resolution, long printing process, a smaller-sized nozzle that affects cell printing by flow resistance, and shear stress within the bio-ink.

Therefore, a printing procedure with high resolution and acceptable bio-ink materials must be improved and created, respectively, to preserve cell viability and ensure the bioprinting's long-term success [165]. Another challenge in bio-printing is cell aggregation and sedimentation in the print cartridge reservoir and syringe [165].

The major limitation for clinical translational of bio-printed large-scale organoids is vascularisation. To regenerate revascularized organoids, 3D printed structures should mediate oxygen and nutrients. Many attempts are being made to overcome the limitations of vascularisation, while a few studies reported bio-printed vascular structures of a few millimeters. However, vascularization of tissue and organ constructions has not yet been accomplished [165]. Bio-ink optimization is another challenge in bio-printing. The materials that will provide high resolution in printing, maintain the overall structure of the construct, and preserve the cell morphology and cell viability, with the creation and maintenance of cellular microenvironment, are highly desired [163].

The ultimate aim of bio-printed organoids is to make patient-specific tissues or organs to reduce patient morbidity and mortality and enhance the quality of life. With the advancement of bio-printing, further developments of simple tissues like skin to complex tissues like lungs and kidneys are also underway. Still, the field is growing, and with constant improvements, the successful clinical translation of bio-printed organoids is not far from reality. The major hurdle in organ bioprinting is the complexity of biological structures that make the particular organ or tissue [163]. Further advancements in morphogenesis, including complex cells, biochemical and biophysical clues, will shape the complex structures and organoids. The use of multiheaded, parallel bioprinters and continuous liquid interface production could overcome the long duration time of printing [46]. Bioprinting has a significant role in the future growth and development of organoids and complex organs. 3D bioprinted organoids have a vital role in drug development, developmental biology, toxicity testing, disease profiling, and repair or regeneration of the various tissues and organs. Thus, bioprinted organoids have massive potential in tissue engineering, regenerative medicine, and healthcare applications.

ABBREVIATIONS

2D	Two Dimensional
3D	Three Dimensional
ARDS	Acute Respiratory Distress Syndrome
ASC	Adult Stem Cells
AM	Additive manufacturing

ALM	Additive layer manufacturing
AFM	Atomic Force Microscopy
ASTM	American Society for Testing and Materials
ASC	Adult stem cells
AM	Additive Manufacturing
ALM	Additive Layer Manufacturing
CIJ	Continuous Inkjet Printing
CAD	Computer-Aided Design
CAM	Computer-Aided Manufacturing
COPD	Chronic obstructive pulmonary Disease
DMLS	Direct Metal Laser Sintering
DOD	Drop On Demand
DED	Direct Energy Deposition
DLP	Digital Light processing
DECM	Decellularised Extracellular Matrix
DM	Diabetes Melitus
EHDJ	Electrohydrodynamic Jetting
ESC	Embryonic Stem Cells
EPC	Endothelial Progenitor Cells
EC	Endothelial Cells
ECM	Extra Cellular Matrix
EBM	Electron Beam Matching
FDM	Fused Deposition Modeling
FFF	Fused Filament Fabrication
FLM	Fused Layer Deposition
GI	Gastro Intestinal
GelMA	Gelatine Methacryloyl
HECFC	Human Endothelial Colony Forming Cells
HUVEC	Human Umbilical Vein Endothelial Cells
HLC	Hepatocyte Like Cells
IEC	Intestinal epithelial cells
ISO	International Organization for Standardization
IPS	Induced Pluripotent Cells
IBD	Inflamatory Bowel Disease
LD	Laser Deposition

LENS	Laser Engineered Net Shaping
LOM	Laminated Object Manufacturing
LGDW	Laser Guided Direct Writing
LIFT	Laser Induced Forward Transfer
LAB	Laser-Assisted Bioprinting
MPP	Multi Photon Polymerization
PCL	PolyCaproLactone
PEG	Poly Ethylene Glycol
PLA	Poly (L-lactic) acid
PLGA	Poly (DL-lactic-co-glycolic) acid
PGA	Poly Glycolic acid
RBC	Red Blood Cells
SLA	Stereolithography
SLM	Selective Laser Melting
SLS	Selective Laser Sintering
UAM	Ultrasound additive manufacturing
USPTO	United States Patent and Trademark Office
UC	Ultrasound Consolidation
UAM	Ultrasonic Additive Manufacturing

ACKNOWLEDGEMENT

We acknowledge intramural research grant DYPES/DU/R&D2021/275 from D. Y. Patil Education Society (Institution Deemed to be University), Kolhapur-416006 (MS), India, for a research fellowship to Shahbaj Mujawar.

REFERENCES

[1]   Gibson I, Cheung LK, Chow SP, *et al.* The use of rapid prototyping to assist medical applications. Rapid Prototyping J 2006; 12(1): 53-8.
[http://dx.doi.org/10.1108/13552540610637273]

[2]   Sachs E, Cima M, Cornie J, Brancazio D, Bredt J, Curodeau A, *et al.* Dimensional Printing : Rapid Tooling and Prototypes Directly from CAD Representation. Int Solid Free Fabr Symp. 27-47.

[3]   Reed C, Grasso F. Recent advances in computational models of natural argument. Int J Intell Syst 2007; 22(1): 1-15.
[http://dx.doi.org/10.1002/int.20187]

[4]   Attaran M. The rise of 3-D printing: The advantages of additive manufacturing over traditional manufacturing. Bus Horiz 2017; 60(5): 677-88.
[http://dx.doi.org/10.1016/j.bushor.2017.05.011]

[5]   Drizo A, Pegna J. Environmental impacts of rapid prototyping: an overview of research to date. Rapid Prototyping J 2006; 12(2): 64-71.

[http://dx.doi.org/10.1108/13552540610652393]

[6]   Matos F, Godina R, Jacinto C, Carvalho H, Ribeiro I, Peças P. Additive manufacturing: Exploring the social changes and impacts. Sustain 2019; 11: 01-18.

[7]   Gebler M, Schoot Uiterkamp AJM, Visser C. A global sustainability perspective on 3D printing technologies. Energy Policy 2014; 74: 158-67.
[http://dx.doi.org/10.1016/j.enpol.2014.08.033]

[8]   Mehrpouya M, Dehghanghadikolaei A, Fotovvati B, Vosooghnia A, Emamian SS, Gisario A. The potential of additive manufacturing in the smart factory industrial 4.0: A review. Appl Sci (Basel) 2019; 9(18): 3865.
[http://dx.doi.org/10.3390/app9183865]

[9]   Ahangar P, Cooke ME, Weber MH, Rosenzweig DH. Current biomedical applications of 3D printing and additive manufacturing. Appl Sci (Basel) 2019; 9(8): 1713.
[http://dx.doi.org/10.3390/app9081713]

[10]   Bourell DL, Beaman JJ. Chronology and Current Processes for Freeform Fabrication. Funtai Oyobi Fummatsu Yakin/Journal Japan Soc Powder. Powder Metall 2003; 50: 981-91.

[11]   Leinster M. Things Pass By, in The Earth In Peril" (D. Wollheim ed.). J Chem Inf Model 2013; 53: 1689-99.

[12]   David jones 3D printing. Mar. Sci. Bull 1984; pp. 1-2.

[13]   Hartary PEW. Hartary PEW Liquid metal recorder United states: United states patent;. 1971.

[14]   Kodama H. Automatic method for fabricating a three-dimensional plastic model with photo-hardening polymer. Rev Sci Instrum 1981; 52(11): 1770-3.
[http://dx.doi.org/10.1063/1.1136492]

[15]   Breinan M, Bernard H. United States Patent 1982; (19): 1-4.

[16]   Ruggiero PE. 4665492 Computer automated manufacturing process and system. Robot Comput-Integr Manuf 1987; 1-2.

[17]   Hull C. On Stereolithography. Virtual Phys Prototyp 2012; 7(3): 177.
[http://dx.doi.org/10.1080/17452759.2012.723409]

[18]   Snyder TJ, Andrews M, Weislogel M, *et al.* 3D systems' technology overview and new applications in manufacturing, engineering, science, and education. 3D Print. 3D Printing and Additive Manufacturing 2014; 1(3): 169-76.
[http://dx.doi.org/10.1089/3dp.2014.1502] [PMID: 28473997]

[19]   Zhao C, Fezzaa K, Cunningham RW, *et al.* Real-time monitoring of laser powder bed fusion process using high-speed X-ray imaging and diffraction. Sci Rep 2017; 7(1): 3602.
[http://dx.doi.org/10.1038/s41598-017-03761-2] [PMID: 28620232]

[20]   Campbell I, Bourell D, Gibson I. Additive manufacturing: rapid prototyping comes of age. Rapid Prototyping J 2012; 18(4): 255-8.
[http://dx.doi.org/10.1108/13552541211231563]

[21]   Ye D, Hong GS, Zhang Y, Zhu K, Fuh JYH. Defect detection in selective laser melting technology by acoustic signals with deep belief networks. Int J Adv Manuf Technol 2018; 96(5-8): 2791-801.
[http://dx.doi.org/10.1007/s00170-018-1728-0]

[22]   Frueh FS, Menger MD, Lindenblatt N, Giovanoli P, Laschke MW. Current and emerging vascularization strategies in skin tissue engineering. Crit Rev Biotechnol 2017; 37(5): 613-25.
[http://dx.doi.org/10.1080/07388551.2016.1209157] [PMID: 27439727]

[23]   Allhoff F, Lin P, Steinberg J. Ethics of human enhancement: an executive summary. Sci Eng Ethics 2011; 17(2): 201-12.
[http://dx.doi.org/10.1007/s11948-009-9191-9] [PMID: 20094921]

[24] Li P, Faulkner A. 3D Bioprinting Regulations: a UK/EU Perspective. Eur J Risk Regul 2017; 8(2): 441-7.
[http://dx.doi.org/10.1017/err.2017.19]

[25] Carvalho C. 3D-Bioplotter® for Bioprinting – There's Only One 3D-Bioplotter® https://envisionteccom/3d-printers/3d-bioplotter/ 2021.

[26] Regenhu- history https://wwwregenhucom/about-us/#history 2021.

[27] Thomas DS. Economics of additive manufacturing. Laser-Based Addit Manuf Met Parts Model Optim Control Mech Prop 2017; 2: 285-320.

[28] Geert DE. Vision of 3D printing with concrete- technical, economic and environmental potentials. 2018; pp. 1-34.

[29] 3D printer pricing how much does a 3D printer cost?. fusion3 https://wwwfusion3designcom/how-much-does-a-3d-printer-cost/ 2021.

[30] Stereolithography The First 3D Printing Technology. Am Soc Mech Eng https://wwwasmeorg/wwwasmeorg/media/resourcefiles/aboutasme/who we are/engineering history/landmarks/261-stereolithographypdf 2016.

[31] Mazzoli A. Selective laser sintering in biomedical engineering. Med Biol Eng Comput 2013; 51(3): 245-56.
[http://dx.doi.org/10.1007/s11517-012-1001-x] [PMID: 23250790]

[32] Wohlers T, Gornet T. History of Additive Manufacturing 2014. Wohlers Rep 2014 - 3D Print Addit Manuf State Ind 2014; 1-34.

[33] Bechtold S. 3d printing and the intellectual property system 2015.

[34] Boland T. Patent US 7051654 B2 2003.

[35] Jones R, Haufe P, Sells E, *et al.* RepRap – the replicating rapid prototyper. Robotica 2011; 29(1): 177-91.
[http://dx.doi.org/10.1017/S026357471000069X]

[36] Gu Z, Fu J, Lin H, He Y. Development of 3D bioprinting: From printing methods to biomedical applications. Asian Journal of Pharmaceutical Sciences 2020; 15(5): 529-57.
[http://dx.doi.org/10.1016/j.ajps.2019.11.003] [PMID: 33193859]

[37] Choudhury D, Anand S, Naing MW. The arrival of commercial bioprinters – Towards 3D bioprinting revolution! International Journal of Bioprinting 2018; 4(2): 1-20.
[http://dx.doi.org/10.18063/ijb.v4i2.139] [PMID: 33102917]

[38] Noor N, Shapira A, Edri R, Gal I, Wertheim L, Dvir T. 3D Printing of Personalized Thick and Perfusable Cardiac Patches and Hearts. Adv Sci (Weinh) 2019; 6(11)1900344
[http://dx.doi.org/10.1002/advs.201900344] [PMID: 31179230]

[39] Bisht B, Hope A, Mukherjee A, Paul MK. Advances in the Fabrication of Scaffold and 3D Printing of Biomimetic Bone Graft. Ann Biomed Eng 2021; 49(4): 1128-50.
[http://dx.doi.org/10.1007/s10439-021-02752-9] [PMID: 33674908]

[40] Gibson I, Rosen D, Stucker B. Generalized Additive Manufacturing Process Chain Addit Manuf Technol. 2nd ed. Springer-Verlag New York 2015; pp. 1-498.

[41] Azman AH, Vignat F, Villeneuve F. Cad tools and file format performance evaluation in designing lattice structures for additive manufacturing. J Teknol 2018; 80(4): 87-95.
[http://dx.doi.org/10.11113/jt.v80.12058]

[42] Bernardini F, Rushmeier H. The 3D model acquisition pipeline. Comput Graph Forum 2002; 21(2): 149-72.
[http://dx.doi.org/10.1111/1467-8659.00574]

[43]   Satyanarayana B, Prakash KJ. Component Replication Using 3D Printing Technology. Procedia Materials Science 2015; 10: 263-9. [Internet].
[http://dx.doi.org/10.1016/j.mspro.2015.06.049]

[44]   Jiménez M, Romero L, Domínguez IA, Espinosa MM, Domínguez M. Additive Manufacturing Technologies: An Overview about 3D Printing Methods and Future Prospects. Complexity 2019; 2019: 1-30.
[http://dx.doi.org/10.1155/2019/9656938]

[45]   Tofail SAM, Koumoulos EP, Bandyopadhyay A, Bose S, O'Donoghue L, Charitidis C. Additive manufacturing: scientific and technological challenges, market uptake and opportunities. Mater Today 2018; 21(): 22-37.
[http://dx.doi.org/10.1016/j.mattod.2017.07.001]

[46]   Rezvani Ghomi E, Khosravi F, Neisiany RE, Singh S, Ramakrishna S. Future of additive manufacturing in healthcare. Curr Opin Biomed Eng 2021; 17: 100255.
[http://dx.doi.org/10.1016/j.cobme.2020.100255]

[47]   Gibson I. D W Rosen BS Rapid Prototyping to Direct Digital Manufacturing Addit Manuf Technol. Springer 2010; pp. 128-45.
[http://dx.doi.org/10.1007/978-1-4419-1120-9]

[48]   Gibson I, Rosen DW, Stucker B. Additive manufacturing technologies. Rapid Prototyp to Direct Digit Manuf 2010; pp. 1-459.
[http://dx.doi.org/10.1007/978-1-4419-1120-9]

[49]   Ziaee M, Crane NB. Binder jetting: A review of process, materials, and methods. Addit Manuf 2019; 28: 781-801. [Internet].
[http://dx.doi.org/10.1016/j.addma.2019.05.031]

[50]   Rezvani Ghomi E, Khosravi F, Neisiany RE, Singh S, Ramakrishna S. Future of additive manufacturing in healthcare. Curr Opin Biomed Eng Elsevier Inc. 2021; 17.
[http://dx.doi.org/10.1016/j.cobme.2020.100255]

[51]   Tofail SAM, Koumoulos EP, Bandyopadhyay A, Bose S, O'Donoghue L, Charitidis C. Additive manufacturing: scientific and technological challenges, market uptake and opportunities. 2018.
[http://dx.doi.org/10.1016/j.mattod.2017.07.001]

[52]   Jiang J, Fu YF. A short survey of sustainable material extrusion additive manufacturing. 2020.
[http://dx.doi.org/10.1080/14484846.2020.1825045]

[53]   Kristensen FK, Sverre JM, Bustad S. Pdb21 a Cost-Utility Analysis of Insulin Glargine (Lantus®) in the Treatment of Patients With Type 1 Diabetes. Value Heal. Value Health 2003; 6(6): 682. [Internet]. [ISPOR].
[http://dx.doi.org/10.1016/S1098-3015(10)61744-5]

[54]   Li W, Mille LS, Robledo JA, Uribe T, Huerta V, Zhang YS. Recent Advances in Formulating and Processing Biomaterial Inks for Vat Polymerization-Based 3D Printing. Adv Healthc Mater 2020; 9(15)2000156
[http://dx.doi.org/10.1002/adhm.202000156] [PMID: 32529775]

[55]   Yue Z, Liu X, Coates PT, Wallace GG. Advances in printing biomaterials and living cells. Curr Opin Organ Transplant 2016; 21(5): 467-75.
[http://dx.doi.org/10.1097/MOT.0000000000000346] [PMID: 27517507]

[56]   Derakhshanfar S, Mbeleck R, Xu K, Zhang X, Zhong W, Xing M. 3D bioprinting for biomedical devices and tissue engineering: A review of recent trends and advances. Bioact Mater 2018; 3: 144-56.
[http://dx.doi.org/10.1016/j.bioactmat.2017.11.008]

[57]   Kang HW, Lee SJ, Ko IK, Kengla C, Yoo JJ, Atala A. A 3D bioprinting system to produce human-scale tissue constructs with structural integrity. Nat Biotechnol 2016; 34(3): 312-9.
[http://dx.doi.org/10.1038/nbt.3413] [PMID: 26878319]

[58] Moroni L, Burdick JA, Highley C, *et al.* Biofabrication strategies for 3D *in vitro* models and regenerative medicine. Nat Rev Mater 2018; 3(5): 21-37.
[http://dx.doi.org/10.1038/s41578-018-0006-y] [PMID: 31223488]

[59] Xue W, Krishna BV, Bandyopadhyay A, Bose S. Processing and biocompatibility evaluation of laser processed porous titanium. Acta Biomater 2007; 3(6): 1007-18.
[http://dx.doi.org/10.1016/j.actbio.2007.05.009] [PMID: 17627910]

[60] Kim JH, Yoo JJ, Lee SJ. Three-dimensional cell-based bioprinting for soft tissue regeneration. Tissue Eng Regen Med 2016; 13(6): 647-62.
[http://dx.doi.org/10.1007/s13770-016-0133-8] [PMID: 30603446]

[61] Murphy SV, Atala A. 3D bioprinting of tissues and organs. Nat Biotechnol 2014; 32(8): 773-85.
[http://dx.doi.org/10.1038/nbt.2958] [PMID: 25093879]

[62] Derby B. Printing and prototyping of tissues and scaffolds. Science 2012; 338: 921-6.
[http://dx.doi.org/10.1126/science.1226340]

[63] Ozbolat IT, Hospodiuk M. Current advances and future perspectives in extrusion-based bioprinting 2016.
[http://dx.doi.org/10.1016/j.biomaterials.2015.10.076]

[64] Pham DT, Dimov SS. Rapid prototyping and rapid tooling—the key enablers for rapid manufacturing. Proc Inst Mech Eng, C J Mech Eng Sci 2003; 217(1): 1-23.
[http://dx.doi.org/10.1243/095440603762554569]

[65] Hutmacher DW, Sittinger M, Risbud MV. Scaffold-based tissue engineering: rationale for computer-aided design and solid free-form fabrication systems. Trends Biotechnol 2004; 22(7): 354-62.
[http://dx.doi.org/10.1016/j.tibtech.2004.05.005] [PMID: 15245908]

[66] Kesti M, Eberhardt C, Pagliccia G, *et al.* Bioprinting Complex Cartilaginous Structures with Clinically Compliant Biomaterials. Adv Funct Mater 2015; 25(48): 7406-17.
[http://dx.doi.org/10.1002/adfm.201503423]

[67] Di Bella C, Duchi S, O'Connell CD, *et al. In situ* handheld three-dimensional bioprinting for cartilage regeneration. J Tissue Eng Regen Med 2018; 12(3): 611-21.
[http://dx.doi.org/10.1002/term.2476] [PMID: 28512850]

[68] Marchioli G, van Gurp L, van Krieken PP, *et al.* Fabrication of three-dimensional bioplotted hydrogel scaffolds for islets of Langerhans transplantation. Biofabrication 2015; 7(2)025009
[http://dx.doi.org/10.1088/1758-5090/7/2/025009] [PMID: 26019140]

[69] Gu Q, Tomaskovic-Crook E, Lozano R, *et al.* Functional 3D Neural Mini-Tissues from Printed Gel-Based Bioink and Human Neural Stem Cells. Adv Healthc Mater 2016; 5(12): 1429-38.
[http://dx.doi.org/10.1002/adhm.201600095] [PMID: 27028356]

[70] Liu F, Liu C, Chen Q, *et al.* Progress in organ 3D bioprinting. International Journal of Bioprinting 2018; 4(1): 1-15.
[http://dx.doi.org/10.18063/ijb.v4i1.128] [PMID: 33102911]

[71] Xie Z, Gao M, Lobo AO, Webster TJ. 3D bioprinting in tissue engineering for medical applications: The classic and the hybrid. Polymers (Basel) 2020; 12(8): 1717.
[http://dx.doi.org/10.3390/polym12081717] [PMID: 32751797]

[72] Wang S, Lee JM, Yeong WY. Smart hydrogels for 3D bioprinting. International Journal of Bioprinting 2015; 1(1): 3-14.
[http://dx.doi.org/10.18063/IJB.2015.01.005]

[73] Naing MW, Chua CK, Leong KF. Computer Aided Tissue Engineering Scaffolds. Virtual Prototyp. Bio Manuf Med Appl 2021.
[http://dx.doi.org/10.1007/978-3-030-35880-8_4]

[74] Jiang T, Munguia-Lopez JG, Flores-Torres S, Kort-Mascort J, Kinsella JM. Extrusion bioprinting of

soft materials: An emerging technique for biological model fabrication. Appl Phys Rev 2019; 6(1)011310
[http://dx.doi.org/10.1063/1.5059393]

[75]   You F, Eames BF, Chen X. Application of extrusion-based hydrogel bioprinting for cartilage tissue engineering. Int J Mol Sci 2017; 18(7): 1597.
[http://dx.doi.org/10.3390/ijms18071597] [PMID: 28737701]

[76]   Peltola SM, Melchels FPW, Grijpma DW, Kellomäki M. A review of rapid prototyping techniques for tissue engineering purposes. Ann Med 2008; 40(4): 268-80.
[http://dx.doi.org/10.1080/07853890701881788] [PMID: 18428020]

[77]   Boularaoui S, Al Hussein G, Khan KA, Christoforou N, Stefanini C. An overview of extrusion-based bioprinting with a focus on induced shear stress and its effect on cell viability. Bioprinting 2020.
[http://dx.doi.org/10.1016/j.bprint.2020.e00093]

[78]   Visser J, Peters B, Burger TJ, Boomstra J, Dhert WJA, Melchels FPW, *et al.* Biofabrication of multi-material anatomically shaped tissue constructs. Biofabrication 2013; 5: 04-10.
[http://dx.doi.org/10.1088/1758-5082/5/3/035007]

[79]   Li X, Liu B, Pei B, *et al.* Inkjet Bioprinting of Biomaterials. Chem Rev 2020; 120(19): 10793-833.
[http://dx.doi.org/10.1021/acs.chemrev.0c00008] [PMID: 32902959]

[80]   Salve RK, Sanklecha SM. An Overview of Inkjet Bioprinting. Int J Futur Gener Commun Netw 2020; 9: 1707-12.

[81]   Bishop ES, Mostafa S, Pakvasa M, Luu HH, Lee MJ, Wolf JM, *et al.* 3-D bioprinting technologies in tissue engineering and regenerative medicine: Current and future trends. Genes Dis . 2017; 4: pp. 185-95.
[http://dx.doi.org/10.1016/j.gendis.2017.10.002]

[82]   Li J, Chen M, Fan X, Zhou H. Recent advances in bioprinting techniques: Approaches, applications and future prospects. J Transl Med. BioMed Central 2016; 14: 1-15.

[83]   Guillotin B, Ali M, Ducom A, Catros S, Keriquel V, Souquet A, *et al.* Laser-Assisted Bioprinting for Tissue Engineering. 2013.
[http://dx.doi.org/10.1016/B978-1-4557-2852-7.00006-8]

[84]   Dobos A, Van Hoorick J, Steiger W, *et al.* Thiol–Gelatin–Norbornene Bioink for Laser-Based High-Definition Bioprinting. Adv Healthc Mater 2020; 9(15)1900752
[http://dx.doi.org/10.1002/adhm.201900752] [PMID: 31347290]

[85]   Kingsley DM, Roberge CL, Rudkouskaya A, *et al.* Laser-based 3D bioprinting for spatial and size control of tumor spheroids and embryoid bodies. Acta Biomater 2019; 95: 357-70.
[http://dx.doi.org/10.1016/j.actbio.2019.02.014] [PMID: 30776506]

[86]   Barron JA, Wu P, Ladouceur HD, Ringeisen BR. Biological laser printing: a novel technique for creating heterogeneous 3-dimensional cell patterns. Biomed Microdevices 2004; 6(2): 139-47.
[http://dx.doi.org/10.1023/B:BMMD.0000031751.67267.9f] [PMID: 15320636]

[87]   Mironov V, Visconti RP, Kasyanov V, Forgacs G, Drake CJ, Markwald RR. Organ printing: Tissue spheroids as building blocks. Biomaterials 2009; 30(12): 2164-74. http://spraakdata.gu.se/svedd/pub/vol46-dana.pdf
[http://dx.doi.org/10.1016/j.biomaterials.2008.12.084] [PMID: 19176247]

[88]   Rezaei FS, Khorshidian A, Beram FM, Derakhshani A, Esmaeili J, Barati A. 3D printed chitosan/polycaprolactone scaffold for lung tissue engineering: hope to be useful for COVID-19 studies. RSC Advances 2021; 11(32): 19508-20.
[http://dx.doi.org/10.1039/D1RA03410C] [PMID: 35479204]

[89]   Gupta A, Kashte S, Gupta M, Rodriguez HC, Gautam SS, Kadam S. Mesenchymal stem cells and exosome therapy for COVID-19: current status and future perspective. Hum Cell 2020; 33(4): 907-18.
[http://dx.doi.org/10.1007/s13577-020-00407-w] [PMID: 32780299]

[90]   Rezaei FS, Khorshidian A, Beram FM, Derakhshani A, Esmaeili J, Barati A. 3D printed chitosan/polycaprolactone scaffold for lung tissue engineering: hope to be useful for COVID-19 studies. RSC Advances 2021; 11(32): 19508-20.
[http://dx.doi.org/10.1039/D1RA03410C] [PMID: 35479204]

[91]   Huang L, Yuan W, Hong Y, *et al.* 3D printed hydrogels with oxidized cellulose nanofibers and silk fibroin for the proliferation of lung epithelial stem cells. Cellulose 2021; 28(1): 241-57.
[http://dx.doi.org/10.1007/s10570-020-03526-7] [PMID: 33132545]

[92]   Wilkinson DC, Mellody M, Meneses LK, Hope AC, Dunn B, Gomperts BN. Development of a Three-Dimensional Bioengineering Technology to Generate Lung Tissue for Personalized Disease Modeling. Curr Protoc Stem Cell Biol 2018; 46(1)e56
[http://dx.doi.org/10.1002/cpsc.56] [PMID: 29927098]

[93]   Grigoryan B, Paulsen SJ, Corbett DC, Sazer DW, Fortin CL, Zaita AJ, *et al.* Multivascular networks and functional intravascular topologies within biocompatible hydrogels. Science 2019; 364: 458-64.
[http://dx.doi.org/10.1126/science.aav9750]

[94]   Galliger Z, Vogt CD, Panoskaltsis-Mortari A. 3D bioprinting for lungs and hollow organs. Transl Res 2019; 211: 19-34.
[http://dx.doi.org/10.1016/j.trsl.2019.05.001] [PMID: 31150600]

[95]   Addario G, Djudjaj S, Farè S, Boor P, Moroni L, Mota C. Microfluidic bioprinting towards a renal *in vitro* model. Bioprinting 2020; 20e00108
[http://dx.doi.org/10.1016/j.bprint.2020.e00108]

[96]   Homan KA, Kolesky DB, Skylar-Scott MA, Herrmann J, Obuobi H, Moisan A, *et al.* Bioprinting of 3D Convoluted Renal Proximal Tubules on Perfusable Chips. Sci Rep. Nature Publishing Group 2016; 6: 1-13.

[97]   Tröndle K, Rizzo L, Pichler R, *et al.* Scalable fabrication of renal spheroids and nephron-like tubules by bioprinting and controlled self-assembly of epithelial cells. Biofabrication 2021; 13(3)035019
[http://dx.doi.org/10.1088/1758-5090/abe185] [PMID: 33513594]

[98]   Higgins JW, Chambon A, Bishard K, Hartung A, Arndt D, Brugnano J, *et al.* Bioprinted pluripotent stem cell-derived kidney organoids provide opportunities for high content screening. bioRxiv 2018.505396
[http://dx.doi.org/10.1101/505396]

[99]   Lawlor KT, Vanslambrouck JM, Higgins JW, *et al.* Cellular extrusion bioprinting improves kidney organoid reproducibility and conformation. Nat Mater 2021; 20(2): 260-71.
[http://dx.doi.org/10.1038/s41563-020-00853-9] [PMID: 33230326]

[100] Goulart E, de Caires-Junior LC, Telles-Silva KA, *et al.* 3D bioprinting of liver spheroids derived from human induced pluripotent stem cells sustain liver function and viability *in vitro*. Biofabrication 2019; 12(1)015010
[http://dx.doi.org/10.1088/1758-5090/ab4a30] [PMID: 31577996]

[101] Kashte S, Maras JS, Kadam S. Bioinspired engineering for liver tissue regeneration and development of bioartificial liver: A review. Crit Rev Biomed Eng 2018; 46(5): 413-27.
[http://dx.doi.org/10.1615/CritRevBiomedEng.2018028276] [PMID: 30806261]

[102] Mao Q, Wang Y, Li Y, Juengpanich S, Li W, Chen M, *et al.* Fabrication of liver microtissue with liver decellularized extracellular matrix (dECM) bioink by digital light processing (DLP) bioprinting. Mater Sci Eng C. 2020.
[http://dx.doi.org/10.1016/j.msec.2020.110625]

[103] Seol YJ, Yoo JJ, Atala A. Bioprinting of three-dimensional tissues and organ constructs Essentials 3D Biofabrication Transl. Elsevier Inc. 2015; pp. 283-92. Internet
[http://dx.doi.org/10.1016/B978-0-12-800972-7.00016-5]

[104] Mao Q, Wang Y, Li Y, Juengpanich S, Li W. Materials Science Engineering C Fabrication of liver

microtissue with liver decellularized extracellular matrix (dECM) bioink by digital light processing (DLP) bioprinting. Mater Sci Eng C. 2020.

[105] Faulkner-Jones A, Fyfe C, Cornelissen DJ, *et al.* Bioprinting of human pluripotent stem cells and their directed differentiation into hepatocyte-like cells for the generation of mini-livers in 3D. Biofabrication 2015; 7(4)044102
[http://dx.doi.org/10.1088/1758-5090/7/4/044102] [PMID: 26486521]

[106] Lee JW, Choi Y-J, Yong W-J, Pati F, Shim J-H, Kang KS, *et al.* Development of a 3D cell printed construct considering angiogenesis for liver tissue engineering. Biofabrication. IOP Publishing 2016; 8: 15007.

[107] Nguyen DG, Funk J, Robbins JB, *et al.* Bioprinted 3D primary liver tissues allow assessment of organ-level response to clinical drug induced toxicity in vitro. PLoS One 2016; 11(7)e0158674
[http://dx.doi.org/10.1371/journal.pone.0158674] [PMID: 27387377]

[108] Bhise NS, Manoharan V, Massa S, *et al.* A liver-on-a-chip platform with bioprinted hepatic spheroids. Biofabrication 2016; 8(1)014101
[http://dx.doi.org/10.1088/1758-5090/8/1/014101] [PMID: 26756674]

[109] Gori M, Giannitelli SM, Torre M, *et al.* Biofabrication of Hepatic Constructs by 3D Bioprinting of a Cell-Laden Thermogel: An Effective Tool to Assess Drug-Induced Hepatotoxic Response. Adv Healthc Mater 2020; 9(21)2001163
[http://dx.doi.org/10.1002/adhm.202001163] [PMID: 32940019]

[110] Wu Y, Wenger A, Golzar H, Tang XS. 3D bioprinting of bicellular liver lobule - mimetic structures *via* microextrusion of cellulose nanocrystal - incorporated shear - thinning bioink. Sci Rep. Nature Publishing Group UK 2020; 10: 1-12.

[111] Kang D, Hong G, An S, *et al.* Bioprinting of Multiscaled Hepatic Lobules within a Highly Vascularized Construct. Small 2020; 16(13)1905505
[http://dx.doi.org/10.1002/smll.201905505] [PMID: 32078240]

[112] Yang H, Sun L, Pang Y, *et al.* Three-dimensional bioprinted hepatorganoids prolong survival of mice with liver failure. Gut 2021; 70(3): 567-74.
[http://dx.doi.org/10.1136/gutjnl-2019-319960] [PMID: 32434830]

[113] Vig K, Chaudhari A, Tripathi S, *et al.* Advances in skin regeneration using tissue engineering. Int J Mol Sci 2017; 18(4): 789.
[http://dx.doi.org/10.3390/ijms18040789] [PMID: 28387714]

[114] Atiyeh B, Hayek S. An Update on Management of Acute and Chronic Open Wounds: The Importance of Moist Environment in Optimal Wound Healing. Med Chem Rev Online 2004; 1(2): 111-21.
[http://dx.doi.org/10.2174/1567203043480304]

[115] Sen CK, Gordillo GM, Roy S, *et al.* Human skin wounds: A major and snowballing threat to public health and the economy. Wound Repair Regen 2009; 17(6): 763-71.
[http://dx.doi.org/10.1111/j.1524-475X.2009.00543.x] [PMID: 19903300]

[116] Jones I, Currie L, Martin R. A guide to biological skin substitutes. Br J Plast Surg 2002; 55(3): 185-93.
[http://dx.doi.org/10.1054/bjps.2002.3800]

[117] Albanna M, Binder KW, Murphy SV, Kim J, Qasem SA, Zhao W, *et al.* In Situ Bioprinting of Autologous Skin Cells Accelerates Wound Healing of Extensive Excisional Full-Thickness Wounds. 2019.
[http://dx.doi.org/10.1038/s41598-018-38366-w]

[118] Albanna M, Binder KW, Murphy SV, Kim J, Qasem SA, Zhao W, *et al.* In Situ Bioprinting of Autologous Skin Cells Accelerates Wound Healing of Extensive Excisional Full-Thickness Wounds. 2019.
[http://dx.doi.org/10.1038/s41598-018-38366-w]

[119] Stojic M, López V, Montero A, Quílez C, de Aranda Izuzquiza G, Vojtova L, *et al.* Skin tissue

engineering. Biomater Ski Repair Regen woodhead publishing;. 2019; pp. 59-99.

[120]   Tarassoli SP, Jessop ZM, Al-Sabah A, Gao N, Whitaker S, Doak S, *et al.* Skin tissue engineering using 3D bioprinting: An evolving research field. J Plast Reconstr Aesthetic Surg 2018; 71: 615-23.
[http://dx.doi.org/10.1016/j.bjps.2017.12.006]

[121]   Lee V, Singh G, Trasatti JP, *et al.* Design and fabrication of human skin by three-dimensional bioprinting. Tissue Eng Part C Methods 2014; 20(6): 473-84.
[http://dx.doi.org/10.1089/ten.tec.2013.0335] [PMID: 24188635]

[122]   Ramasamy S, Davoodi P, Vijayavenkataraman S, *et al.* Optimized construction of a full thickness human skin equivalent using 3D bioprinting and a PCL/collagen dermal scaffold. Bioprinting 2021; 21e00123 [Internet].
[http://dx.doi.org/10.1016/j.bprint.2020.e00123]

[123]   Desanlis A, Albouy M, Rousselle P, *et al.* Validation of an implantable bioink using mechanical extraction of human skin cells: First steps to a 3D bioprinting treatment of deep second degree burn. J Tissue Eng Regen Med 2021; 15(1): 37-48.
[http://dx.doi.org/10.1002/term.3148] [PMID: 33170542]

[124]   Baltazar T, Merola J, Catarino C, *et al.* Three Dimensional Bioprinting of a Vascularized and Perfusable Skin Graft Using Human Keratinocytes, Fibroblasts, Pericytes, and Endothelial Cells. Tissue Eng Part A 2020; 26(5-6): 227-38.
[http://dx.doi.org/10.1089/ten.tea.2019.0201] [PMID: 31672103]

[125]   Shapiro AMJ, Pokrywczynska M, Ricordi C. Clinical pancreatic islet transplantation. Nat Rev Endocrinol 2017; 13(5): 268-77.
[http://dx.doi.org/10.1038/nrendo.2016.178] [PMID: 27834384]

[126]   Rickels MR, Robertson RP. Pancreatic islet transplantation in humans: Recent progress and future directions. Endocr Rev 2019; 40(2): 631-68.
[http://dx.doi.org/10.1210/er.2018-00154] [PMID: 30541144]

[127]   Gamble A, Pepper AR, Bruni A, Shapiro AMJ. The journey of islet cell transplantation and future development 2018.
[http://dx.doi.org/10.1080/19382014.2018.1428511]

[128]   Peiris H, Bonder CS, Coates PTH, Keating DJ, Jessup CF. The β-cell/EC axis: how do islet cells talk to each other? Diabetes 2014; 63(1): 3-11.
[http://dx.doi.org/10.2337/db13-0617] [PMID: 24357688]

[129]   Salg GA, Poisel E, Munoz MN, Cebulla D, Vieira V, Bludszuweit-Philipp C, *et al.* Towards 3D-Bioprinting of an Endocrine Pancreas: A Building-Block Concept for Bioartificial Insulin-Secreting Tissue. bioRxiv 2021.433164 https://www.biorxiv.org/content/10.1101/2021.02.27.433164v1%0Ahttps://www.biorxiv.org/content/10.1101/2021.02.27.433164v1 Internet abstract
[http://dx.doi.org/10.1101/2021.02.27.433164]

[130]   Kim J, Kang K, Drogemuller CJ, Wallace GG, Coates PT. Bioprinting an Artificial Pancreas for Type 1 Diabetes. Curr Diab Rep 2019; 19(8): 53.
[http://dx.doi.org/10.1007/s11892-019-1166-x]

[131]   Duin S, Schütz K, Ahlfeld T, *et al.* 3D Bioprinting of Functional Islets of Langerhans in an Alginate/Methylcellulose Hydrogel Blend. Adv Healthc Mater 2019; 8(7)1801631
[http://dx.doi.org/10.1002/adhm.201801631] [PMID: 30835971]

[132]   Nikolova G, Jabs N, Konstantinova I, *et al.* The vascular basement membrane: a niche for insulin gene expression and β cell proliferation. Dev Cell 2006; 10(3): 397-405.
[http://dx.doi.org/10.1016/j.devcel.2006.01.015] [PMID: 16516842]

[133]   Lammert E, Thorn P. The Role of the Islet Niche on Beta Cell Structure and Function. J Mol Biol 2020; 432: 1407-8.
[http://dx.doi.org/10.1016/j.jmb.2019.10.032]

[134] Liu X, Carter SSD, Renes MJ, *et al.* Development of a Coaxial 3D Printing Platform for Biofabrication of Implantable Islet-Containing Constructs. Adv Healthc Mater 2019; 8(7)1801181
[http://dx.doi.org/10.1002/adhm.201801181] [PMID: 30633852]

[135] Sefano Guandalini M. 13 Most Common Gastrointestinal Conditions and What to Do About Them i am aware 2020.

[136] Bijlsma MF, Sadanandam A, Tan P, Vermeulen L. Molecular subtypes in cancers of the gastrointestinal tract. Nat Rev Gastroenterol Hepatol 2017; 14(6): 333-42.
[http://dx.doi.org/10.1038/nrgastro.2017.33] [PMID: 28400627]

[137] Graham DY. History of *Helicobacter pylori*, duodenal ulcer, gastric ulcer and gastric cancer. World J Gastroenterol 2014; 20(18): 5191-204.
[http://dx.doi.org/10.3748/wjg.v20.i18.5191] [PMID: 24833849]

[138] Silvio Danese claudio fiocchi. Ulcerative colitis. N Engl J Med 2011; 365: 1713-25.

[139] Eichele DD, Kharbanda KK. Dextran sodium sulfate colitis murine model: An indispensable tool for advancing our understanding of inflammatory bowel diseases pathogenesis. World J Gastroenterol 2017; 23(33): 6016-29.
[http://dx.doi.org/10.3748/wjg.v23.i33.6016] [PMID: 28970718]

[140] Gonzalez LM, Moeser AJ, Blikslager AT. Porcine models of digestive disease: The future of large animal translational research. Transl Res 2015.
[http://dx.doi.org/10.1016/j.trsl.2015.01.004]

[141] Yin L, Yang H, Li J, *et al.* Pig models on intestinal development and therapeutics. Amino Acids 2017; 49(12): 2099-106.
[http://dx.doi.org/10.1007/s00726-017-2497-z] [PMID: 28986749]

[142] Zhao X, Pack M. Modeling intestinal disorders using zebrafish Methods Cell Biol. Elsevier Ltd 2017; pp. 241-70. Internet
[http://dx.doi.org/10.1016/bs.mcb.2016.11.006]

[143] Sato T, Clevers H. Primary mouse small intestinal epithelial cell cultures. Methods Mol Biol 2012; 945: 319-28.
[http://dx.doi.org/10.1007/978-1-62703-125-7_19] [PMID: 23097115]

[144] Zhang M, Liu Y, Chen YG. Generation of 3D human gastrointestinal organoids: principle and applications. Cell Regen (Lond) 2020; 9(1): 6.
[http://dx.doi.org/10.1186/s13619-020-00040-w] [PMID: 32588198]

[145] Wang X, Yamamoto Y, Wilson LH, *et al.* Cloning and variation of ground state intestinal stem cells. Nature 2015; 522(7555): 173-8.
[http://dx.doi.org/10.1038/nature14484] [PMID: 26040716]

[146] Post Y, Clevers H. Defining Adult Stem Cell Function at Its Simplest: The Ability to Replace Lost Cells through Mitosis. Cell Stem Cell. 2019.
[http://dx.doi.org/10.1016/j.stem.2019.07.002]

[147] Qu M, Xiong L, Lyu Y, Zhang X, Shen J, Guan J, *et al.* Establishment of intestinal organoid cultures modeling injury-associated epithelial regeneration. 2021.
[http://dx.doi.org/10.1038/s41422-020-00453-x]

[148] Lancaster MA, Knoblich JA. Organogenesisin a dish: Modeling development and disease using organoid technologies. Science. 2014; 345.

[149] Sato T, Clevers H. Growing self-organizing mini-guts from a single intestinal stem cell: Mechanism and applications. Science . 2013; 340: pp. 1190-4.

[150] Sato T, Vries RG, Snippert HJ, *et al.* Single Lgr5 stem cells build crypt-villus structures in vitro without a mesenchymal niche. Nature 2009; 459(7244): 262-5.
[http://dx.doi.org/10.1038/nature07935] [PMID: 19329995]

[151] Min S, Kim S, Cho SW. Gastrointestinal tract modeling using organoids engineered with cellular and microbiota niches. 2020.
[http://dx.doi.org/10.1038/s12276-020-0386-0]

[152] Bhatia SN, Ingber DE. Microfluidic organs-on-chips. Nat Biotechnol 2014; 32(8): 760-72.
[http://dx.doi.org/10.1038/nbt.2989] [PMID: 25093883]

[153] Kim W, Kim G. Intestinal Villi Model with Blood Capillaries Fabricated Using Collagen-Based Bioink and Dual-Cell-Printing Process. ACS Appl Mater Interfaces 2018; 10(48): 41185-96.
[http://dx.doi.org/10.1021/acsami.8b17410] [PMID: 30419164]

[154] Creff J, Courson R, Mangeat T, Foncy J, Souleille S, Thibault C, *et al.* Fabrication of 3D scaffolds reproducing intestinal epithelium topography by high-resolution 3D stereolithography. Biomaterials. 2019.
[http://dx.doi.org/10.1016/j.biomaterials.2019.119404]

[155] Chen EP, Toksoy Z, Davis BA, Geibel JP. 3D Bioprinting of Vascularized Tissues for *in vitro* and *in vivo* Applications. Front Bioeng Biotechnol 2021; 9664188
[http://dx.doi.org/10.3389/fbioe.2021.664188] [PMID: 34055761]

[156] Sarker MD, Naghieh S, Sharma NK, Ning L, Chen X. Bioprinting of vascularized tissue scaffolds: Influence of biopolymer, cells, growth factors, and gene delivery. J Healthc Eng 2019; 2019: 1-20.
[http://dx.doi.org/10.1155/2019/9156921] [PMID: 31065331]

[157] Tomasina C, Bodet T, Mota C, Moroni L, Camarero-espinosa S. Despite the great advances that the tissue engineering field has. Materials (Basel) 2019; 12: 2701.
[http://dx.doi.org/10.3390/ma12172701] [PMID: 31450791]

[158] Datta P, Ayan B, Ozbolat IT. Bioprinting for vascular and vascularized tissue biofabrication. Acta Biomater 2017; 51: 1-20.
[http://dx.doi.org/10.1016/j.actbio.2017.01.035] [PMID: 28087487]

[159] Inglis S, Christensen D, Wilson DI, Kanczler JM, Oreffo ROC. Human endothelial and foetal femur-derived stem cell co-cultures modulate osteogenesis and angiogenesis. Stem Cell Res Ther 2016; 7(1): 13.
[http://dx.doi.org/10.1186/s13287-015-0270-3] [PMID: 26781715]

[160] Park TY, Yang YJ, Ha DH, Cho DW, Cha HJ. Marine-derived natural polymer-based bioprinting ink for biocompatible, durable, and controllable 3D constructs. Biofabrication 2019; 11(3)035001
[http://dx.doi.org/10.1088/1758-5090/ab0c6f] [PMID: 30831562]

[161] Mazurek R, Dave JM, Chandran RR, Misra A, Sheikh AQ, Greif DM. Vascular Cells in Blood Vessel Wall Development and Disease Adv Pharmacol. 1st ed. Elsevier Inc. 2017; pp. 323-50. Internet
[http://dx.doi.org/10.1016/bs.apha.2016.08.001]

[162] Weinberg CB, Bell E. A blood vessel model constructed from collagen and cultured vascular cells. 1986.
[http://dx.doi.org/10.1126/science.2934816]

[163] Yilmaz B, Al Rashid A, Mou YA, Evis Z, Koç M. Bioprinting: A review of processes, materials and applications. Bioprinting 2021; 23e00148
[http://dx.doi.org/10.1016/j.bprint.2021.e00148]

[164] Abdulhameed O, Al-Ahmari A, Ameen W, Mian SH. Additive manufacturing: Challenges, trends, and applications. Adv Mech Eng 2019; 11(2)
[http://dx.doi.org/10.1177/1687814018822880]

[165] Seol Y, Yoo JJ, Atala A. Bioprinting of Three-Dimensional Tissues and Organ Constructs Essentials 3D Biofabrication Transl. Academic Press 2015; pp. 283-92.
[http://dx.doi.org/10.1016/B978-0-12-800972-7.00016-5]

CHAPTER 10

Large-Scale Organoid Culture for High Throughput Drug Screening

Shraddha Gautam[1], Atul Kumar Singh[2] and Sachin Kadam[3,*]

[1] *Advancells Group, A 102, Sector 5, NOIDA, Uttar Pradesh 201301, India*

[2] *Central Research Facility, IIT-Delhi Sonipat Campus, Rajiv Gandhi Education City, Sonipat, Haryana 131029, India*

[3] *Manipal Center for Biotherapeutics Research, Manipal Academy of Higher Education (Institute of Eminence Deemed to be University), Manipal 576104, India*

Abstract: Despite several limitations, two-dimensional cell culture has been widely used in drug and drug-related compound selection and screening studies. A more recent approach of using three-dimensional (3D) organoid culture enables researchers with a more robust and accurate model for drug screening. Numerous studies have reported the successful use of stem cells, including induced pluripotent stem cells (iPSCs) and adult stem cells, for organoid generation to predict therapy response in various disease conditions, including cancer. The development of high-throughput drug screening and organoids-on-a-chip technology can advance the use of patient-derived organoids in clinical practice and facilitate therapeutic decision-making. Although organoids are in complaisant with high-throughput screenings, extensive manipulation studies are required by current methods.

Keywords: High-throughput drug screening, Organoid culture, Stem cells, 3D culture.

INTRODUCTION

Today, the world has acknowledged the fundamental power of clinical research. This global health emergency has highlighted the significance of global health and its impact on the socioeconomic status by reminding us that health risks can even transcend national boundaries. With this global pandemic, researchers, scientific policymakers, and regulatory authorities have understood the importance of rapid decisions to initiate clinical studies and rapid implementation of drugs in regular practice. Moreover, as opposed to the current scenario where almost 10% of the total drugs were implemented in routine practice [1], it is essential to develop and

* **Corresponding Author Sachin Kadam**: Manipal Center for Biotherapeutics Research, Manipal Academy of Higher Education (Institute of Eminence Deemed to be University), Manipal 576104, India; E-mail: kadamsachin@gmail.com

Manash K. Paul (Ed.)

standardize methodologies to reduce high costs and minimize time to identify suitable drug candidates. At the same time, current research methods require two-dimensional (2D) cell culture applications and animal models, mainly for drug screening and other toxicological studies [2, 3]. However, researchers reported several challenges in drug discovery applications; like maintenance of species-specific cell lines for a longer duration without contamination, the phenotypical and genotypical difference in several cell lines obtained from different interrelated species, as well as the inability to control the microenvironment, thereby regulating the epigenetic status of various cell lines, *etc.* These challenges are further responsible for failure in predicted outcomes, unanticipated toxicity issues that can be translated to humans, and further increasing time and cost in developing new drug candidates. Studies have acknowledged that drug development can be bettered by mimicking the three-dimensional (3D) structure of the cellular microenvironment to regulate its complex dynamics and identify effective target sites [3 - 5]. These studies have also proposed that the science of cell culture in current practice does not account for these proposals, leading to the failure of new drug implementation. It should be noted that advances in technology and cell culture science have developed such a 3D microenvironment *In vitro* [6]. Accordingly, the table below represents the comparative analysis of the advantages of using 3D cell culture over its 2D counterparts.

Table 1. Comparative Analysis of issues encountered in 2D cell culture vis-a-vis 3D organoid development

Type	Actual Representation of Human Tissue	Ethical Issues	Cost of Treatment	Maintenance	Risk	Assay Development
Animal Model	Low	Low	High	Medium	Low	Possible
2D Culture	Low	Low	Low	High	High	Not Possible
Organoid	High	Low	Low	High	Medium	Possible

With the advent of technology and recent scientific development, it is possible to develop different organs on a microfabricated chip. In simpler language, these chips are designed with the help of different primary cells obtained from human tissues to form 3D cellular aggregates incorporated in a microfluidic device [3, 4]. However, many issues need to be sorted out, further limiting the potential commercial use of these systems, which can further be noted as a lack of exact representation of human tissues due to the presence of specific matrix proteins [8, 9]. Moreover, certain studies also confirmed that the organ-on-a-chip (OOAC) models developed with the help of primary cells exhibit a limited capacity for self-regeneration and organization capacity to mimic tissue of interest.

In this regard, recently, the generation of OOAC models with the help of human induced pluripotent stem cells (hiPSCs) received tremendous acknowledgment; due to the self-regenerating capacity of stem cells and, importantly, their ability to copy exact structural and functional aspects of human tissues in a 3D cell culture setup [10]. Studies have also proposed that integrating these stem cells with microfluidic devices is an added advantage for the more realistic development of healthy and diseased models (Fig. **1**). Regarding organ-on-a-chip development, some of the characteristic features of stem cells, like the ability to transform into any tissue-specific cell lineage, self-renewal, indefinite proliferation, *etc.*, are advantageous to tackle challenges encountered previously with primary cells Fig. (**2**) [11, 12]. Certain studies have revealed the mechanism of fetal development by generating strong patterning signals that are released through a special signaling mechanism and have inductive effects on a wide range of tissue-specific cells. The same mechanism has been detected in hiPSCs, promoting them as the most suitable candidates for organoid generation [13]. Scientists are thus hopeful about the ability of hiPSCs to generate disease-specific models, further helping in strategizing personalized medicines as the foundation of the 21st-century healthcare [14].

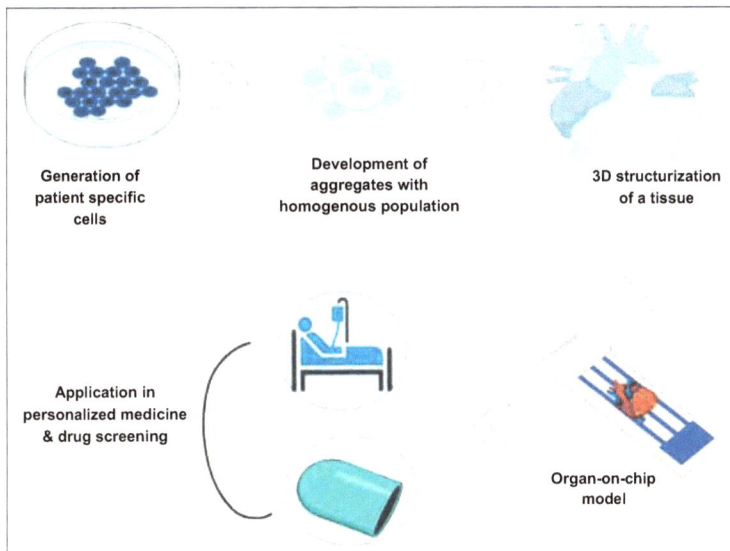

Fig. (1). Development of organ-on-a-chip model with induced pluripotent stem cells as crucial tools in personalized medicine and drug discovery.

Accordingly, the current chapter is intended to review the science behind the production of the organ-on-a-chip model with the help of hiPSCs, their regulated differentiation to specifically committed spheroids, and the controlled development of aggregates to provide adequate cues with homogenous cellular makeup [14, 15]. We shall also discuss some of the organ-on-a-chip models that have been used for drug screening required for different ailments like cardiovascular disorders, liver diseases, and different types of cancers [15, 16].

HUMAN-INDUCED PLURIPOTENT STEM CELLS: WELL-CONTROLLED SPHEROIDAL AGGREGATES

Today, the scientific community has witnessed many advancements in regenerative medicine, and the current decade is all about exploiting the unlimited potential of stem cells to produce cells of choice [13]. The aim cannot be fulfilled without incorporating hiPSCs. However, in many of these applications, large-scale expansion of pluripotent stem cells for clinical use, along with authenticated purification systems, can be pretty challenging [17-19, 20].

As discussed earlier, due to practical limitations encountered in using 2D cell culture techniques in drug screening, scientists are rapidly approaching aggregate-based cell culture systems. These systems work on the basic principle of mass-scaled development of homogenous cell types isolated and purified using microfluidic separation models to form spheroids and/or aggregates with a group of cells [21]. Essentially researchers have further raised a concern regarding the regulation of the inner diameter of aggregates with increasing culture uniformity of isolated cells [22]. Recently developed methodologies like micropatterning techniques can address this challenge. These techniques help analyze the effect of the inner diameter of the aggregates and their impact on cellular pluripotency and differentiation trajectories. Another study also supported the idea of a micropatterning system to promote cellular aggregation and the generation of a homogenous cellular population. Apart from the same, some relevant studies have also highlighted the role of chemical inducers like dextran sulphates for inducing aggregate formation in human pluripotent stem cells.

REGULATION OF AGGREGATE SIZE: ESSENTIAL TOOL TOWARDS LINEAGE-SPECIFIC DIFFERENTIATION

Cell Density

As we all know, initial seeding density plays a crucial role in monolayer 2D cell culture techniques. At the same time, the same rule applies to hiPSCs culture and their committed differentiation towards lineage-specific cells [23, 24]. Studies have confirmed that when cultured with higher initial cellular densities and subjected to neurological differentiation using dual SMAD inhibition techniques, hiPSCs cells were able to differentiate into neural retinal precursor cells and epithelial cells with Rx+ and Pax6+ markers [25]. Contrary to this, lower initial cellular densities can give rise to cells of the central nervous system. Similarly, studies have also suggested that the differentiation commitment of hiPSCs towards brain microvascular lineages is more potent than cells seeded at higher densities [25 - 27]. Thus, seeding density should be considered as one of the essential factors to control the commitment of hiPSCs toward cell-specific lineages.

Colony Formation Capacity

Certain studies have indicated that primary cells' well-controlled colony formation capacity encourages a higher degree of pluripotency and maintenance of the same for the long term [28, 29]. The direct impact of colony size has been proposed to be essential for pluripotency maintenance [30]. This was further elaborated by variation in the OCT-4 expression due to special arrangements, further connected to the activation of local signaling factors [31]. Some fundamental studies have further explained that this spatial rearrangement is largely controlled by signaling pathways like Activin-Nodal, BMP, and WNT [32, 33]. The in-depth analysis in the same direction helped researchers retain three germ layers' spatial organization *In vitro* using colonies of induced pluripotent stem cells [34]. Later, it was confirmed that density-dependent receptors are mainly responsible for spatial regulation and self-controlled differentiation patterns; through the regulation of BMP4 [35]. These outcomes have provoked many scientists to apply the same findings to 3D culture systems by having complete control over cell numbers and the diameter of each colony formed by hiPSCs. After the initial success, significant work is done with very positive findings [36, 37]. Studies confirmed that cellular colonies with a smaller diameter could easily differentiate into the population of cells with characteristics similar to that of endoderm-enriched cells [38]. At the same time, it was observed that cellular aggregates with very high diameters could form cells enriched with neuronal origin. Certain studies also demonstrated that cellular aggregates exhibit the capacity to transdifferentiate into cells of different origins, ranging from cardiomyocytes to cells of different neuro-ectodermal populations [39 - 42].

However, the outcome of many such studies could not explain the diffusional limitations associated with the 3D cell culture system, directly impacting the expansion and differentiation potential of human induced pluripotent stem cells. At the same time, it was confirmed that the regulation of the entire cell population and their fate is primarily dependent upon the regulatory pathways [41]. Accordingly, it is essential to discuss in detail how these regulatory pathways are crucial for the expansion and differentiation potential of cells and the influence of the diameter of the colonies formed to differentiate into cells of different origins. One of the studies showed that in colonies with less than 100 μm diameters, the rate of hypoxia was minimal compared to that of colonies with higher diameters, further leading to the necrotic stage. With this, it was then proposed that it is essential to maintain and modulate the diameter of colonies formed for a higher degree of differentiation [42].

In the subsequent sections of this chapter, we have discussed how crucial it is to regulate different signaling pathways. We shall also discuss the orientation of hiPSCs towards different lineages, like neuroectodermal, mesoderm, and/or endoderm [43].

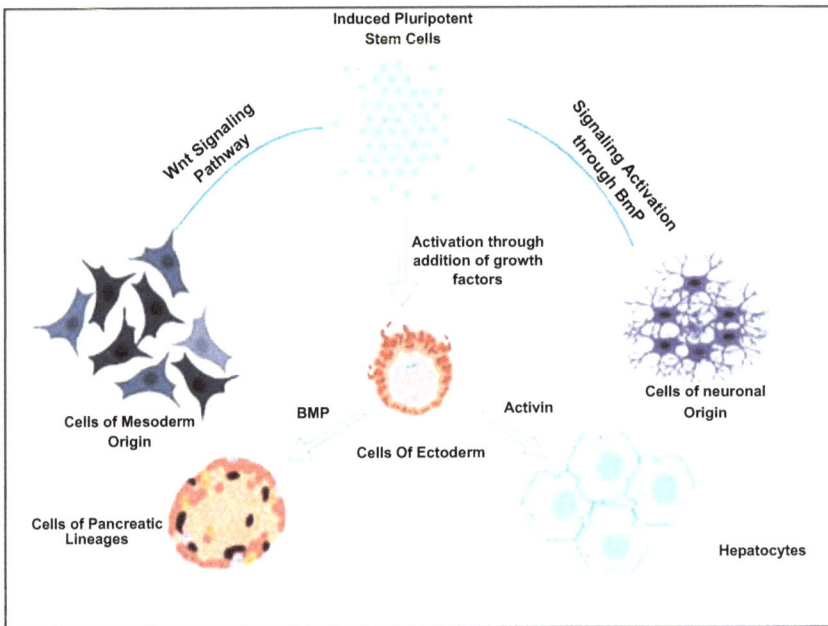

Fig. (2). Development of organ-on-a-chip model with induced pluripotent stem cells as crucial tools in personalized medicine and drug discovery.

Differentiation of Induced Pluripotent Stem Cells Towards Mesodermal Origin and its Modern-day Application

Due to the increased morbidity and mortality in developed countries like the USA due to cardiovascular disorders, hiPSCs are highly analyzed in pre-clinical and clinical setups due to their ability to differentiate into mesodermal lineages [41]. However, researchers are still struggling with many prerequisites; one is the requirement of high cell numbers of similar lineages. Additionally, cellular toxicity is another issue that has been raised recently, causing a higher intensity of drug failure during the process of drug screening. Several studies have reported that with the help of chemically defined conditions to regulate different signaling pathways, suspension aggregates can maintain cardiac differentiation potential to a certain extent [42, 43]. The method was further improved with the help of advanced technologies. The usage of microwell plates is turning out to be the golden tool for organ regeneration. With a similar intent, an experiment was conducted to regulate signaling pathways and modulate cellular communication; where hiPSCs were placed in V-shaped microwell culture plates with the optimized concentration of cardiac lineage-specific growth factors [44, 45]. The method proved to be successful, with almost 89% differentiation efficacy. Interestingly, the beating phenotype was detected by the end of the 7th day of incubation with specific growth factors, including BMP4, polyvinyl alcohol, insulin, and serum [46]. Furthermore, quality control procedures like rotary agitation were observed with 100% contractility of cardio-spheres when grown in 3D suspension cultures in microwells [47]. Various researchers are contributing to these outcomes in different ways; some have reported an average efficiency of 90% with the production of higher cardiomyocytes in suspension cultures [48]; while some are proposing reasonable control over the diameter of aggregates formed in suspension culture to increase the efficacy of differentiation [49, 50]. These analyses further propose an excellent start to initiate therapeutic applications of hiPSCs in the near future.

Differentiation of hiPSCs into the Endodermal Origin and Modern-day Applications

Previously 2D approaches had taught us that it is possible to redirect the cellular fate with the help of inhibition of specific signaling pathways [51]. Contrary to this, modern-day science of induced pluripotent stem cells has confirmed that not only the regulation of signaling pathways and the communication potential of cells but also their spatial rearrangements and maintenance of the physicochemical conditions of 3D aggregates are essential factors that also need to be considered when it comes to the organoid formation [52, 53]. To confirm the differentiation potential of hiPSCs, it was demonstrated that 3D suspension aggregates of hiPSCs can be redirected to specific differentiated cells with the exposure of the right concentration of growth factors, creating right environmental conditions like microwells and regulation of physicochemical condition of the culture, *i.e.*, diameters of colony-forming units, *etc* [54 - 57]. This was confirmed by demonstrating that clusters of beta cells increased insulin secretion rate better than a single cell. Another study has shown how colonies of hiPSCs with regulated diameter, when incubated with specific growth factors at the optimized concentration for a prescribed period, can initiate differentiation [58]. Thus, when hiPSCs were exposed to keratinocyte growth factors, researchers could observe tube-like structures in the culture, qualifying the quality control tests of cells of endodermal lineages. With a similar method yet with different growth factors, hiPSCs could demonstrate the potential to differentiate into cells of pancreatic lineage and/or hepatic origin, confirming their ability to differentiate into ectodermal lineage for therapeutic applications [59 - 62].

Differentiation of hiPSCs into Neuroectodermal Lineage and Applications in Modern Medicine

With loads of research and scientific data, it has also been confirmed that hiPSCs could differentiate into cells of neuronal origin, serving as the best tool for therapeutic applications in case of neurodegenerative disorders, the prevalence of which is increasing day by day [63]. Studies have confirmed that first-generation hiPSCs, when incubated with specific neuronal growth factors, were able to differentiate into neuronal precursor cells [64]. When further exposed to different growth factors, these cells could activate proper signaling pathways to form neurons, astrocytes, and oligodendrocytes [65 - 67]. In certain other relative studies, the importance of fibroblast growth factor-2 [68] was also demonstrated for maintenance of neuronal precursor cells for quite a long, further omitting the necrotic stage.

While studying the importance of different communication pathways, studies have shed light on the fact that signaling pathways like Activin/Nodal inhibition are crucial for differentiation into cells of neuronal origin [69]. When incubated with pluripotent cells, regulating certain factors like BMP-4 antagonists and Noggin can promote differentiation into neuronal precursor cells [70]. Moreover, maintaining pluripotency in the culture can also be achieved with the help of TGF-beta addition [71]. Commonly, out of many relevant studies mentioned herewith, most scientists acknowledge the double inhibition method, also widely known as the SMAD inhibition method [72]; wherein both BMP-4 and Activin/Nodal pathways are inhibited to block cellular communication from other lineages and only promote differentiation towards the neuroectodermal origin [73 - 75]. However, there are also certain variables for this commonly acknowledged differentiation protocol. It was demonstrated in one interesting study carried out with similar intent and confirmed how high-density cellular culture has a tremendous impact on the differentiation ability of cells [76]. The study demonstrated that at higher cell density with several cells close to each other, they get differentiated into the central nervous system cells after inhibiting differentiation pathways from other lineages. At low density, they differentiate into peripheral nervous system cells [77]. A similar concept was applied to demonstrate the differentiation potential of hiPSCs under 3D suspension conditions [78]. The report confirmed that suspension cells under a 3D setting differentiated into cells of neuronal origin when Noggin pathways were initiated, and the other pathways were blocked [79]. Quality control reports confirmed the presence of neuronal markers PSA-NCAM+ with 90% differentiation efficacy.

Thus, cells of neuroectodermal origin can be easily generated using optimized culture conditions in a bioreactor system using hiPSCs, with more than 90% efficacy in generating PAX6-NCAM positive cells for therapeutic applications [80]. Although 3D colonies of hiPSCs under suspension conditions showed great potential in processing differentiation into precursor cells of a particular lineage, terminal differentiation of precursors into tissue-specific cells required 2D setup only, further limiting the recapitulation of the tissue microenvironment [81 - 84]. Thus, there has always been a need for the sturdy development of a 3D microenvironment, mimicking the original environment and communication specificities.

A ROADMAP AHEAD FROM COLONIES TO ORGANOIDS

Although many studies have demonstrated the differentiation potential of progenitor precursors with 2D cell culture techniques, cellular complexity should be primarily maintained for better results and higher differentiation efficacy [85]. Contrary to the age-old practice, scientists have diverted their attention to the 3D culture to form organoids of different colonies with the help of stem cells; without manipulating their inherent traits like the capacity to mimic the exact cellular complexity, replicate the functional attributes of the original organ and promote differentiation into organ-specific cellular lineages with a higher degree of efficacy [86]. In-depth analysis of the same was done by designing organoids from different cellular lineages, including but not limited to lungs, brains, gut, and an optic cup [87 - 89].

Thus, the generation of organoids from specific cellular raw material represents a new tool of modern science to enhance all facets of cellular therapy, including drug discovery, toxicity analysis, and manipulating genetic disorders [89]. Although the discovery of organoids has brought scientists one step closer to a cure, several challenges must be overcome. One such challenge is scaling up stem cell expansion, developing cellular aggregates, and maintaining the size and diameter to facilitate differentiation into the desired outcome [90]. In the following sections, we shall focus on generating organoids from different types of pluripotent stem cells.

Modelling Neurodegenerative Disorders with CNS Organoids

With almost 100 billion neurons and many trillion *synapses*, the human central nervous system is the most complex system, and the modeling of which *In vitro* is somewhat tricky. However, since the generation of the first organoids from the brain, the scientific world has witnessed an exponential rise in the number of studies exploiting 3D culture technology as a platform supporting therapeutic applications [91 - 93]. Many scientists have focused their attention on the derivation of 3D cell culture using induced pluripotent stem cells. Most of them have used bioreactor systems for the mass-scale production of cells. It was demonstrated that induced pluripotent stem cells, when incubated with growth factors relevant to the neuroectodermal lineage, can give rise to cells of neurogenic precursors, which, when further encapsulated and cultured with a suitable substrate like matrigel in many cases, can form cells of cerebral organoids [91]. With controlled spatial arrangements, regulation of cytoskeletal contraction, and nuclear expansion, scientists could generate different types of brain organoids, like forebrain, midbrain, hypothalamus organoids, *etc*. Interestingly, these organoids passed all the quality control parameters, confirming them to recapitulate the exact characteristics of *In vivo* systems [91]. These location-specific organoids helped us get an in-depth idea about the inside-out pattern of cell production, communication, and overall complexities. Another such report also confirmed that with the help of induced pluripotent stem cells and a 3D cell culture system, it is possible to produce polarised epithelium that further helped create electrophysiologically active Purkinje fibers [92].

Today, researchers and the medical fraternity hope that induced pluripotent stem cells and organoids derived from them will provide a robust platform yielding potential therapeutic opportunities. Consequently, using brain organoids, it is possible to treat several orphan neurodegenerative disorders, including but not limited to autism spectrum disorders, *etc.*, where the connection between the central nervous system and the rest of the body requires further investigation [93]. Recently, during the Zika outbreak in Brazil, scientists were able to exploit the same region-specific brain organoids to understand the pathophysiology of the disease [93]. It was then clear that the virus was destroying neural precursor cells, due to which people were experiencing microcephaly. Another study employed these region-specific brain organoids to understand the direct and/or indirect effect of foreign substances, like lead, nicotine, *etc.*, on the functional and structural aspects of the brain. The study confirmed that exposure to toxic substances like nicotine in the early stage of life could lead to shunted neuronal development [94]. Currently, researchers cannot provide a proper supply of oxygen and other nutrients to the brain organoid due to the lack of adequate vasculature. Scientists are trying to address this challenge by co-culturing brain organoids with endothelial cells that give rise to a network of tube-like structures within the organoids. When engrafted in the mouse brain, the vascularized organoids gave rise to functional vasculature.

Modelling Kidney Disorders with Organoids

Effective generation of kidney organoids facilitates exact mimicking of the *In vivo* production of metabolic by-products during the process of excretion. Researchers are trying to understand renal damage associated with toxicity, drug overdose, *etc.* Various studies have explained how organoids with different components of nephrotic systems can be produced with the help of hiPSCs [95]. Interestingly, these organoid systems displayed exact functional attributes of human kidneys with uretic epithelium, nephrotic progenitor cells, stromal progenitor cells, fibroblasts, and endothelial cells [96]. The protocol, commonly acknowledged for organoid generation, was followed in two steps. In the first step, the Wnt singling pathway in hiPSCs was regulated with the help of CHIR99021 to induce the posterior primitive streak [97]. In the second stage of differentiation, cells were incubated with FGF-9 and heparin at a specific concentration to induce mesodermal lineage. The nephrogenesis was induced on day 7 with the addition of a pulse of CHIR99021, which then turned into functional kidney mimics upon further maturation [97].

It is essential to note that upon transplantation to the host, kidney organoids were able to promote host-derived vascularisation; when co-cultured with endothelial cells [97]. Studies have collected relevant data to show that co-culturing activity with endothelial cells has increased the survival rate of organoids due to a better supply of oxygen and other nutrient factors [98]. The successful derivation of kidney organoids has paved the way for the potential application of regenerative medicine in modeling renal diseases and various other applications like drug screening, toxicity analysis, and testing of nephrotoxicity in different compounds [98]. However, more deep diving is essential to improve the efficacy of differentiation protocol and functional attributes of kidney organoids [98].

However, it should be noted that there are still numerous challenges to overcome before kidney organoids may be used directly in therapeutic applications. Out of many concerns, the safety of the use of kidney organoids is a significant concern. Fortunately, no study has so far reported an event with dangerous signs and symptoms; all induced pluripotent stem cells need to undergo thorough quality control procedures to avoid unwanted side effects.

Modelling of Liver Disorders using Liver Organoids

As explained above, along with kidney organoids, liver organoids also gained tremendous applause recently from the scientific community. The pharmaceutical industry mainly relies on liver organoids for many of its applications, like hepatotoxicity analysis. Like many other endodermal lineages-derived organs, the liver can also be easily mimicked with the help of hiPSCs by converting them into liver bud progenitor cells as well as HLC [99]. For information, liver buds are condensed masses of liver tissues that are delaminated by the foregut and further vascularised using endothelial cells and angioblasts to promote improved blood circulation with a higher oxygen supply [99]. This system is beneficial in the complex system through a higher degree of self-regularization. Traditionally, 2D cell culture techniques have confirmed that a combination of human mesenchymal stem cells and human endothelial cells can create a good platform for generating a stable assembly of liver buds [99]. Advances in the same direction have demonstrated that co-culturing of hiPSCs, mesenchymal stem cells, and endothelial cells can enhance vascular network formation through specific signaling profiles and allow the formation of liver buds [99]. When infused directly into chimeric mouse models, these co-culture systems promoted the survival rate of differentiated liver organoids by the rapid vascularisation in the integrated tissue [100]. On the contrary to this, recently, scientists have successfully created liver organoids with the help of a pure population of hiPSCs using 3D cell culture techniques; and have confirmed that they show similar characteristics both *In vivo* and in vitro. The maturation levels of liver organoids were confirmed by biomarker analysis and their potency to screen different drugs at various concentrations [100].

Primary liver cancer has been one of the major causes of cancer-related deaths, and the prevalence of the same is increasing exponentially over the period. Some of the most important liver cancers with high detection rates are hepatocyte carcinoma (HCC), intrahepatic cholangiocarcinomas, *etc.* Researchers face many issues *in* developing new cell-based therapies due to the lack of appropriate disease models, both *In vivo* and *In vitro* setups [100]. Previously, patient-derived xenograft models were utilized to analyze HCC by transplanting HCC tissue into immunodeficient mice [101]. These models sorted some of the issues encountered to a certain extent; however, creating and maintaining such models for the long term can be costly and labor-intensive. Moreover, these models were further confirmed to be non-amendable to high-throughput screenings. Researchers have reported that organoid technologies can overcome challenges encountered in traditional 2D setups and PDX models [101].

With the combined approach of using healthy liver-derived organoids and cancer organoids from cells obtained from patients suffering from gastrointestinal cancers, scientists were able to recover liver cancer organoids [102]. Another such study demonstrated the combined use of HCC, CCA, and hepato-cholangiocarcinoma for the development of liver cancer models [102]. Recently, researchers have acknowledged that tissue biopsy itself can be used to create liver cancer organoids. The histopathological analysis performed on the tissue biopsies compared with that of the cancer organoid models confirmed that they closely resemble the histology of the primary tumors, indicating higher efficacy of these organoids in mimicking the *In vivo* environment. Thus, liver-derived organoid models have rapidly been used and refined for many biomedical research and liver cancer models.

CONCLUSION

Thus, scientists have diverted their attention to exploiting the unique regeneration and self-organization properties of hiPSCs in multiple applications like drug screening, toxicity analysis, and therapies. Traditionally, 2D cultures adherent to the surface were used for these applications. Still, they fail to give the exact details of pathophysiology and an in-depth understanding of the disease modeling. However, 3D organoid formation generated using hiPSCs gave better insights into organ complexities, signaling pathways, and communication channels. Understanding these facts helps scientists to regulate specific cellular pathways for intended differentiation.

REFERENCES

[1]   Seruga B, Ocana A, Amir E, Tannock IF. Failures in Phase III. Clin Cancer Res 2015; 21(20): 4552-60.
[http://dx.doi.org/10.1158/1078-0432.CCR-15-0124] [PMID: 26473191]

[2]   Wang YI, Abaci HE, Shuler ML. Microfluidic blood–brain barrier model provides in vivo-like barrier properties for drug permeability screening. Biotechnol Bioeng 2017; 114(1): 184-94.
[http://dx.doi.org/10.1002/bit.26045] [PMID: 27399645]

[3]   Lee H, Cho DW. One-step fabrication of an organ-on-a-chip with spatial heterogeneity using a 3D bioprinting technology. Lab Chip 2016; 16(14): 2618-25.
[http://dx.doi.org/10.1039/C6LC00450D] [PMID: 27302471]

[4]   Skardal A, Devarasetty M, Soker S, Hall AR. *In situ* patterned micro 3D liver constructs for parallel toxicology testing in a fluidic device. Biofabrication 2015; 7(3)031001
[http://dx.doi.org/10.1088/1758-5090/7/3/031001] [PMID: 26355538]

[5]   Skardal A, Murphy SV, Devarasetty M, *et al.* Multi-tissue interactions in an integrated three-tissue organ-on-a-chip platform. Sci Rep 2017; 7(1): 8837.
[http://dx.doi.org/10.1038/s41598-017-08879-x] [PMID: 28821762]

[6]   Sasai Y. Next-generation regenerative medicine: organogenesis from stem cells in 3D culture. Cell Stem Cell 2013; 12(5): 520-30.
[http://dx.doi.org/10.1016/j.stem.2013.04.009] [PMID: 23642363]

[7]   Warmflash A, Sorre B, Etoc F, Siggia ED, Brivanlou AH. A method to recapitulate early embryonic spatial patterning in human embryonic stem cells. Nat Methods 2014; 11(8): 847-54.
[http://dx.doi.org/10.1038/nmeth.3016] [PMID: 24973948]

[8]   Takahashi K, Tanabe K, Ohnuki M, *et al.* Induction of pluripotent stem cells from adult human fibroblasts by defined factors. Cell 2007; 131(5): 861-72.
[http://dx.doi.org/10.1016/j.cell.2007.11.019] [PMID: 18035408]

[9]   Fernandes TG, Duarte ST, Ghazvini M, *et al.* Neural commitment of human pluripotent stem cells under defined conditions recapitulates neural development and generates patient-specific neural cells. Biotechnol J 2015; 10(10): 1578-88.
[http://dx.doi.org/10.1002/biot.201400751] [PMID: 26123315]

[10] Badenes SM, Fernandes TG, Cordeiro CSM, *et al.* Defined Essential 8™ Medium and Vitronectin Efficiently Support Scalable Xeno-Free Expansion of Human Induced Pluripotent Stem Cells in

Stirred Microcarrier Culture Systems. PLoS One 2016; 11(3)e0151264
[http://dx.doi.org/10.1371/journal.pone.0151264] [PMID: 26999816]

[11] Bauwens CL, Peerani R, Niebruegge S, *et al.* Control of human embryonic stem cell colony and aggregate size heterogeneity influences differentiation trajectories. Stem Cells 2008; 26(9): 2300-10.
[http://dx.doi.org/10.1634/stemcells.2008-0183] [PMID: 18583540]

[12] Miranda CC, Fernandes TG, Pascoal JF, *et al.* Spatial and temporal control of cell aggregation efficiently directs human pluripotent stem cells towards neural commitment. Biotechnol J 2015; 10(10): 1612-24.
[http://dx.doi.org/10.1002/biot.201400846] [PMID: 25866360]

[13] Kinney MA, Saeed R, McDevitt TC. Systematic analysis of embryonic stem cell differentiation in hydrodynamic environments with controlled embryoid body size. Integr Biol 2012; 4(6): 641-50.
[http://dx.doi.org/10.1039/c2ib00165a] [PMID: 22609810]

[14] Fernandes TG, Rodrigues CAV, Diogo MM, Cabral JMS. Stem cell bioprocessing for regenerative medicine. J Chem Technol Biotechnol 2014; 89(1): 34-47.
[http://dx.doi.org/10.1002/jctb.4189]

[15] Rodrigues GMC, Rodrigues CAV, Fernandes TG, Diogo MM, Cabral JMS. Clinical-scale purification of pluripotent stem cell derivatives for cell-based therapies. Biotechnol J 2015; 10(8): 1103-14.
[http://dx.doi.org/10.1002/biot.201400535] [PMID: 25851544]

[16] Rodrigues GMC, Matos AFS, Fernandes TG, *et al.* Integrated platform for production and purification of human pluripotent stem cell-derived neural precursors. Stem Cell Rev 2014; 10(2): 151-61.
[http://dx.doi.org/10.1007/s12015-013-9482-z] [PMID: 24221956]

[17] Amit M, Chebath J, Margulets V, *et al.* Suspension culture of undifferentiated human embryonic and induced pluripotent stem cells. Stem Cell Rev 2010; 6(2): 248-59.
[http://dx.doi.org/10.1007/s12015-010-9149-y] [PMID: 20431964]

[18] Zweigerdt R, Olmer R, Singh H, Haverich A, Martin U. Scalable expansion of human pluripotent stem cells in suspension culture. Nat Protoc 2011; 6(5): 689-700.
[http://dx.doi.org/10.1038/nprot.2011.318] [PMID: 21527925]

[19] Olmer R, Haase A, Merkert S, *et al.* Long term expansion of undifferentiated human iPS and ES cells in suspension culture using a defined medium. Stem Cell Res (Amst) 2010; 5(1): 51-64.
[http://dx.doi.org/10.1016/j.scr.2010.03.005] [PMID: 20478754]

[20] Peerani R, Rao BM, Bauwens C, *et al.* Niche-mediated control of human embryonic stem cell self-renewal and differentiation. EMBO J 2007; 26(22): 4744-55.
[http://dx.doi.org/10.1038/sj.emboj.7601896] [PMID: 17948051]

[21] Bauwens CL, Song H, Thavandiran N, *et al.* Geometric control of cardiomyogenic induction in human pluripotent stem cells. Tissue Eng Part A 2011; 17(15-16): 1901-9.
[http://dx.doi.org/10.1089/ten.tea.2010.0563] [PMID: 21417693]

[22] Ungrin MD, Joshi C, Nica A, Bauwens C, Zandstra PW. Reproducible, ultra high-throughput formation of multicellular organization from single cell suspension-derived human embryonic stem cell aggregates. PLoS One 2008; 3(2)e1565
[http://dx.doi.org/10.1371/journal.pone.0001565] [PMID: 18270562]

[23] Miranda CC, Fernandes TG, Pinto SN, Prieto M, Diogo MM, Cabral JMS. A scale out approach towards neural induction of human induced pluripotent stem cells for neurodevelopmental toxicity studies. Toxicol Lett 2018; 294: 51-60.
[http://dx.doi.org/10.1016/j.toxlet.2018.05.018] [PMID: 29775723]

[24] Bauwens CL, Song H, Thavandiran N, *et al.* Geometric control of cardiomyogenic induction in human pluripotent stem cells. Tissue Eng Part A 2011; 17(15-16): 1901-9.
[http://dx.doi.org/10.1089/ten.tea.2010.0563] [PMID: 21417693]

[25] Lipsitz YY, Tonge PD, Zandstra PW. Chemically controlled aggregation of pluripotent stem cells.

Biotechnol Bioeng 2018; 115(8): 2061-6.
[http://dx.doi.org/10.1002/bit.26719] [PMID: 29679475]

[26] Wilson HK, Canfield SG, Hjortness MK, Palecek SP, Shusta EV. Exploring the effects of cell seeding density on the differentiation of human pluripotent stem cells to brain microvascular endothelial cells. Fluids Barriers CNS 2015; 12(1): 13.
[http://dx.doi.org/10.1186/s12987-015-0007-9] [PMID: 25994964]

[27] Chambers SM, Fasano CA, Papapetrou EP, Tomishima M, Sadelain M, Studer L. Highly efficient neural conversion of human ES and iPS cells by dual inhibition of SMAD signaling. Nat Biotechnol 2009; 27(3): 275-80.
[http://dx.doi.org/10.1038/nbt.1529] [PMID: 19252484]

[28] D'Amour KA, Agulnick AD, Eliazer S, Kelly OG, Kroon E, Baetge EE. Efficient differentiation of human embryonic stem cells to definitive endoderm. Nat Biotechnol 2005; 23(12): 1534-41.
[http://dx.doi.org/10.1038/nbt1163] [PMID: 16258519]

[29] Etoc F, Metzger J, Ruzo A, *et al.* A Balance between Secreted Inhibitors and Edge Sensing Controls Gastruloid Self-Organization. Dev Cell 2016; 39(3): 302-15.
[http://dx.doi.org/10.1016/j.devcel.2016.09.016] [PMID: 27746044]

[30] Kattman SJ, Witty AD, Gagliardi M, *et al.* Stage-specific optimization of activin/nodal and BMP signaling promotes cardiac differentiation of mouse and human pluripotent stem cell lines. Cell Stem Cell 2011; 8(2): 228-40.
[http://dx.doi.org/10.1016/j.stem.2010.12.008] [PMID: 21295278]

[31] Fonoudi H, Ansari H, Abbasalizadeh S, *et al.* A universal and robust integrated platform for the scalable production of Human Cardiomyocytes from Pluripotent Stem Cells. Stem Cells Transl Med 2015; 4(12): 1482-94.
[http://dx.doi.org/10.5966/sctm.2014-0275] [PMID: 26511653]

[32] Burridge PW, Anderson D, Priddle H, *et al.* Improved human embryonic stem cell embryoid body homogeneity and cardiomyocyte differentiation from a novel V-96 plate aggregation system highlights interline variability. Stem Cells 2007; 25(4): 929-38.
[http://dx.doi.org/10.1634/stemcells.2006-0598] [PMID: 17185609]

[33] Nazareth EJP, Ostblom JEE, Lücker PB, *et al.* High-throughput fingerprinting of human pluripotent stem cell fate responses and lineage bias. Nat Methods 2013; 10(12): 1225-31.
[http://dx.doi.org/10.1038/nmeth.2684] [PMID: 24141495]

[34] Gassmann M, Fandrey J, Bichet S, *et al.* Oxygen supply and oxygen-dependent gene expression in differentiating embryonic stem cells. Proc Natl Acad Sci USA 1996; 93(7): 2867-72.
[http://dx.doi.org/10.1073/pnas.93.7.2867] [PMID: 8610133]

[35] Sen A, Kallos MS, Behie LA. Effects of Hydrodynamics on Cultures of Mammalian Neural Stem Cell Aggregates in Suspension Bioreactors. Ind Eng Chem Res 2001; 40(23): 5350-7.
[http://dx.doi.org/10.1021/ie001107y]

[36] Itskovitz-Eldor J, Schuldiner M, Karsenti D, *et al.* Differentiation of human embryonic stem cells into embryoid bodies compromising the three embryonic germ layers. Mol Med 2000; 6(2): 88-95.
[http://dx.doi.org/10.1007/BF03401776] [PMID: 10859025]

[37] Zhang SC, Wernig M, Duncan ID, Brüstle O, Thomson JA. *In vitro* differentiation of transplantable neural precursors from human embryonic stem cells. Nat Biotechnol 2001; 19(12): 1129-33.
[http://dx.doi.org/10.1038/nbt1201-1129] [PMID: 11731781]

[38] Reubinoff BE, Itsykson P, Turetsky T, *et al.* Neural progenitors from human embryonic stem cells. Nat Biotechnol 2001; 19(12): 1134-40.
[http://dx.doi.org/10.1038/nbt1201-1134] [PMID: 11731782]

[39] Itsykson P, Ilouz N, Turetsky T, *et al.* Derivation of neural precursors from human embryonic stem cells in the presence of noggin. Mol Cell Neurosci 2005; 30(1): 24-36.

[http://dx.doi.org/10.1016/j.mcn.2005.05.004] [PMID: 16081300]

[40]   Pera MF, Tam PPL. Extrinsic regulation of pluripotent stem cells. Nature 2010; 465(7299): 713-20.
        [http://dx.doi.org/10.1038/nature09228] [PMID: 20535200]

[41]   Gerrard L, Rodgers L, Cui W. Differentiation of human embryonic stem cells to neural lineages in
        adherent culture by blocking bone morphogenetic protein signaling. Stem Cells 2005; 23(9): 1234-41.
        [http://dx.doi.org/10.1634/stemcells.2005-0110] [PMID: 16002783]

[42]   Sanvitale CE, Kerr G, Chaikuad A, *et al.* A new class of small molecule inhibitor of BMP signaling.
        PLoS One 2013; 8(4)e62721
        [http://dx.doi.org/10.1371/journal.pone.0062721] [PMID: 23646137]

[43]   Shi Y, Kirwan P, Smith J, Robinson HPC, Livesey FJ. Human cerebral cortex development from
        pluripotent stem cells to functional excitatory synapses. Nat Neurosci 2012; 15(3): 477-486, S1.
        [http://dx.doi.org/10.1038/nn.3041] [PMID: 22306606]

[44]   Liu C, Sun Y, Arnold J, Lu B, Guo S. Synergistic contribution of SMAD signaling blockade and high
        localized cell density in the differentiation of neuroectoderm from H9 cells. Biochem Biophys Res
        Commun 2014; 452(4): 895-900.
        [http://dx.doi.org/10.1016/j.bbrc.2014.08.137] [PMID: 25218470]

[45]   Steiner D, Khaner H, Cohen M, *et al.* Derivation, propagation and controlled differentiation of human
        embryonic stem cells in suspension. Nat Biotechnol 2010; 28(4): 361-4.
        [http://dx.doi.org/10.1038/nbt.1616] [PMID: 20351691]

[46]   Zhou S, Szczesna K, Ochalek A, *et al.* Neurosphere Based Differentiation of Human iPSC Improves
        Astrocyte Differentiation. Stem Cells Int 2016; 2016: 1-15.
        [http://dx.doi.org/10.1155/2016/4937689] [PMID: 26798357]

[47]   Miranda CC, Fernandes TG, Diogo MM, Cabral JMS. Scaling up a chemically-defined aggregate-
        based suspension culture system for neural commitment of human pluripotent stem cells. Biotechnol J
        2016; 11(12): 1628-38.
        [http://dx.doi.org/10.1002/biot.201600446] [PMID: 27754603]

[48]   Simão D, Pinto C, Piersanti S, *et al.* Modeling human neural functionality in vitro: three-dimensional
        culture for dopaminergic differentiation. Tissue Eng Part A 2015; 21(3-4): 654-68.
        [http://dx.doi.org/10.1089/ten.tea.2014.0079] [PMID: 25257211]

[49]   Choi YJ, Park J, Lee SH. Size-controllable networked neurospheres as a 3D neuronal tissue model for
        Alzheimer's disease studies. Biomaterials 2013; 34(12): 2938-46.
        [http://dx.doi.org/10.1016/j.biomaterials.2013.01.038] [PMID: 23369217]

[50]   Kimbrel EA, Lanza R. Current status of pluripotent stem cells: moving the first therapies to the clinic.
        Nat Rev Drug Discov 2015; 14(10): 681-92.
        [http://dx.doi.org/10.1038/nrd4738] [PMID: 26391880]

[51]   Fox IJ, Daley GQ, Goldman SA, Huard J, Kamp TJ, Trucco M. Use of differentiated pluripotent stem
        cells in replacement therapy for treating disease. Science 2014; 345(6199)1247391
        [http://dx.doi.org/10.1126/science.1247391] [PMID: 25146295]

[52]   Onakpoya IJ, Heneghan CJ, Aronson JK. Post-marketing withdrawal of 462 medicinal products
        because of adverse drug reactions: a systematic review of the world literature. BMC Med 2016; 14(1):
        10.
        [http://dx.doi.org/10.1186/s12916-016-0553-2] [PMID: 26843061]

[53]   Yang L, Soonpaa MH, Adler ED, *et al.* Human cardiovascular progenitor cells develop from a KDR+
        embryonic-stem-cell-derived population. Nature 2008; 453(7194): 524-8.
        [http://dx.doi.org/10.1038/nature06894] [PMID: 18432194]

[54]   Burridge PW, Thompson S, Millrod MA, *et al.* A universal system for highly efficient cardiac
        differentiation of human induced pluripotent stem cells that eliminates interline variability. PLoS One
        2011; 6(4)e18293

[http://dx.doi.org/10.1371/journal.pone.0018293] [PMID: 21494607]

[55]   Nguyen DC, Hookway TA, Wu Q, *et al.* Microscale generation of cardiospheres promotes robust enrichment of cardiomyocytes derived from human pluripotent stem cells. Stem Cell Reports 2014; 3(2): 260-8.
[http://dx.doi.org/10.1016/j.stemcr.2014.06.002] [PMID: 25254340]

[56]   Kempf H, Kropp C, Olmer R, Martin U, Zweigerdt R. Cardiac differentiation of human pluripotent stem cells in scalable suspension culture. Nat Protoc 2015; 10(9): 1345-61.
[http://dx.doi.org/10.1038/nprot.2015.089] [PMID: 26270394]

[57]   Chng Z, Teo A, Pedersen RA, Vallier L. SIP1 mediates cell-fate decisions between neuroectoderm and mesendoderm in human pluripotent stem cells. Cell Stem Cell 2010; 6(1): 59-70.
[http://dx.doi.org/10.1016/j.stem.2009.11.015] [PMID: 20074535]

[58]   Funa NS, Schachter KA, Lerdrup M, *et al.* β-Catenin Regulates Primitive Streak Induction through Collaborative Interactions with SMAD2/SMAD3 and OCT4. Cell Stem Cell 2015; 16(6): 639-52.
[http://dx.doi.org/10.1016/j.stem.2015.03.008] [PMID: 25921273]

[59]   McLean AB, D'Amour KA, Jones KL, *et al.* Activin a efficiently specifies definitive endoderm from human embryonic stem cells only when phosphatidylinositol 3-kinase signaling is suppressed. Stem Cells 2007; 25(1): 29-38.
[http://dx.doi.org/10.1634/stemcells.2006-0219] [PMID: 17204604]

[60]   Teo AKK, Ali Y, Wong KY, *et al.* Activin and BMP4 synergistically promote formation of definitive endoderm in human embryonic stem cells. Stem Cells 2012; 30(4): 631-42.
[http://dx.doi.org/10.1002/stem.1022] [PMID: 22893457]

[61]   McLean AB, D'Amour KA, Jones KL, *et al.* Activin a efficiently specifies definitive endoderm from human embryonic stem cells only when phosphatidylinositol 3-kinase signaling is suppressed. Stem Cells 2007; 25(1): 29-38.
[http://dx.doi.org/10.1634/stemcells.2006-0219] [PMID: 17204604]

[62]   Johannesson M, Ståhlberg A, Ameri J, Sand FW, Norrman K, Semb H. FGF4 and retinoic acid direct differentiation of hESCs into PDX1-expressing foregut endoderm in a time- and concentration-dependent manner. PLoS One 2009; 4(3)e4794
[http://dx.doi.org/10.1371/journal.pone.0004794] [PMID: 19277121]

[63]   Nostro MC, Sarangi F, Ogawa S, *et al.* Stage-specific signaling through TGFβ family members and WNT regulates patterning and pancreatic specification of human pluripotent stem cells. Development 2011; 138(5): 861-71.
[http://dx.doi.org/10.1242/dev.055236] [PMID: 21270052]

[64]   Jaques F, Jousset H, Tomas A, *et al.* Dual effect of cell-cell contact disruption on cytosolic calcium and insulin secretion. Endocrinology 2008; 149(5): 2494-505.
[http://dx.doi.org/10.1210/en.2007-0974] [PMID: 18218692]

[65]   Brereton HC, Carvell MJ, Asare-Anane H, *et al.* Homotypic cell contact enhances insulin but not glucagon secretion. Biochem Biophys Res Commun 2006; 344(3): 995-1000.
[http://dx.doi.org/10.1016/j.bbrc.2006.03.214] [PMID: 16643853]

[66]   Van Hoof D, Mendelsohn AD, Seerke R, Desai TA, German MS. Differentiation of human embryonic stem cells into pancreatic endoderm in patterned size-controlled clusters. Stem Cell Res (Amst) 2011; 6(3): 276-85.
[http://dx.doi.org/10.1016/j.scr.2011.02.004] [PMID: 21513906]

[67]   Vosough M, Omidinia E, Kadivar M, *et al.* Generation of functional hepatocyte-like cells from human pluripotent stem cells in a scalable suspension culture. Stem Cells Dev 2013; 22(20): 2693-705.
[http://dx.doi.org/10.1089/scd.2013.0088] [PMID: 23731381]

[68]   Lancaster MA, Knoblich JA. Organogenesis in a dish: Modeling development and disease using organoid technologies. Science 2014; 345(6194)1247125

[http://dx.doi.org/10.1126/science.1247125] [PMID: 25035496]

[69]   Arora N, Alsous JI, Guggenheim JW, *et al.* A process engineering approach to increase organoid yield. Development 2017; 144(6)dev.142919
[http://dx.doi.org/10.1242/dev.142919] [PMID: 28174251]

[70]   Eiraku M, Takata N, Ishibashi H, *et al.* Self-organizing optic-cup morphogenesis in three-dimensional culture. Nature 2011; 472(7341): 51-6.
[http://dx.doi.org/10.1038/nature09941] [PMID: 21475194]

[71]   Lancaster MA, Renner M, Martin CA, *et al.* Cerebral organoids model human brain development and microcephaly. Nature 2013; 501(7467): 373-9.
[http://dx.doi.org/10.1038/nature12517] [PMID: 23995685]

[72]   Dye BR, Dedhia PH, Miller AJ, *et al.* A bioengineered niche promotes *in vivo* engraftment and maturation of pluripotent stem cell derived human lung organoids. eLife 2016; 5e19732
[http://dx.doi.org/10.7554/eLife.19732] [PMID: 27677847]

[73]   Spence JR, Mayhew CN, Rankin SA, *et al.* Directed differentiation of human pluripotent stem cells into intestinal tissue in vitro. Nature 2011; 470(7332): 105-9.
[http://dx.doi.org/10.1038/nature09691] [PMID: 21151107]

[74]   Boj SF, Hwang CI, Baker LA, *et al.* Organoid models of human and mouse ductal pancreatic cancer. Cell 2015; 160(1-2): 324-38.
[http://dx.doi.org/10.1016/j.cell.2014.12.021] [PMID: 25557080]

[75]   van de Wetering M, Francies HE, Francis JM, *et al.* Prospective derivation of a living organoid biobank of colorectal cancer patients. Cell 2015; 161(4): 933-45.
[http://dx.doi.org/10.1016/j.cell.2015.03.053] [PMID: 25957691]

[76]   Lu YC, Fu DJ, An D, *et al.* Scalable Production and Cryostorage of Organoids Using Core–Shell Decoupled Hydrogel Capsules. Adv Biosyst 2017; 1(12)1700165
[http://dx.doi.org/10.1002/adbi.201700165] [PMID: 29607405]

[77]   Karzbrun E, Kshirsagar A, Cohen SR, Hanna JH, Reiner O. Human brain organoids on a chip reveal the physics of folding. Nat Phys 2018; 14(5): 515-22.
[http://dx.doi.org/10.1038/s41567-018-0046-7] [PMID: 29760764]

[78]   Qian X, Jacob F, Song MM, Nguyen HN, Song H, Ming G. Generation of human brain region–specific organoids using a miniaturized spinning bioreactor. Nat Protoc 2018; 13(3): 565-80.
[http://dx.doi.org/10.1038/nprot.2017.152] [PMID: 29470464]

[79]   Paşca AM, Sloan SA, Clarke LE, *et al.* Functional cortical neurons and astrocytes from human pluripotent stem cells in 3D culture. Nat Methods 2015; 12(7): 671-8.
[http://dx.doi.org/10.1038/nmeth.3415] [PMID: 26005811]

[80]   Muguruma K, Nishiyama A, Kawakami H, Hashimoto K, Sasai Y. Self-organization of polarized cerebellar tissue in 3D culture of human pluripotent stem cells. Cell Rep 2015; 10(4): 537-50.
[http://dx.doi.org/10.1016/j.celrep.2014.12.051] [PMID: 25640179]

[81]   Garcez PP, Loiola EC, Madeiro da Costa R, *et al.* Zika virus impairs growth in human neurospheres and brain organoids. Science 2016; 352(6287): 816-8.
[http://dx.doi.org/10.1126/science.aaf6116] [PMID: 27064148]

[82]   Gabriel E, Ramani A, Karow U, *et al.* Recent Zika Virus Isolates Induce Premature Differentiation of Neural Progenitors in Human Brain Organoids. Cell Stem Cell 2017; 20(3): 397-406.e5.
[http://dx.doi.org/10.1016/j.stem.2016.12.005] [PMID: 28132835]

[83]   Wang Y, Wang L, Zhu Y, Qin J. Human brain organoid-on-a-chip to model prenatal nicotine exposure. Lab Chip 2018; 18(6): 851-60.
[http://dx.doi.org/10.1039/C7LC01084B] [PMID: 29437173]

[84]   Shen Q, Goderie SK, Jin L, *et al.* Endothelial cells stimulate self-renewal and expand neurogenesis of

neural stem cells. Science 2004; 304(5675): 1338-40.
[http://dx.doi.org/10.1126/science.1095505] [PMID: 15060285]

[85]   Yin X, Mead BE, Safaee H, Langer R, Karp JM, Levy O. Engineering Stem Cell Organoids. Cell Stem Cell 2016; 18(1): 25-38.
[http://dx.doi.org/10.1016/j.stem.2015.12.005] [PMID: 26748754]

[86]   Mansour AA, Gonçalves JT, Bloyd CW, *et al.* An in vivo model of functional and vascularized human brain organoids. Nat Biotechnol 2018; 36(5): 432-41.
[http://dx.doi.org/10.1038/nbt.4127] [PMID: 29658944]

[87]   Takasato M, Er PX, Chiu HS, *et al.* Kidney organoids from human iPS cells contain multiple lineages and model human nephrogenesis. Nature 2015; 526(7574): 564-8.
[http://dx.doi.org/10.1038/nature15695] [PMID: 26444236]

[88]   Taguchi A, Nishinakamura R. Higher-Order Kidney Organogenesis from Pluripotent Stem Cells. Cell Stem Cell 2017; 21(6): 730-746.e6.
[http://dx.doi.org/10.1016/j.stem.2017.10.011] [PMID: 29129523]

[89]   van den Berg CW, Ritsma L, Avramut MC, *et al.* Renal Subcapsular Transplantation of PSC-Derived Kidney Organoids Induces Neo-vasculogenesis and Significant Glomerular and Tubular Maturation In Vivo. Stem Cell Reports 2018; 10(3): 751-65.
[http://dx.doi.org/10.1016/j.stemcr.2018.01.041] [PMID: 29503086]

[90]   Fuchs TC, Hewitt P. Biomarkers for drug-induced renal damage and nephrotoxicity-an overview for applied toxicology. AAPS J 2011; 13(4): 615-31.
[http://dx.doi.org/10.1208/s12248-011-9301-x] [PMID: 21969220]

[91]   Materne EM, Tonevitsky AG, Marx U. Chip-based liver equivalents for toxicity testing – organotypicalness versus cost-efficient high throughput. Lab Chip 2013; 13(18): 3481-95.
[http://dx.doi.org/10.1039/c3lc50240f] [PMID: 23722971]

[92]   Ang LT, Tan AKY, Autio MI, *et al.* A Roadmap for Human Liver Differentiation from Pluripotent Stem Cells. Cell Rep 2018; 22(8): 2190-205.
[http://dx.doi.org/10.1016/j.celrep.2018.01.087] [PMID: 29466743]

[93]   Zhao R, Duncan SA. Embryonic development of the liver. Hepatology 2005; 41(5): 956-67.
[http://dx.doi.org/10.1002/hep.20691] [PMID: 15841465]

[94]   Takebe T, Sekine K, Enomura M, *et al.* Vascularized and functional human liver from an iPSC-derived organ bud transplant. Nature 2013; 499(7459): 481-4.
[http://dx.doi.org/10.1038/nature12271] [PMID: 23823721]

[95]   Camp JG, Sekine K, Gerber T, *et al.* Multilineage communication regulates human liver bud development from pluripotency. Nature 2017; 546(7659): 533-8.
[http://dx.doi.org/10.1038/nature22796] [PMID: 28614297]

[96]   Takebe T, Sekine K, Kimura M, *et al.* Massive and Reproducible Production of Liver Buds Entirely from Human Pluripotent Stem Cells. Cell Rep 2017; 21(10): 2661-70.
[http://dx.doi.org/10.1016/j.celrep.2017.11.005] [PMID: 29212014]

[97]   Fordham RP, Yui S, Hannan NRF, *et al.* Transplantation of expanded fetal intestinal progenitors contributes to colon regeneration after injury. Cell Stem Cell 2013; 13(6): 734-44.
[http://dx.doi.org/10.1016/j.stem.2013.09.015] [PMID: 24139758]

[98]   Forster R, Chiba K, Schaeffer L, *et al.* Human intestinal tissue with adult stem cell properties derived from pluripotent stem cells. Stem Cell Reports 2014; 2(6): 838-52.
[http://dx.doi.org/10.1016/j.stemcr.2014.05.001] [PMID: 24936470]

[99]   Sato T, Vries RG, Snippert HJ, *et al.* Single Lgr5 stem cells build crypt-villus structures *in vitro* without a mesenchymal niche. Nature 2009; 459(7244): 262-5.
[http://dx.doi.org/10.1038/nature07935] [PMID: 19329995]

[100] Watson CL, Mahe MM, Múnera J, *et al.* An *in vivo* model of human small intestine using pluripotent stem cells. Nat Med 2014; 20(11): 1310-4.
[http://dx.doi.org/10.1038/nm.3737] [PMID: 25326803]

[101] Nadkarni RR, Abed S, Cox BJ, *et al.* Functional Enterospheres Derived *In Vitro* from Human Pluripotent Stem Cells. Stem Cell Reports 2017; 9(3): 897-912.
[http://dx.doi.org/10.1016/j.stemcr.2017.07.024] [PMID: 28867347]

[102] Roerink SF, Sasaki N, Lee-Six H, *et al.* Intra-tumour diversification in colorectal cancer at the single-cell level. Nature 2018; 556(7702): 457-62.
[http://dx.doi.org/10.1038/s41586-018-0024-3] [PMID: 29643510]

[103] Ronaldson-Bouchard K, Vunjak-Novakovic G. Organs-on-a-Chip: A Fast Track for Engineered Human Tissues in Drug Development. Cell Stem Cell 2018; 22(3): 310-24.
[http://dx.doi.org/10.1016/j.stem.2018.02.011] [PMID: 29499151]

SUBJECT INDEX

A

Acetaminophen 238
Acid 10, 20, 30, 31, 88, 114, 116, 118, 119, 120, 139, 149, 151, 165, 243
  hyaluronic 20, 114, 118
  linoleic 88
  mycophenolic 31, 149, 151
  polyglycolic 120, 243
  polylactic 116, 119
  retinoic 10, 139, 165
  zoledronic 30
Activin-Nodal 261, 264
  inhibition 264
  pathways 264
Additive layer manufacturing 221
Adenocarcinomas 12, 29, 70, 210
  mouse ductal 12
  pancreatic ductal 70
  primary colorectal 29
Adenoviruses 211
Adrenergic stimulation 139
Airway organoids 7, 180, 181, 182
Alginate 115, 118, 235
  beads 235
  bioink 118
  hydrogels 115
Alkaline phosphatase activity 112, 168
Alveolar 7, 145, 148, 204, 205, 206, 207, 235
  domain 7
  ducts 205
  epithelial progenitor cells (AEPCs) 205
  epithelial type 7, 205, 207
  sacs 205
  scaffolds 235
  stem cell niches 148
  tissue 145
  type 204, 206
Anterior foregut 8, 10
  endoderm 8
  organoids 10
Arrhythmogenic syndromes 139

Artificial 113, 117, 240
  bone 113
  matrices 177
  skin models 240

B

Barrier 26, 85, 91, 150, 239
  epithelial 91
  epithelial-mucosal 85
  physical 239
  selective air-blood 150
  simulate blood-brain 26
Bicalcium phosphates 113
Bio- 2, 20, 21, 123, 131, 159, 221, 243, 244
  banking 2, 123
  fabrication 20
  ink 20, 21, 243
  medical industries 221
  mimetic 131
  printing 2, 159, 244
Bioengineering 1, 2 3, 5, 7, 9, 11, 13, 15, 17, 19, 21, 23, 25, 27, 29, 34, 40, 41, 78, 243
  advances 34
  approaches 7, 78
  organoids 1, 3, 5, 7, 9, 11, 13, 15, 17, 19, 21, 23, 25, 27, 29
Biomarker 202, 268
  analysis 268
  studies 202
Biomaterial 8, 9, 133
  matrix 9
  scaffolds 8
  science 133
Biophysical 6, 94, 130, 237, 244
  clues 244
  factors 6
  parameters 237
  processes 130
  properties 94
Biopsies 70, 90, 175, 177, 179
  fine-needle 70

W

Whole genome sequencing 175
Wilson's disease 166

Y

Yamanaka factors 4

Z

Zika virus 64, 174
Zirconia 117

www.ingramcontent.com/pod-product-compliance
Lightning Source LLC
Chambersburg PA
CBHW050814220326
41598CB00006B/208